ACS SYMPOSIUM SERIES **732**

Modeling NMR Chemical Shifts

Gaining Insights into Structure and Environment

Julio C. Facelli, EDITOR
University of Utah

Angel C. de Dios, EDITOR
Georgetown University

American Chemical Society, Washington, DC

Library of Congress Cataloging-in-Publication Data

Modeling NMR chemical shifts : gaining insights into structure and environment / Julio C. Facelli, Angel C. de Dios, editors.

p. cm.—(ACS symposium series ; 732)

Developed from the Second International Symposium on NMR Chemical Shifts held at the 216th National Meeting of the American Chemical Society, Boston, Mass., Aug. 23–26, 1998.

Includes bibliographical references and index.

ISBN 0–8412–3622–4

1. Nuclear magnetic resonance spectroscopy—Congresses. 2. Molecular Structure—Congresses.

I. Facelli, Julio C. II. Dios, Angel C. de. III. American Chemical Society. Meeting (216th : 1998 : Boston, Mass.). IV. International Symposium on NMR Chemical Shifts (2nd : 1998 : Boston, Mass.) V. Series.

QD96.N8M59 1999
543′.0877—dc21

99-24229
CIP

The paper used in this publication meets the minimum requirements of American National Standard for Information Sciences—Permanence of Paper for Printed Library Materials, ANSI Z39.48–1984.

Distributed by Oxford University Press

PRINTED IN THE UNITED STATES OF AMERICA

Advisory Board

ACS Symposium Series

Foreword

THE ACS SYMPOSIUM SERIES was first published in 1974 to provide a mechanism for publishing symposia quickly in book form. The purpose of the series is to publish timely, comprehensive books developed from ACS sponsored symposia based on current scientific research. Occasionally, books are developed from symposia sponsored by other organizations when the topic is of keen interest to the chemistry audience.

Before agreeing to publish a book, the proposed table of contents is reviewed for appropriate and comprehensive coverage and for interest to the audience. Some papers may be excluded in order to better focus the book; others may be added to provide comprehensiveness. When appropriate, overview or introductory chapters are added. Drafts of chapters are peer-reviewed prior to final acceptance or rejection, and manuscripts are prepared in camera-ready format.

As a rule, only original research papers and original review papers are included in the volumes. Verbatim reproductions of previously published papers are not accepted.

ACS BOOKS DEPARTMENT

Contents

INDEXES

Preface

In the past fifty years, the NMR Chemical Shielding have evolved from "corrections to the measurement of nuclear magnetic moments" as quoted from Ramsey's 1950 original papers (*Phys. Rev.* *77,* 567 and *78,* 699), to one of the most important tools for structural elucidation in many branches of chemistry. There are no simple relationships between molecular structure and chemical shifts. Their dependence on the molecular electronic and geometrical structure can be derived via complex quantum mechanical equations.

Since the first Symposium on theoretical and physical aspects of the NMR chemical shielding organized in 1992 by Professor Tossel, at a NATO workshop entitled "Nuclear Magnetic Shielding and Molecular Structure", a number of significant findings, new methodologies, and novel applications have been reported. In addition, the number of practitioners using NMR chemical shifts to gain information on molecular structure and function has dramatically increased given impetus to the idea of organizing a second meeting in 1998.

This volume was developed from the second international symposium on NMR chemical shifts. This meeting was organized by the editors at the 216th National Meeting of the American Chemical Society, Boston, Massachusetts, August 23–26, 1998, and was cosponsored by the ACS Divisions of Computers in Chemistry and Physical Chemistry. The symposium included four extended lectures by Jameson, Ando, Oldfield, and Nicholas, which are included in this volume as Chapters 1–4, respectively. These lectures provide a convenient review of the current state of the art in the calculation of the NMR chemical shielding (Jameson, Chapter 1), and its application to different areas of chemistry: polymers (Ando), biomolecules (Oldfield), and catalysis (Nicholas).

The rest of the papers included in the volume provide a "snap shot" of the field at the time of the symposium. The papers by Ferraro, Sternberg, Schreckenbach, and Webb address new methodologies to calculate NMR chemical shifts. The papers by Kuroki, Kurosu, Sakurai, and Facelli present novel calculations of chemical shifts in molecular systems of biological interest. The contributions by McDermott, Case, and Martin report important advances in the understanding of ^{1}H chemical shifts, and the contribution from de Dios emphasizes the importance of the local geometry on the determination of the NMR chemical shifts. The modeling of chemical shifts in inorganic compounds is discussed in detail in the contributions from Bühl, Moore, Wasylishen, Henry, Tossell, Alam, and Jameson (Chapter 23).

Acknowledgments

We thank the contributing authors and the ACS editorial staff for their timely contributions in making this volume a success. Financial support from the ACS Computers in Chemistry Division in supporting the symposium is gratefully appreciated.

JULIO C. FACELLI
Center for High Performance Computing
University of Utah
405 INSCC
Salt Lake City, UT 84112

ANGEL C. DE DIOS
Department of Chemistry
Georgetown University
37th and O Streets, N.W.
Washington, DC 20057

Chapter 1

Recent Advances in Nuclear Magnetic Shielding Theory and Computational Methods

Cynthia J. Jameson

Department of Chemistry, University of Illinois at Chicago, Chicago, IL 60607

As the well-established methods employing multiple origins, GIAO, IGLO, LORG, IGAIM, and common origin CHF methods are applied to diverse molecular systems, using ever larger basis sets up to near Hartree-Fock limit, the limitations of calculations at the SCF level in accurately predicting the nuclear shielding becomes more evident. The challenge of applying ab initio methods and density functional methods to heavier nuclei or to nuclei having neighboring halogens and other atoms with large spin orbit coupling constants demanded the development of theoretical methods for improved description of shielding in systems where electron correlation is important, and systems where relativistic effects are important. At the same time, the calculations on small molecules have become sufficiently accurate that comparisons with experiment have to include accurate rovibrational averaging. This overview of the recent successes, new challenges, and new directions for the theory of nuclear magnetic shielding, provides only a hint of the new and exciting results that are reported in this Symposium.

The first conference devoted to the NMR chemical shift was held in Maryland, July 20-24, 1992, as a NATO Advanced Scientific Workshop on Nuclear Magnetic Shielding and Molecular Structure (*1*). Experimentalists measuring fundamental quantities such as the chemical shift tensor and its dependence on geometry and intermolecular interactions and theoreticians involved in computations of the nuclear magnetic shielding exchanged information on the state of the subject. At that time, the following was already known:

1. The use of distributed origins (IGLO, LORG, GIAO) leads to dramatic improvement in damping the errors arising from incomplete cancellation of long-range contributions due to basis set inadequacies.

2. At the CHF level, common origin calculations with large basis sets can yield results indistinguishable from distributed origins calculations with modest basis sets.
3. The full shielding tensor provides more complete information and more stringent tests of calculations than the isotropic shielding. Ingenious experimental techniques for single crystals in systems of up to 24 distinct ^{13}C nuclear sites are complemented by ab initio calculations of the shielding tensor, which are nearly indispensable in interpreting experimental data.
4. The dependence of the nuclear shielding on molecular geometry defines shielding surfaces. The shielding surface determines the experimentally observed thermal average shielding, the temperature dependence of the shielding, and the isotopic chemical shifts. Rovibrational averaging over the shielding surface are required for good comparisons of theoretical values with experimental values.
5. The intermolecular shielding surface and the effects of electric fields on shielding (shielding polarizabilities and hyperpolarizabilities) are useful in the interpretation of long-range intramolecular effects and in intermolecular effects on shielding.
6. Electron correlation is important for multiple bonded systems, especially when lone pairs are also involved, but the magnitude of the contribution from electron correlation is clouded by the use of modest basis sets in correlated calculations.
There are many unanswered questions about nuclei in the 3rd row and below in the Periodic Table, for transition as well as representative elements. Basis set development for such atoms are required before quantitative results for σ may be expected. The possible importance of relativistic effects, the unknown geometries (especially of complex ions) in solution, and the lack of absolute shielding scales for such nuclei makes any "good agreement" of small basis set uncorrelated calculations with chemical shifts observed in solution very suspect.

At the end of the conference the following directions for future work were discussed.
1. We need to optimize basis sets for 3rd and 4th row atoms in the Periodic Table.
2. How does one do relativistic calculations? Although two formalisms had been in the literature outlining relativistic approaches to shielding calculations, no calculations had been carried out for molecules.
3. Absolute shielding scales are not available for many interesting nuclei. In a relativistic system, is the relation between the spin-rotation tensor and the paramagnetic shielding derived in the nonrelativistic limit preserved? Instead, C. J. Jameson suggested using a linear molecule: measurement of $(\sigma_{\|}-\sigma_{\perp})$ together with theoretical $\sigma_{\|}^{d}$ gives absolute shielding tensor components.
4. There is a need of more gas phase data for accurate comparisons of ab initio calculations with experiments since the intermolecular effects present in condensed phases are especially large for nuclei having large numbers of electrons.
5. There is a need for rovibrational averaging of theoretical shielding prior to comparison of calculated values with experiment. This becomes even more important as the theoretical methods begin to yield more accurate results.
6. There is a need for shielding calculations including electron correlation at higher levels than have already been reported at the conference (MBPT2, SOPPA/SOLO) to really find out how large the electron correlations effects are in molecules such as N_2, HCN, CO.

7. Experimentalists are very interested in interpretation of the NMR chemical shifts observed in complex systems. How does one deal with computations of NMR chemical shifts in complex systems such as proteins, zeolites, and other extended networks?

These directions for future work have correctly predicted the multiple thrusts of theoretical computations of the NMR shielding from the summer of 1992 to the present time. There have been major theoretical developments in the following areas: (a) New treatments of the gauge origin problem, (b) development and widespread implementation of density functional methods for shielding calculations, (c) treatment of scalar and spin orbit relativistic effects on σ, (d) high-level electron correlated calculations of σ, (e) more efficient computations by using parallelized code in multiple processors, (f) an analysis of and demonstration of basis set convergence, (g) beginning calculations of σ for some heavy nuclei, (h) accurate rovibrational averaging of σ for small molecules, (i) supermolecule calculations of intermolecular shielding functions, other approaches to intermolecular effects, and ensemble averages including intermolecular effects on σ, (j) theoretical approach to calculations of shielding in proteins, (k) model calculations in zeolites and inorganic solids, and (l) fine geometry effects on shielding: NMR refinement of diffraction results. Probably the most important developments from the point of view of widespread applications of shielding calculations to new areas were the calculations of shielding surfaces for proteins and the calculations of average chemical shifts in adsorbed Xe in zeolites. With these applications, theoretical calculations of NMR chemical shifts have entered the mainstream of chemical tools, no longer a highly specialized activity for a selective few practitioners. The papers included in this Symposium proceedings, describing some of the recent developments listed above, demonstrate quite clearly that theoretical computation of the NMR chemical shift has finally reached a level of accuracy and efficiency that it has become a generally recognized tool in the elucidation of molecular structure and interactions.

New Treatments of the Gauge Origin Problem.

The CSGT (continuous set of gauge transformations) method introduced by Keith and Bader (*2*), more accurately renamed CTOCD (continuous transformations of the origin of the current density) by Lazzeretti et al. (*3-5*), uses a different gauge origin in real space for each point **r**; by using a shift function, one can empirically shift the gauge origin towards the nucleus nearest to the point **r**, for example. The paper by Paolo Lazzeretti in this Symposium demonstrates that one can choose to rigorously annihilate either the diamagnetic or the paramagnetic contribution to the first order induced current density. The first, annihilation of the diamagnetic contribution, turns out to be equivalent to Geertsen's polarization propagator-based procedure. Both methods have been applied to small molecules. Gauge-invariance tests of all the components of the shielding tensor is an important objective criterion for the robustness of a method of theoretical calculation of magnetic properties. The adequacy of the basis set used can be tested separately from the extent of electron correlation contributions. Such purely theoretical tests (as opposed to comparisons against experiments) are a particular emphasis in the work of Lazzeretti and co-workers.

Density Functional Methods for Shielding Calculations

Since the Maryland meeting, density functional theory has been developed as a major tool for shielding calculations (*6*). The pioneering studies by Malkin and Salahub (*7*) on one hand, were followed by independently developed methods by Handy et al. (*8*), Schreckenbach and Ziegler (*9*), Pulay et al. (*10*), and Cheeseman et al. (*11*) on the other hand. The practical utility of this method compared to ab initio methods is that correlation is intrinsically included in the exchange-correlation energy functional used, and once the functional is chosen, there are only basis set levels to choose for describing the electron density. Unlike the perturbation expansion approaches to electron correlation, MPn or MBPTn, the level of electron correlation in the DFT method is not a variable choice. The method is intrinsically efficient in incorporating correlation as the size of the system increases. The disadvantage is that results depend on which functional is used in the calculation. Functionals used in calculations of magnetic properties should have an explicit dependence on the electron current density as well as the electron density itself, but neglect of this may be a minor problem (*8*).

Existing field-free exchange correlation (XC) functionals can be classified as local density approximations (LDA), or generalized gradient approximations, or hybrid functionals that incorporate part of the Hartree-Fock exchange. There are gradient corrected exchange functionals by Becke without (B) or with Hartree-Fock exchange (B3) which can be combined with correlation functionals by Lee, Yang, and Parr (LYP) or by Perdew (P86) or Perdew and Wang (PW91). Thus DFT calculations may use LDA, BP86, BLYP, or BPW91 or the hybrid B3LYP functionals. It has been shown that the simple LDA is insufficient for chemical shifts (*6*). There is yet no universal functional that works uniformly well for all shielding calculations. For example, hybrid functionals have been found to give inferior results compared to various generalized gradient approximations in the study of ^{17}O shielding in $[MO_4]^{n-}$ complexes (*12*), but hybrid functionals are clearly superior to generalized gradient functionals in recovering the full range of the ^{57}Fe and ^{103}Rh chemical shifts (*13*), although not for ^{95}Mo (Bühl, M. to be published). The main problem is that the functionals presently in use do not provide the correct asymptotic behavior in the immediate vicinity of the nucleus and are therefore intrinsically deficient for calculation of magnetic properties that involve the r^{-3} operators. Nevertheless, when differences in shielding (chemical shifts) rather than absolute shielding values are being compared with experiment, these intrinsic deficiencies appear not to be a significant problem. Earlier DFT calculations by Malkin et al. used IGLO local origins, but more recent DFT implementations use GIAOs. The latter present no ambiguities related to the localization methods used.

While DFT may or may not be more accurate than MP2 for absolute shielding calculations is debatable, the strength of the DFT method in calculations of shieldings is in the ability of DFT to provide a consistent picture over a wide range of chemical systems, since calculations can be done at a very modest computational cost compared to MP2. Among the successes of the method is in ligand chemical shifts in transition metal complexes. For example, ^{13}C, ^{17}O, ^{31}P and ^{1}H chemical shifts for oxo (*12,14,15*), carbonyl (*16-19*), interstitial carbide (*20*), phosphine (*21,22*), hydride (*23*), and other ligands have been successfully reproduced to within tens of ppm in most cases. Other successes include ^{77}Se (*24*), and ^{125}Te (*25*) shieldings in a large number of chemical bonding situations. DFT methods have also been used to include relativistic corrections, as we see below.

Scalar and Spin Orbit Relativistic Effects on σ.

Fully relativistic methods based on four-component wavefunctions and the complete Dirac equation would be impractical. Several approaches have been used. One may neglect the spin orbit operator completely and find only "scalar" or "spin-free" relativistic effects. Since the mass-velocity and Darwin contributions may induce large perturbations in the electronic structure close to the nuclei, perturbation theory may not be adequate. To circumvent this difficulty, techniques such as the relativistic zeroth order Hamiltonian, quasirelativistic effective core potentials, or the frozen-core approximation have been used. Within the DFT method, one approach is a quasirelativistic theory in which the highly relativistic core is described with 4-component wavefunctions, from which are extracted the core electron density and potential (effective core potential, ECP), and these are used in molecular calculations (*15,26*). Another is a combination of a first order quasirelativistic approach with the frozen core approximation (*16,27*). The scalar relativistic effects have a large contribution to the absolute shieldings. The atomic core contributions to the diamagnetic part of the shielding are increased. When relative chemical shifts are considered however, the shielding contributions from the inner cores are quite similar in the various molecular environments and tend to cancel to a large extent in the chemical shift values. The valence MOs are orthogonalized against all core MOs; this ensures their correct asymptotic behavior near the nucleus. What gives rise to the scalar relativistic effects on the chemical shifts is that the core tails of the valence orbitals are shaped differently from one molecule to another.

The first attempts to calculate spin orbit contributions to the shielding with nonrelativistic wavefunctions used a third order perturbation treatment within a semiempirical theory. Recently, Nakatsuji et al. developed an approach using ab initio unrestricted Hartree-Fock wavefunctions and the finite field perturbation method with a common gauge origin (*28*). They applied this method to the calculation of halogen substitution-induced chemical shifts of ^{1}H, ^{13}C, ^{27}Al, ^{29}Si, ^{119}Sn and other nuclei (*28-31*). Later on, they included the scalar relativistic corrections, still using a common origin and no electron correlation (*32*). With this approach, significant cross terms between scalar relativistic effects and spin orbit coupling have been found for ^{199}Hg and ^{183}W chemical shifts (*33,34*). The halogen spin orbit contributions to the chemical shifts of neighboring nuclei are in the direction of increased shielding and is responsible for the so-called normal halogen dependence, in which the central atom shielding increases when going from Cl to Br to I substituents.

A DFT-based third order perturbation theory approach includes the FC term by FPT. Based on the perturbed nonrelativistic Kohn-Sham orbitals spin polarized by the FC operator, a sum over states treatment (SOS-DFPT) calculates the spin orbit corrections (*35-37*). This approach, in contrast to that of Nakatsuji et al., includes both electron correlation and local origins in the calculations of spin orbit effects on chemical shifts. In contrast to these approaches that employed the finite perturbation method the SO corrections to NMR properties can be calculated analytically from quadratic response functions (*38,39*). The latter method using SCF and MCSCF wavefunctions have been used to calculate the ^{1}H shieldings in the hydrogen halides and the ^{13}C shielding in the methyl halides, including only SO corrections, without the scalar relativistic effects (*39*).

An analogy of the role of the Fermi contact (FC) operator in the spin orbit (SO) shifts to the FC role in spin spin coupling constants indicates that the higher the s character in the bond to the halogen, the greater the SO shifts (*37*). This provides understanding of experimentally observed facts: the large SO shifts for nuclei having a large s orbital participation in the bond (in ^{1}H, ^{13}C, ^{29}Si, in main group elements in high oxidation states, and in d^{10} transition metals). In contrast, a low s character of the bond leads to an inefficient FC mechanism, to small SO shifts, and thus to inverse halogen dependence, the usual situation when SO effects are not included, such as for early transition metals in high oxidation states, or for p-block main group elements in low oxidation states.

Electron Correlated Calculations of σ in Small Molecules.

What is the current state of shielding calculations for small molecules? This question can be asked in the context of state of the art calculations of shielding in molecules

Table I. ^{13}C Shieldings, ppm Relative to $^{13}CH_4$

^{13}C in Molecule	*SCF*[a]	*MBPT(2)*[a]	*Expt.*[b]
CH_3CH_3	11.7	13.5	14.2
$H_2C{=}CH_2$	135.8	130.3	130.6
$HC{\equiv}CH$	81.8	78.2	77.9
CH_3F	71.6	79.7	78.3
CH_3OH	52.0	59.3	58.5
CH_3NH_2	31.9	36.6	36.8
$\mathit{CH_3}CHO$	33.5	38.7	37.9
$(\mathit{CH_3})_2CO$	32.2	37.0	37.1
$\mathit{CH_3}CN$	4.8	7.9	7.4
CO	224.9	190.4	194.1
CO_2	147.9	138.0	136.3
H_2CO	205.0	194.8	
$CH_3\mathit{C}HO$	211.3	200.3	201.8
$(CH_3)_2\mathit{C}O$	218.8	207.3	208.2
HCN	127.5	114.2	113.0
$CH_3\mathit{C}N$	135.1	125.4	121.3
$CH_2{=}\mathit{C}{=}CH_2$	240.0	227.5	224.4
$\mathit{CH_2}{=}C{=}\mathit{CH_2}$	81.7	80.6	79.9
CF_4	116.4	137.1	130.6
C_6H_6	140.6	137.5	137.9

[a] Ref. (*42*).
[b] Ref. (*41*).

Table II. Calculated Shieldings in Small Molecules

METHOD	*^{13}C in CH$_4$*	*^{13}C in CO*	*^{13}C in HCN*	*^{15}N in NH$_3$*	*^{15}N in N$_2$*	*^{15}N in HCN*	*Ref.*
DFT	184.33	-15.35			-84.82		(*8*)
DFT	187.5	-17.7		259.2	-87.9		(*11*)
DFT	187.80	-12.27	71.74	259.42	-80.55	-43.47	(*10*)
DFT	191.2	-9.3	91.5	262.0	-72.9	8.4	(*9*)
SCF	194.8	-25.5	70.9	262.3	-112.4	-50.7	(*44*)
MBPT(2)	201.0	10.6	87.6	276.5	-41.6	-0.3	(*44*)
MBPT(3)	198.8	-4.2	80.0	270.1	-72.2	-26.2	(*44*)
MBPT(4)	198.6	4.1	84.3	269.9	-60.1	-14.9	(*44*)
MCSCF	198.2	8.22	86.76		-52.2	2.63	(*47,48*)
CCSD	198.7	0.8	84.1	269.7	-63.9	-16.7	(*46*)
CCSD(T)	198.9	5.6	86.3	270.7	-58.1	-13.6	(*46*)
Expt. σ_e [b]	198.7	3.2	84	273.3	-59.6	-15	(*51*)
	±1.2	±0.9	±1	±0.2	±0.2	±1	
Expt. σ_0 [a]	195.1	1.0	82.1	264.5	-61.6	-20.4	(*51*)
	±1.2	±0.9	±0.9	±0.2	±0.2	±0.2	

[a] These are absolute isotropic shielding values σ_0 which are the thermal averages in the gas at the zero pressure limit. The error bars are associated with the determination of the absolute scale based on the spin rotation constants for specific molecules (^{13}CO, ^{15}NH$_3$, H^{19}F). References for the experimental values σ_0 and the calculations of the rovibrational corrections that lead from σ_0 to σ_e are given in Ref. (*51*).

[b] These are absolute isotropic shielding values σ_e for a molecule fixed at its equilibrium molecular geometry, with which the calculations are to be compared.

containing only 1-2 second row atoms. In another context, with modest effort, how closely can computations predict ^{13}C chemical shifts in simple organic compounds? Gauss's answer is, within 1.7% at MBPT(2) level or 0.8 ppm, with exceptions (*40*), in calculations for carbon in molecules where comparison of the isotropic shielding against either the absolute shielding scale or at least gas phase experiments (*41*) is possible. These are shown in Table I (*42*).

Gauss has derived shielding theory with electron correlated calculations in the Møller-Plesset expansion (*43,44*) and also in the coupled cluster approach (*45*). He has provided benchmark calculations on a set of small molecules shown in Tables II and III at the CCSD(T) level (*46*).

In these molecules the sign and magnitude of the contribution of electron correlation to ^{13}C, ^{15}N, ^{17}O, ^{19}F are found to be positive (except for F_2), as little as 4 ppm, as much as 54 ppm. How well can we do with including intermediate levels of electron correlation? Gauss finds that MBPT(2) or MP2 overcorrects, except in the case of F_2. For F_2 molecule the full CCSD(T) calculations and also rovibrational averaging have to be included before comparison with experiment is possible. In Tables II and III the various methods are compared with one another. All have used GIAOs as basis functions in the computations, so as not to cloud the comparisons. The signs and magnitudes of the electron correlation contribution to shielding in these molecules are finally settled. In 1992, the discrepancies were still large in the calculated shieldings of the various heavy nuclei in the set of N_2, NH_3, HCN, NNO molecules or in the set of CH_4, CO, HCN molecules when compared with experimental absolute shieldings. It was suggested that something might be wrong with absolute shielding scales for ^{13}C and ^{15}N. It is obvious now that MBPT(2) level of theory is insufficient for some of these systems (*44*). At this time, the consensus is that the absolute shielding scales for ^{13}C and ^{15}N are fine. On the other hands, the ^{17}O absolute shielding is thought to be at the far edge of the rather large error bar associated with the reference molecule ($C^{17}O$) (*49*). How well did density functional theory do for these benchmark systems? Tables II and III show that, where there is little electron correlation, DFT did not do as well as SCF. Different exchange-correlation functionals give different answers. In some cases, DFT provided absolute shielding results that are not yet as accurate as MP2-level calculations.

Table III. Calculated Shieldings in Small Molecules

METHOD	^{17}O *in* H_2O	^{17}O *in CO*	^{19}F *in HF*	^{19}F *in* F_2	*Ref.*
DFT	317.86	-77.14	405.05	-271.70	(*8*)
DFT	324.8	-80.7			(*11*)
DFT	326.37	-73.60	410.87	-277.14	(*10*)
DFT	331.5	-68.4	412.5	-282.7	(*9*)
SCF	328.1	-87.7	413.6	-167.9	(*44*)
MBPT(2)	346.1	-46.5	424.2	-170.0	(*44*)
MBPT(3)	336.7	-68.3	417.8	-176.9	(*44*)
MBPT(4)	337.5	-52.0	418.7	-174.0	(*44*)
MCSCF	335.3	-38.92	419.6	-136.6	(*47*)
CCSD	336.9	-56.0	418.1	-171.1	(*46*)
CCSD(T)	337.9	-52.9	418.6	-186.5	(*46*)
Expt. σ_e [b]	357.6 ±17	-36.7 ±17	419.7 ±6	-192.8 ±6	(*51*)
Expt. σ_0 [a]	344.0 ±17.2	-42.3 ±17.2	410.0 ±6	-232.8 ±6	(*51*)

[a, b] The same footnotes as in Table II.

Full configuration interaction (CI) calculations for the shieldings in H_2 (*50*) and BH molecules have demonstrated that the CCSD(T) results are sufficiently close to the full CI results, so that for practical purposes, the CCSD(T) calculations may be considered as the benchmark calculations for each molecule, against which other methods such as density functional methods may be tested.

More Efficient Computations by Using Parallelized Code in Multiple Processors.

Single processor calculations of nuclear shielding at the SCF level are limited by practical computation times in most hardware to about 800 basis functions with no symmetry or 1600 with high symmetry. Thus, the obvious solution of the problem is parallel processing using an array of inexpensive workstations or PCs. In a significant breakthrough, Peter Pulay et al. have implemented the first parallel computation of shielding (*52*). The program by Wolinski, Haacke, Hinton and Pulay uses the new integral program of Wolinski that calculates integrals in vectorized batches, increases efficiency, and simplifies parallelization. The self consistent field step includes a multi-Fock algorithm, i.e., the simultaneous calculation of several Fock matrices with the same integral set, increasing the efficiency of the direct SCF procedure. Replacing the SCF diagonalization by pseudo-diagonalization, a method widely used in semiempirical programs, provides further time savings. Including higher angular momentum functions in the basis sets is an important factor in calculations of shielding of heavier elements. By concentrating all logic in integer arrays, the program has in principle been made open ended with respect to angular momentum integrals. The results are very promising, presently a scaling of computational time to $N^{2.8}$, N being the number of basis functions, has been achieved toward the ultimate goal of N^2 scaling. The paper presented by Peter Pulay in this Symposium proceedings provides details and comparisons.

Basis Set Convergence.

A case study of the basis set dependence of the anisotropy and the isotropic shielding has been presented by Enevoldsen and Oddershede (*53*) for ^{19}F in the CH_3F molecule. The RPA (SCF level) results are shown in Tables IV and V.

Table IV. Calculation of $\Delta\sigma = (\sigma_{\parallel} - \sigma_{\perp})$ for ^{19}F in the CH_3F molecule.[a,b]

Basis set size	*LORG*	*Common origin*	*Geertsen method*	*CTOCD*	*GIAOs*
4-31G, (24)	29.34	5.91	53.61	19.38	-75.34
6-31G, (24)	21.21	-2.78	44.88	11.26	-74.14
D95, (26)	17.81	7.79	39.38	10.98	-80.84
6-31G+, (36)	-44.89	-65.47	-40.00	-47.78	-68.95
6-31G++,(45)	-49.03	-65.85	-42.48	-52.18	-67.42
75	20.51	10.22	40.03	-0.51	-86.11
132	-69.75	-70.17	-66.02	-71.90	-67.28
135	-67.44	-67.63	-63.52	-69.38	-67.78
141	-67.76	-67.96	-63.84	-69.71	-68.67
158	-67.69	-67.95	-66.82	-67.72	-68.58

[a]From Ref. (*53*).
[b]Experimental value is -60.86±15 ppm (*54*).

The conclusions are as follows: The basis set must include polarization functions. GIAOs converge fastest, better than LORG or IGLO. The inclusion of polarization functions in standard basis sets (6-31G+ and 6-31G++) do not guarantee convergence. The basis set may be considered converged (for 132 functions or more) when, using any method, only small basis set dependences are observed. These trends are the same both in RPA (equivalent to coupled Hartree-Fock) and SOPPA (second order polarization propagation approximation), although SOPPA results are not shown here. As expected, the Geertsen method is gauge invariant at all basis set levels. The full tensor is more sensitive to deficiencies in the basis set than the isotropic average shielding.

Table V. Calculation of σ_{iso} for ^{19}F in the CH_3F Molecule.[a,b]

Basis set	*LORG*	*Common origin*	*CTOCD and Geertsen*	*GIAOs*
4-31G, (24)	404.11	420.55	-46.67	488.82
6-31G, (24)	405.58	422.24	-47.08	489.35
D95, (26)	407.05	425.69	-55.29	492.96
6-31G+, (36)	453.29	467.06	100.34	488.32
6-31G++, (45)	458.66	469.25	106.76	488.52
75	405.16	411.60	235.25	493.99
132	480.12	480.22	466.62	483.75
135	482.31	482.39	468.81	483.26
141	482.31	482.40	475.31	483.65
158	482.27	482.44	480.99	483.39

[a] From Ref. (*53*).
[b] Experimental value is 471 ppm (*55*).

Calculations of σ for Heavy Nuclei.

What is the current state for shielding calculations of nuclei in the third row of the Periodic Table? We illustrate the current state with two nuclei for which absolute shielding scales exist (*56,57*), ^{77}Se and ^{71}Ga, in Tables VI and VII. It has been found that, in addition to complications arising from relativistic corrections, electron correlation contributions can be large in heavy atoms, so the DFT method should have an advantage in large systems. For ^{77}Se the best calculations, which can be considered benchmarks, are the CCSD calculations by Gauss et al. (*58*). There are few experimental gas phase results for comparisons. For the O=C=Se molecule the experimental value of the absolute ^{77}Se shielding has been derived (see Table 8 of Ref. (*59*) from the measured ^{77}Se spin rotation tensor and the equilibrium molecular geometry. No error estimates were provided with the spin rotation tensor. On the optimistic side, we assign an uncertainty of about ± 60 ppm. For a linear molecule such as the Se=C=Se molecule, the absolute isotropic shielding can be obtained from the shielding anisotropy by a simple relation: $\sigma = \sigma_{\parallel}{}^{d} - (2/3)\Delta\sigma$. The nonrelativistic free atom value 2998 ppm (*61*), used in the Flygare approximation (*62*), including atom dipole correction terms equal to 9.6 ppm, gives $\sigma_{\parallel}{}^{d} = 3008$ ppm. From the experimental shielding anisotropy measured in liquid crystal solution (2210±120 ppm) (*60*) and the parallel component of the diamagnetic shielding, we estimate the absolute isotropic ^{77}Se shielding in liquid phase SeCSe = 1535±80 ppm, the uncertainty reflecting only that of the shielding anisotropy measurement. With an experimental gas-to-liquid shift of 73.7 ppm, between the liquid at 20°C and the gas at 90°C (*63*), the gas phase isotropic shielding is estimated to be 1610±80 ppm. Table VI shows us that calculated ^{77}Se shieldings are in reasonable agreement with the absolute shielding values. There are still discrepancies which could be attributed to rovibrational effects.

Table VI. Absolute Shielding σ(^{77}Se) in Small Molecules

	H_2C=Se	*SeF_6*	*Me_2Se*	*MeSeH*	*H_2Se*	*O=C=Se*	*Se=C=Se*
DFT[a]		953	1668	1837	2093	2270	1441
CONV[g]			1905.4	1988.0	2156.7		
CONV[b]			1926.3	2015.7	2171.0		
SCF[c]			1927.0	2043.8	2178.5		
MP2[c]			1922.2	2093.0	2275.6		
SCF[d]	-1011.4	1320.6	1897.6	2027.6	2167.6	2261.6	1451.6
MP2[d]	-994.5	1147.5	1874.7	2054.5	2236.5	2406.5	1753.5
SCF[e]	-886		1894.1		2170	2291	1488
MP2[e]	-813		1895.4		2260	2418	1748
CCSD[e]	-741		1877.5		2213	2345	1596
Expt.	-900 ±200[f]	1138 ±64[g]	1756 ±64[h]	1911 ±64[h]	2101 ±64[g]	2348 ±60[i]	1610 ±80[d]

[a] Ref. (*24*).
[b] Ref. (*64*).
[c] Ref .(*65*).
[d] Ref. (*66*).
[e] Ref. (*58*).
[f] From Table 16 in Ref. (*67*), derived from ^{77}Se spin rotation tensor and rotational constants in high resolution microwave data of Ref. (*68*).
[g] From Ref. (*56*), the ^{77}Se absolute shielding scale, but without the relativistic corrections (300 ppm) in the diamagnetic shielding of the Se free atom used in Ref. (*56*).
[h] The chemical shifts between H_2Se vs. MeSeH and Me_2Se (all in the gas phase) from Ref. (*1*) were converted to absolute shieldings by using the absolute shielding σ(^{77}Se, H_2Se, gas) = 2101±64 ppm from Ref. (*56*), again leaving out the relativistic corrections.
[i] See the text for the determination of these experimental values.

Calculations on ^{125}Te shielding have been very successful in reproducing the 4000 ppm range of isotropic chemical shifts in a wide variety of compounds in solutions or neat liquids (*25*).although no corrections of medium effects had been made in comparing with experimental values.

For the ^{71}Ga nucleus, the availability of an absolute shielding scale also makes possible a comparison of calculated absolute shieldings with experimental values. The standard used experimentally, $[Ga(OH_2)_6]^{3+}$ in solution, is unsuitable as a theoretical reference due to the lack of consideration of water molecules beyond the first solvation shell in the calculations. On the other hand, there exists an absolute shielding scale for ^{71}Ga, based on the NMR measurement in an atomic Ga beam at the same time as in the ion at infinite dilution in a D_2O solution (*57*).

By using the calculated nonrelativistic diamagnetic shielding for the free Ga atom = 2638.6 ppm (*61*), it is found that,

$\sigma(^{71}Ga, [Ga(OD_2)_6]^{3+}$ infinitely dilute solution in D_2O) = 1840(45) ppm.

With this value we can convert the experimental chemical shift measurements to absolute shieldings. Table VII shows that within the experimental errors, without taking into account the solvation effects, the agreement of the calculations of ^{71}Ga shielding at the MP2 level with experiment is reasonable. These three examples, ^{77}Se, ^{125}Te, and ^{71}Ga, provide convincing evidence that shielding calculations for heavy nuclei of representative elements are presently feasible.

Table VII. Absolute Shielding $\sigma(^{71}Ga)$

$\sigma(^{71}Ga)$	*MP2 calc*[a]	*exptl. (± 45 ppm)*[b]	*% recovery*
$GaMe_3$	1175.5	1120	105
$[GaCl_4]^-$	1613.2	1593	101
Ga_2Cl_6	1625.0	1598	102
$[Ga(OH)_4]^-$	1629.4	1620	100
$[GaH_4]^-$	1827.3	1730	106
$[Ga(OH_2)_6]^{3+}$	1930.4	1840	105
$Ga(C_5H_5)$	2593.2	2554	101

[a] From Ref. (*69*).

[b] The experimental chemical shifts (compiled by Ref. (*69*)) have been converted to absolute shielding, based on the σ ($^{71}Ga, [Ga(OD_2)_6]^{3+}$ infinitely dilute in D_2O) = 1840(45) ppm, as described in the text.

Rovibrational Averaging of σ for Small Molecules.

How large are the rovibrational corrections? How well can they be calculated? To do the rovibrational corrections, we need both the shielding surface for the nucleus and the potential energy surface for the molecule. The average is obtained from the shielding value at each point on the shielding surface weighted according to the probability of finding the molecular system at that nuclear configuration. There had been, prior to 1992, a large quantity of data on the rovibrational effects on the NMR shielding. Furthermore, the theoretical interpretation of the observed consequences of rovibrational averaging (temperature dependence of the molecule in the limit of zero pressure, and shielding changes upon isotopic substitution) were already well established (*70,71*). However, few accurate shielding surfaces had been calculated, for several diatomic molecules, and only for very few polyatomic molecules: H_2O (*72*), NH_3 (*73*), PH_3 (*74*), and CH_4 (*75*).

The developments since 1992 include very accurate calculations of the shielding surfaces in the immediate vicinity of the potential minimum for diatomic molecules. Rovibrational corrections to shielding calculated with these more accurate shielding surfaces (CCSD(T) level calculations) are shown in Table VIII.

Some of the shielding surfaces are found to have significant curvature at the equilibrium internuclear separation, so that it is important to evaluate higher than first derivatives of the surface at this configuration. In comparing with experimental data,

data from the full range of temperature should be used, rather than making a comparison between theoretical derivatives (σ_e', σ_e'') and experimental ones. For example, the empirical value for σ_e' in F_2 molecule was obtained from a single parameter fitting to the observed temperature dependence of the ^{19}F chemical shift in F_2 gas in the limit of zero pressure (*76*). Thus, the empirical parameter actually stands for the effects of both the first derivative (σ_e') and the second derivative (σ_e'') terms. The single parameter is found to be too large compared to current theoretical values for the first derivative of ^{19}F shielding (*49*), nevertheless, the change in the thermal average ^{19}F shielding over the range 230 to 350 K predicted from the theoretical

Table VIII. Magnitudes of Rovibrational Corrections to Shielding

System	$\langle\sigma\rangle_{v=0} - \sigma_e$ *a*	$\langle\sigma\rangle_{300\,K} - \langle\sigma\rangle_{v=0}$ *a*	$\langle\sigma\rangle_{300\,K} - \sigma_e$ *a*	$\langle\sigma\rangle_{300\,K} - \sigma_e$ *EXPT*
^{1}H in H_2	-0.355	-0.014	-0.369	-0.303 [d,f], -0.375 [d,g]
^{1}H in HF	-0.323	-0.035	-0.358	-0.38 [i]
^{19}F in HF	-10.0	-0.42	-10.42	-9.75 [i], -10.5[d,g]
^{13}C in CO	-2.24	-0.15, (-0.087)[b,c]	-2.39	-1.91 [d,h]
^{17}O in CO	-5.73	-0.35	-6.07	-4.88 [d,h]
^{15}N in N_2	-4.03	-0.24, (-0.255)[b,c]	-4.27	-3.49 [d,h]
^{19}F in F_2	-30.87	-4.69	-35.56	-40 [c,j]
^{17}O in H_2O	-9.85[e]	-0.36[e]	-10.22[e]	-13.0 [d,h]
^{33}S in H_2S			-16.4[k, m]	
^{77}Se in H_2Se			-58.9 [n]	-56.9 [d,h]
^{15}N in NH_3			-8.81[p]	-8.3 [d,h]
^{31}P in PH_3			-12.78 [q]	-10.4 [d,h]
^{13}C in CH_4	-3.591	-0.104	-3.695[r]	-3.3[d,g]

[a] Except where noted, the values are taken from Ref. (*49*).
[b] From Ref. (*77*).
[c] experimental estimate from the temperature dependence of the chemical shift in the gas phase at the zero-pressure limit.
[d] experimental estimate from the isotope shift.
[e] The more accurate values reported by W. T. Raynes at this Symposium. Earlier values were -13.116, -0.457, and -13.573 ppm respectively, from a CHF shielding surface Ref. (*78*).
[f] Ref. (*79*).
[g] Ref. (*80*).
[h] Ref. (*70*).
[i] Ref. (*81*).
[j] Ref. (*82*).
[k] Using shielding derivatives of Ref. (*83*).
[m] Ref. (*84*).
[n] Ref. (*85*), page 23.
[p] Ref. (*73*).
[q] Ref. (*74*).
[r] Ref. (*86*).

derivatives $\sigma_e'$and σ_e'', and the mean bond and mean square displacements, is in good agreement with experiment. The calculations for the H_2 molecule have been carried out with full configuration interaction, and the last remaining minor discrepancy with experiments, including the isotope shifts, have been accounted for (*50*).

For comparison, some polyatomic molecules have been included in Table VIII. The most recent more accurate shielding surface for H_2O is reported by W. T. Raynes in a paper at this Symposium, and an independent calculation by Vaara et al. has recently appeared (*87*). Both calculations use MCSCF wavefunctions, however, the calculations by Raynes et al. have the benefit of more accurate force constants and perhaps a more accurate shielding value at the equilibrium geometry. With increasing accuracy of the shielding calculations for small molecules, rovibrational corrections are increasingly important for comparison with absolute shielding scales. In addition, accurate recovery of the experimental isotope shifts provide a stringent test of the derivatives of the theoretical shielding surface.

Intermolecular Effects on σ

There are several approaches to the calculations of intermolecular effects on shielding. Where the experiment can measure binary collision effects, as in the dilute gas, or in a van der Waals complex in a beam, comparisons with experiment can be achieved by calculations of the binary complex supermolecule shielding surface accompanied by determination of the intermolecular potential surface. In doing supermolecule shielding surface calculations, it is very important to make counterpoise corrections at every configuration (*88*). The second virial coefficients of the shielding, i.e., the density coefficient of the intermolecular chemical shift as a function of temperature can be calculated by doing the integration of the shielding surface over all configurations, weighted by the intermolecular potential function. The experimental work done in gases in the early 1970s were quantitatively interpreted for the first time in 1992 (*88*), in which ab initio calculations on the rare gas atom pairs were scaled to the heavier rare gas atoms by a corresponding states analog. Recently, the intermolecular shielding functions for Xe-CO_2, Xe-CO, Xe-N_2 have been reported (*89*). These are the first complete configuration shielding surfaces involving molecules, providing for the first time, the shape of a atom-linear molecule shielding surface. The Xe-Xe shielding function obtained using very large basis sets on the Xe atom, compared with that obtained from scaling the Ar-Ar shielding function, demonstrated excellent agreement over the range of configurations sampled in the integration for the experimental temperature range $\pm$100 K. The very recent CCSD(T) calculations on the Ne-Ne shielding surface established that there is a negligible effect of correlation on the shielding surface for this rare gas pair (Bühl, M.; Kaupp, M.; Malkin, V. G.; Malkina, O. L. *J. Comp. Chem.*, **1999**, in press.).

For Xe adsorbed in a zeolite, the system configurations involve variable occupancy so that a grand canonical ensemble average must be used in calculating the shielding. Here, one starts with a shielding surface which involves the Xe atom in question, all the atoms of the zeolite cage, plus a variable number of Xe atoms in the same cage. The approach that has been adopted for this system is to consider one rare gas atom in a supermolecule with a rigid fragment of the zeolite cage (*90*). The shielding values obtained at a multitude of configurations of the supermolecule can be expressed in a functional form as pairwise sums of distance dependent atom-atom (Xe-O, Xe-Na^+, etc.) shielding functions. The Xe-Xe contributions for the n Xe atoms in the cage and adjacent cages are likewise treated as pairwise sums of the ab initio σ function of the Xe-Xe distance. By using periodic boundary conditions, the Xe configurations in the entire zeolite can be simulated. Monte Carlo methods are used to determine distributions of configurations to find the grand canonical ensemble average of $\sigma(^{129}Xe)$ in this environment (*91*). In this simulation approach all parts of the medium are treated as atoms with orbitals and electrons, and the framework atoms of the real zeolite are located in the known positions based on diffraction data. The experimental Xe chemical shifts in mixtures of Xe with other molecules within the zeolite cages have been reproduced in this way (*92*).

Other approaches do not consider the overlap/exchange/dispersion interactions of the molecule with the surrounding molecules, but only the electric fields and gradients generated by the molecules of the medium. The linear and quadratic response to external static electric fields **F** and field gradients **∇F** were first introduced by Buckingham (*93*).

$$\sigma_{\alpha\beta}(\mathbf{F},\nabla\mathbf{F}) = \sigma_{\alpha\beta} + \sigma'_{\alpha\beta\gamma}\mathbf{F}_\gamma + (1/2)\sigma''_{\alpha\beta\gamma\delta}\mathbf{F}_\gamma\,\mathbf{F}_\delta + \sigma'_{\alpha\beta\gamma\delta}\mathbf{F}_{\gamma\,\delta} + \ldots \qquad (1)$$

The physical picture is a simple one: a static homogeneous magnetic field and intrinsic nuclear magnetic moments induce stationary currents within the electronic charge distribution, whereas a static electric field polarizes it. Therefore, the distortion induced in the electron clouds by the latter gives rise to additional effects, which can be rationalized in terms of response tensors of higher rank. In one approach to intermolecular effects, Dykstra, Oldfield, and Pearson calculate the average shielding of a molecule in a liquid in a molecular dynamics simulation by using Eq. (1), with the shielding polarizability parameters as constants obtained from a previous ab initio calculation (*94,95*). Following this method, Mikkelsen et al. used molecular dynamics simulations to obtain the ensemble averages of the electric field and field gradient at the ^{17}O and ^{1}H nuclear sites as a function of temperature. In this approach, the surrounding water molecules are represented entirely by fields and field gradients, i.e., they possess no electrons with which to generate exchange and overlap contributions to shielding at the H_2O molecule of interest. Including the effects of the different average geometry in the liquid from that of an isolated H_2O, plus the total contribution calculated from Eq. (1) leads to only 28% and 16% of the experimental gas-to-liquid shift of the ^{17}O and ^{1}H, respectively. Further, the temperature coefficient of the chemical shifts in the liquid is found to have the correct sign, but too small a magnitude compared to experiment (*96*).

Another method of calculating the electric field and field gradient contributions to the shielding in a molecule in a liquid is to use a reaction field model. This idea was first introduced by Buckingham, using the Onsager dipole model (*93*). Mikkelsen et al. present an update of the Buckingham model, considering a molecule within a spherical cavity in a homogeneous isotropic and linear dielectric medium, and include higher multipole terms (*97,98*). This model accounts for only a small portion of the observed gas-to-solution shifts of ^{13}C and ^{1}H in CH_4 in solvents of various dielectric constants. The sign of the gas-to-liquid shift for ^{17}O in water calculated by this method is opposite to that observed experimentally (9.4 ppm increase in shielding rather than the observed 36 ppm decrease in shielding in going to the liquid) (*97*).

Results for ^{1}H improved by introducing the surrounding 4 water molecules into the cavity, but still only leads to 45% of the gas-to-liquid shift for the ^{17}O nucleus (*97*). Likewise, this method fails to account for all of the gas-to-liquid shift of ^{19}F in fluoromethanes (*99*) and of ^{77}Se in H_2Se (*100*). Clearly, medium effects can not be treated accurately by using a reaction field model. The major problem with the above two approaches is that only the electric polarization effects are included in the model.

For associated liquids such as the above examples, the best approach appears to be the quantum cluster equilibrium (QCE) theory of Farrar and Weinhold (*101*). In this method, σ is calculated in various stable clusters (n-mers of the molecule, typically, n = 2-6). Each cluster is geometry optimized via an *ab initio* method and the shielding is calculated at the optimized geometry for each of the clusters. Excellent tests for the geometry and electronic properties of the clusters have been made against spectroscopy of van der Waals complexes. Using standard statistical thermodynamics and employing energies (vibrational frequencies, etc.) obtained from *ab initio* calculations, partition functions are calculated and the distribution of molecules among the various clusters can be obtained. This distribution is then used for determining population-weighted NMR shieldings. This scheme has been applied to ammonia (*102,103*), N-methyl formamide(*104*), CH_3NC (*105*) and N-methyl acetamide (*106*). The success in reproducing the temperature dependence is an indication that most of the temperature dependence is in the changes in the distributions of the molecules among various clusters.

Ab Initio Calculations of Shielding in Proteins

The nuclei in the same amino acid residue in different parts of a protein exhibit uniquely different chemical shifts. Secondary structural information is therefore encoded in the NMR shifts. Herein lies the challenge: how to predict chemical shifts for given secondary (and tertiary) structures. Then, there is the even more difficult challenge of how to unencode the experimental chemical shifts to deduce or refine secondary structures of proteins. Earlier work on intermolecular effects had focused on electrostatic effects and magnetic anisotropy, and considered dispersion contributions to be paramount, using the Raynes, Buckingham, and Bernstein framework (*107*). Within this framework, research groups working on chemical shifts in proteins had been largely concerned with proton chemical shifts (*108*), particularly electrostatic effects, ring currents and magnetic anisotropy using parameterized empirical models. These magnetic anisotropy effects are of very little consequence for chemical shifts of nuclei other than protons. The earlier interpretations of protein chemical shifts reported at the Maryland conference were highly invested in the electric polarization idea, that the way to calculate these was by calculating shielding polarizabilities of the observed nuclei coupled with calculations of local electric fields and electric field gradients generated by all the atoms in the surroundings of the nucleus of interest, as in Eq. (1) (*94*).

An alternative approach is to consider the shielding surface, i.e., the nuclear magnetic shielding as a function of nuclear coordinates of a molecule (distances, bond angles, torsion angles, ...) and to calculate traces on the shielding hypersurfaces to elicit the main factors influencing chemical shifts in proteins. For a ^{13}C nucleus, for example, there was a hint already in experiments that the torsion angles were important in distinguishing helix from sheet (*109*). Therefore, the model system had to permit exploration of the dependence of ^{13}C, ^{15}N, ^{17}O shielding on local geometry, especially torsion angles, as well as on hydrogen bonding. Furthermore, the effects of the more distant parts of the protein had to be taken into account.

Since the Maryland meeting, de Dios, Oldfield, and others have established that ab initio calculations of ^{13}C and ^{15}N (and ^{19}F) chemical shifts in proteins can be done successfully using a model system, specifically, a molecular fragment having the essential attributes of local geometry and short range interactions, including hydrogen-bonding, calculated at a sufficiently high ab initio level while surrounded by the rest of the protein represented with a simpler description such as point charges at atomic locations (*110-115*). In this approach to the calculation of a property, the local geometry and electronic structure (including a hydrogen bonding partner) are represented in the greatest detail by a judiciously chosen molecular fragment with locally enhanced basis sets, and good ab initio calculations at various local geometries (in the case of proteins the torsion angles phi and psi are paramount, but also chi has been considered). The remainder of the protein (or the remainder of the crystal in the case of amino acid shielding tensor calculations) is represented at a lower level of detail, the long range intermolecular effects treated by point charges at each atom of the residues located within a sphere, for example. All short range intermolecular effects are of course treated correctly ab initio within the fragment. Others who have later adopted the model originated by de Dios et al. have given this approach the name "embedded cluster", and of course, the model can and has been generalized to permit the surrounding framework to be treated by molecular mechanics, and to include even further out, the solvent molecules under molecular dynamics. Thus, the initial work in this area has influenced the direction of research in this and other fields, with other researchers adopting the approach and embellishing further. This approach to calculating chemical shifts in very large molecules (or networks) such as proteins is discussed with new examples in the paper by Oldfield in this Symposium proceedings.

The prediction that the component of the ^{13}C(alpha) shielding tensor in a peptide along the C-H bond has a larger range of values depending on the torsion angles phi, psi than the range of the isotropic shieldings for the same nucleus (*116*) has been corroborated experimentally by Bax et al. (*117*). This is an important development. First of all, theoretical calculations (*116*) have demonstrated that the experimental method of Bax et al. which depends on the cross relaxation terms involving chemical shift anisotropy and dipolar coupling does indeed lead to good values for the shielding component along the C-H bond direction. Second, the theoretical calculations demonstrate that this is more sensitively associated with the local geometry than is the isotropic shielding of the same ^{13}C nuclear site, thus providing complementary and more sensitive information for deducing structure.

Model Calculations in Zeolites and Inorganic Solids.

One of the most difficult challenges for theoretical treatments of the NMR chemical shift is in extended networks such as inorganic solids. Indeed, while the immediate future approaches to isolated small molecule chemical shifts were being laid out even at the time of the Maryland conference, the theoreticians attending did not have any words of comfort for the experimentalists who brought fascinating trends and puzzles in covalent solids.

One approach to solids that are not molecular crystals is to carve out a fragment of the solid as a model system, with appropriate terminations of what would otherwise be dangling bonds This is a logical approach for an ionic solid, where replacing the distant neighbors by point charges is an excellent approximation. The paper by de Dios in this Symposium provides a very nice example for the theoretical predictions of trends in ^{87}Rb shielding in various Rb doped into alkali ion sites in various alkali halide crystals, using the full complement of nearest neighbors in the model fragment. In aluminosilicates the size of the model fragment used in the calculations of the ^{29}Si shielding tensor has varied from the very early 3 heavy atoms (one O and two T atoms) to the larger fragments containing up to 15 heavy atoms (*118-120*). There is a controversy about the extent to which geometry optimization of the model fragment should be carried out, or even whether it makes sense to use anything different than accurate experimental T-O distances and T-O-T bond angles in the actual aluminosilicate. Similarly, for the calculation of the ^{13}C shielding tensor of a small organic molecule (acetone) adsorbed in a zeolite, a model fragment that has been used to represent the HZSM5 zeolite has one central oxygen bridging the Si and Al moieties each of which is terminated with OSiH3 groups, i.e., the $(H_3SiO)_3SiOHAl(OSiH_3)_3$ fragment (*121*). In contrast to the more usual practice of using smaller terminating groups of the type $OSiH_3$, $OSi(OH)_3$ terminating groups have been used by Farrar et al. for aluminosilicate glasses (*122*). In the model fragment used in this work, the central Si or Al is surrounded by four groups of the type $OSi(OH)_3$ or $OAl(OH)_3$ or OH, and counter ions to balance the charge. The calculated ^{29}Si and ^{27}Al shielding in various clusters representative of the Q^n species (Q is the central tetrahedral Al or Si and *n* is the number of oxygens that bridge the central atom to tetrahedral Al or Si sites) in aluminosilicate glasses can accurately predict experimentally observed trends in these glasses (*122*).

In inorganic solids like carbides and nitrides, Tossell has found that the theoretical shielding results are sensitive to the second nearest neighbors in the solid (*123*), i. e., these systems demand a larger fragment that includes not only the nearest neighbors but also the second nearest neighbors. Tossell has calculated the effects arising from second nearest neighbors on the ^{15}N shielding in crystalline α- and β-Si_3N_4 are as large as 80 ppm (*123*). The shielding calculated with the largest fragment, $Si_9N_9H_{21}$, is still 50 ppm more shielded than the experimental shielding in β-Si_3N_4, indicating that the this size fragment is still an inadequate model system.

A completely different approach is to start with basis functions that already have the periodic nature of the solid and extend indefinitely, for example, using basis functions of plane waves. This method has been used very successfully for calculations of electric field gradients (*124*) and has only very recently been applied to shielding (*125*).

Fine Geometry Effects on Shielding, NMR Refinement of Diffraction Results.

Complete assignment of ^{13}C shielding tensors in the entire molecule from single crystal studies has been developed to the highest level by D. M. Grant and co-workers. Single crystal NMR measurements provide complete tensor information and the increased resolution in the 2D chemical shift-chemical shift correlation method and the multiple-axis sample reorientation mechanism developed in this group permits the study of crystals containing 50-100 magnetically different nuclei per unit cell. The redundant data provided by the crystal symmetry increases the number of independent observations of a given shift tensor, enabling the calibration of the magnetic field directions in the single crystal sample and improving the precision of the shift measurements to better than 0.5 ppm. These experimental techniques are combined with ab initio calculations which provides a high level of agreement between calculated and experimental tensors for ^{13}C in a large number of polycyclic aromatic compounds and sugars. Only the structural parameters (bond distances and angles) limit the level of agreement. This means that *ab initio* calculations and measurements together can be used to address certain fine details of solid-state structure, surpassing the accuracy of x-ray data. In a dramatic application of ab initio calculations toward the elucidation of the shielding tensor and its dependence on molecular geometry, it has been shown that the departures from D_{2h} symmetry observed in the experimental chemical shift tensor components of naphthalene in the single crystal (leading to the reduced symmetry, C_i) could be accounted for entirely by small geometrical variations that are smaller than the error bars of the x-ray diffraction parameters (*126*). This leads to the conclusion that solid state NMR methods can be used for refining structural data, especially in those cases where imperfections such as translational disorder or occlusion of molecular impurities degrade diffraction data while having no effect on chemical shift data (*127*).

In summary, the challenges in the computations of nuclear magnetic shielding are being met, and the rich and diverse structural and dynamical NMR chemical shift data are beginning to yield to rigorous, accurate interpretation. This Symposium presents some of the most recent breakthroughs in the field.

Literature Cited

1. *Nuclear Magnetic Shieldings and Molecular Structure*; Tossel, J. A., Ed.; NATO ASI Series C; Kluwer: Dordrecht, 1993, Vol. 386.
2. Keith, T. A.; Bader, R. F. W. *Chem. Phys. Lett.* **1993,** *210*, 223.
3. Lazzeretti, P.; Malagoli, M.; Zanasi, R. *Chem. Phys. Lett.* **1994**, *220*, 299.
4. Coriani, S.; Lazzeretti, P.; Malagoli, M.; Zanasi, R. *Theor. Chim. Acta* **1994**, *89*, 181.
5. Zanasi, R.; Lazzeretti, P.; Malagoli, M.; Piccinini, F. *J. Chem. Phys.* **1995**, *102*, 7150.
6. Malkin, V. G.; Malkina, O. L.; Eriksson, L. A.; Salahub, D. R. In *Modern Density Functional Theory: A Tool for Chemistry*; Politzer, P.; Seminario, J. M., Eds.; Theoretical and Computational Chemistry; Elsevier: Amsterdam, 1995, Vol. 2; pp. 273-347.
7. Malkin, V. G.; Malkina, O. L. ; Salahub, D. R. *Chem. Phys. Lett.* **1993**, *204*, 80.
8. Lee, A. M.; Handy, N. C.; Colwell, S. M. *J. Chem. Phys.* **1995**, *103*, 10095.
9. Schreckenbach, G.; Ziegler, T. *J. Phys. Chem.* **1995**, *99*, 606.
10. Rauhut, G.; Puryear, S.; Wolinski, K.; Pulay, P. *J. Phys. Chem.* **1996**, *100*, 6310.

11. Cheeseman, J. R.; Trucks, G. W.; Keith, T. A.; Frisch, M. J. *J. Chem. Phys.* **1996,** *104*, 5497.
12. Kaupp, M.; Malkina, O. L.; Malkin, V. G. *J. Chem. Phys.* **1997,** *106*, 9201.
13. Bühl, M. *Chem. Phys. Lett.* **1997**, *267*, 251.
14. Schreckenbach, G.; Ziegler, T. *Int. J. Quantum Chem.* **1997**, *61*, 899.
15. Kaupp, M.; Malkin, V. G.; Malkina, O. L.; Salahub, D. R. *Chem. Phys.Lett.* **1995,** *235*, 382.
16. Ruiz-Morales, Y.; Schreckenbach, G.; Ziegler, T. *J. Phys. Chem.* **1996**, *100*, 3359.
17. Kaupp, M. *Chem. Eur. J.* **1996**, *2*, 348.
18. Kaupp, M.; Malkin, V. G.; Malkina, O. L.; Salahub, D. R. *Chem. Eur. J.* **1996,** *2*, 24.
19. Salzmann, R.; Kaupp, M.; McMahon, M.; Oldfield, E. *J. Am. Chem. Soc.* **1998,** *120*, 4771.
20. Kaupp, M. *J. Am. Chem. Soc.* **1996**, *118*, 3018.
21. Ruiz-Morales, Y.; Ziegler, T. *J. Phys. Chem.A* **1998**, *102*, 3970.
22. Kaupp, M. *Chem. Ber.* **1996**, *129*, 535.
23. Ruiz-Morales, Y.; Schreckenbach, G.; Ziegler, T. *Organometallics* **1996**, *15*, 3920.
24. Schreckenbach, G.; Ruiz-Morales, Y.; Ziegler, T. *J. Chem. Phys.* **1996**, *104*, 8605.
25. Ruiz-Morales, Y.; Schreckenbach, G.; Ziegler, T. *J. Phys. Chem.* **1997**, *101*, 4121.
26. Kaupp, M.; Malkin, V. G.; Malkina, O. L.; Salahub, D. R. *J. Am. Chem. Soc.* **1995**, *117*, 1851.
27. Schreckenbach, G.; Ziegler, T. *Int. J. Quantum Chem.* **1996**, *60*, 753.
28. Nakatsuji, H.; Takashima, H.; Hada, M. *Chem. Phys. Lett.* **1995**, *233*, 95.
29. Nakatsuji, H.; Nakajima, T.; Hada, M.; Takashima, H.; Tanaka, S. *Chem. Phys. Lett.* **1995**, *247*, 418.
30. Nakatsuji, H.; Hada, M.; Tejima, T.; Nakajima, T.; Sugimoto, M. *Chem. Phys. Lett.* **1996**, *249*, 284.
31. Kaneko, H.; Hada, M.; Nakajima, T.; Nakatsuji, H. *Chem. Phys. Lett.* **1996**, *261*, 1.
32. Ballard, C. C.; Hada, M.; Kaneko, H.; Nakatsuji, H. *Chem. Phys. Lett.* **1996,** *254*, 170.
33. Nakatsuji, H.; Hada, M.; Kaneko, H.; Ballard, C. C. *Chem. Phys. Lett.* **1996,** *255*, 195.
34. Hada, M.; Kaneko, H.; Nakatsuji, H. *Chem. Phys. Lett.* **1996**, *261*, 7.
35. Malkin, V. G.; Malkina, O. L. ; Salahub, D. R. *Chem. Phys. Lett.* **1996**, *261*, 335.
36. Kaupp, M.; Malkina, O. L.; Malkin, V. G. *Chem. Phys. Lett.* **1996,** *265*, 55.
37. Kaupp, M.; Malkina, O. L.; Malkin, V. G.; Pyykko, P. *Chem. Eur. J.* **1998**, *4*, *118.*
38. Kirpekar, S.; Jensen, H. J. A.; Oddershede, J. *Theoret. Chim. Acta* **1997,** *95*, 35.
39. Vaara, J.; Ruud, K.: Vahtras, O.; Ågren, H.; Jokisaari, J. *J. Chem. Phys.* **1998,** *109*, 1212.
40. Gauss, J. *Ber. Bunsenges. Phys. Chem.* **1995,** *99*, 1001.

41. Jameson, A. K.; Jameson, C. J. *Chem. Phys. Lett.* **1987**, *134*, 461.
42. Gauss, J. *J. Chem. Phys.* **1993**, *99*, 3629.
43. Gauss, J. *Chem. Phys. Lett.* **1992**, *191*, 614.
44. Gauss, J. *Chem. Phys. Lett.* **1994**, *229*, 198.
45. Gauss, J.; Stanton, J. F. *J. Chem. Phys.* **1995**, *103*, 3561.
46. Gauss, J.; Stanton, J. F. *J. Chem. Phys.* **1996**, *104*, 2574.
47. Ruud, K.; Helgaker, T.; Kobayashi, R.; Jørgensen, P.; Bak, K. L.; Jensen, H. J. A. *J. Chem. Phys.* **1994**, *100*, 8178.
48. Barszczewicz, A.; Helgaker, T.; Jaszunski, M.; Jørgensen, P.; Ruud, K. *J. Magn. Reson. A* **1995**, *114*, 212.
49. Sundholm, D.; Gauss, J.; Schäfer, A. *J. Chem. Phys.* **1996**, *105*, 11051.
50. Sundholm, D.; Gauss, J.; Ahlrichs, R. *Chem. Phys. Lett.* **1995**, *243*, 264.
51. Jameson, C. J. In *Nuclear Magnetic Resonance*; Webb, G. A., Ed.; Specialist Periodical Reports; The Royal Society of Chemistry: Cambridge, UK, 1997, Vol. 26; pp. 46-87. (p. 55,56)
52. Wolinski K.; Haacke R.; Hinton J. F.; Pulay P. *J. Comput. Chem.* **1997,** *18*, 816
53. Enevoldsen, T.; Oddershede, J. *Mol. Phys.* **1995**, *86*, 235.
54. Wofsy, S. C.; Muenter, J. S.; Klemperer, W. *J. Chem. Phys.* **1971**, *55*, 2014.
55. Jameson, C. J.; Jameson, A. K.; Burrell, P. M. *J. Chem. Phys.* **1980**, *73*, 6013.
56. Jameson, C. J.; Jameson, A. K. *Chem. Phys. Lett.* **1987**, *135*, 254.
57. Kodweiss, J.; Lutz, O.; Messner, W.; Mohn, K. R.; Nolle, A.; Stutz, B.; Zepf, D. *J. Magn. Reson.* **1981**, *43*, 495.
58. Bühl, M.; Gauss, J.; Stanton, J. F. *Chem. Phys. Lett.* **1995**, *241*, 248.
59. Jameson, C. J. In *Nuclear Magnetic Resonanc*; Webb, G. A., Ed.; Specialist Periodical Reports; The Royal Society of Chemistry: London, 1996; Vol. 25, pp. 39-82.
60. Jokisaaari, J.; Lazzeretti, P.; Pyykkö, P. *Chem. Phys.* **1988**, *123*, 339.
61. Malli, G.; Froese, C. *Intl. J. Quantum Chem.* **1967**, *1S*, 95.
62. Gierke, T. D.; Flygare, W. H. *J. Am. Chem. Soc.*, **1972**, *94*, 7277.
63. Chen, Q.; Freeman, D. L.; Odom, J. D.; Ellis, P. D. Poster Abstracts, 36th Experimental NMR Conference, Boston, 1995, p. 333.
64. Higashioji, T.; Hada, M.; Sugimoto, M.; Nakatsuji, H. *Chem. Phys.* **1996**, *203*, 159.
65. Magyarfalvi, G.; Pulay, P. *Chem. Phys. Lett.* **1994**, *225*, 280.
66. Bühl, M.; Thiel, W.; Fleischer, U.; Kutzelnigg, W. *J. Phys. Chem.* **1995**, *99*, 4000.
67. Jameson, C. J. In *Nuclear Magnetic Resonance*; Webb, G. A., Ed.; Specialist Periodical Reports; The Royal Society of Chemistry: Cambridge, UK, 1989, Vol. 18; pp. 1-33. (Table 16)
68. Brown, R. D.; Dyall, K. G.; Godfrey, P. D.; McNaughton, D. *Mol. Phys.*, **1987**, *61*, 783.
69. Bühl, M. *Magn. Reson. Chem.* **1996**, *34*, 782.
70. Jameson, C. J.; Osten, H. J. *Annu. Rep. NMR Spectrosc.* **1986**, *17*, 1.
71. Jameson, C. J. In *Theoretical Models of Chemical Bonding*; Maksic, Z. B., Ed.; Molecular Spectroscopy, Electronic Structure, and Intramolecular Interactions, Part 3; Springer-Verlag: Berlin, 1991, pp. 457-519.
72. Fowler, P. W.; Riley, G.; Raynes, W. T. *Mol. Phys.* **1981,** *42,* 1463.

73. Jameson, C. J.; de Dios, A. C.; Jameson, A. K. *J. Chem. Phys.* **1991**, *95*, 1069.
74. Jameson, C. J.; de Dios, A. C.; Jameson, A. K. *J. Chem. Phys.* **1991**, *95*, 9042.
75. Lazzeretti, P.; Zanasi, R.; Sadlej, A. J.; Raynes, W. T. *Mol. Phys.* **1987**, *62*, 605.
76. Jameson, A. K.; Schuett, K.; Jameson, C. J.; Cohen, S. M.; Parker, H. *J. Chem. Phys.* **1977**, *67*, 2821.
77. Jameson, C. J.; Jameson, A. K.; Wille S.; Burrell, P. M. *J. Chem. Phys.* **1981**, *74*, 853.
78. Fowler, P. W.; Zanasi, R.; Cadioli, B.; Steiner, E. *Chem. Phys. Lett.* **1996**, *251*, 132.
79. Raynes, W. T.; Panteli, N. *Mol. Phys.* **1983**, *48*, 439.
80. Jameson, C. J.; Osten, H. J. *J. Chem. Phys.* **1984**, *81*, 4300.
81. Hindermann, D. K.; Cornwell, C. D. *J. Chem. Phys.* **1968**, *48*, 4148.
82. Jameson, C. J.; Jameson, A. K.; Burrell, P. M. *J. Chem. Phys.* **1980**, *73*, 6013.
83. Chesnut, D. B.; Foley, C. K. *J. Chem. Phys.* **1986**, *85*, 2814.
84. Osten, H. J.; Jameson, C. J. *J. Chem. Phys.* **1985**, *82*, 4595.
85. Jameson, C. J. In *Nuclear Magnetic Resonance*; Webb, G. A., Ed.; Specialist Periodical Reports; The Royal Society of Chemistry: Cambridge, UK, 1989, Vol. 19; pp. 1-33. (page 23)
86. Raynes, W. T.; Fowler, P. W.; Lazzeretti, P.; Zanasi, R.; Grayson, M. *Mol. Phys.* **1988**, *64*, 143.
87. Vaara, J.; Lounila, J.; Ruud, K; Helgaker, T. *J. Chem. Phys.* **1998**, *109*, 8388.
88. Jameson, C. J.; de Dios, A. C. *J. Chem. Phys.* **1992**, *97*, 417.
89. de Dios, A. C.; Jameson, C. J. *J. Chem. Phys.* **1997**, *107*, 4253.
90. Jameson, C. J.; Lim, H.-M. *J. Chem. Phys.* **1995**, *103,* 3885.
91. Jameson, C. J.; Jameson, A. K.; Baello, B. I.; Lim, H. M. *J. Chem. Phys.* **1994**, *100*, 5965.
92. Jameson, C. J.; Jameson, A. K.; Lim, H. M. *J. Chem. Phys.* **1996**, *104*, 1709.
93. Buckingham, A. D. *Can. J. Chem.* **1960**, *38*, 300.
94. Augspurger, J. D.; Dykstra, C. E.; Oldfield, E.; Pearson, J. G. In *Nuclear Magnetic Shieldings and Molecular Structure*; Tossell, J. A. Ed.; NATO ASI Series C Vol. 386, Kluwer: Dordrecht, 1993, pp. 75-94.
95. Pearson, J. G.; Oldfield, E.; Lee, F. S.; Warshel, A. *J. Am. Chem. Soc.* **1993**, *115*, 6851.
96. Augspurger, J. D.; Dykstra, C. E. *J. Phys. Chem.* **1991**, *95*, 9230
97. Mikkelsen, K. V.; Ruud, K.; Helgaker, T. *Chem. Phys. Lett.* **1996**, *253*, 443.
98. Mikkelsen, K. V.; Jorgenson, P.; Ruud, K.; Helgaker, T. *J. Chem. Phys.* **1997**, *106*, 1170.
99. Åstrand, P. O.; Mikkelsen, K. V.; Ruud, K.; Helgaker, T. *J. Phys. Chem.*, **1996**, *100*, 19771.
100. Åstrand, P. O.; Mikkelsen, K. V.; Jørgensen, P.; Ruud, K.; Helgaker, T. *J. Chem. Phys.* **1998,** *108*, 2528.
101. Ludwig, R.; Weinhold, F.; Farrar, T. C.; *J. Chem. Phys.* **1995,** *102*, 5118.
102. Ludwig, R.; Weinhold, F.; Farrar, T. C.; *Ber. Bunsenges. Phys. Chem.* **1998,** *102*, 197
103. Ludwig, R.; Weinhold, F.; Farrar, T. C. *Ber. Bunsenges. Phys. Chem.* **1998,** *102*, 205.
104. Ludwig, R.; Weinhold, F.; Farrar, T. C.; *J. Chem. Phys.* **1997,** *107*, 499.

105. Stringfellow, T. C.; Farrar, T. C. *Spectrochim. Acta* **1997,** *53*, 2425.
106. Ludwig, R.; Weinhold, F.; Farrar, T. C.; *J. Phys. Chem.* **1997,** *101*, 8861.
107. Raynes, W. T.; Buckingham, A. D.; Bernstein, H. J. *J. Chem. Phys.* **1962**, *36*, 3481.
108. Case, D. A. *J. Biomol. NMR* **1995**, *6*, 341.
109. Spera, S.; Bax, A. *J. Am. Chem. Soc.*, **1991**, *113*, 5490.
110. de Dios, A. C.; Pearson, J. G.; Oldfield, E. *Science* **1993**, *260*, 1491.
111. de Dios, A. C.; Oldfield, E. *Chem. Phys. Lett.* **1993**, *205*, 108.
112. de Dios, A. C.; Pearson, J. G.; Oldfield, E. *J. Am. Chem. Soc.* **1993**, *115*, 9768.
113. de Dios, A. C.; Oldfield, E. *J. Am. Chem. Soc.* **1994**, *116*, 5307.
114. de Dios, A. C.; Laws, D. D.; Oldfield, E. *J. Am. Chem. Soc.* **1994**, *116*, 7784.
115. Le, H.; Pearson, J. G.; de Dios, A.C.; Oldfield, E. *J. Am. Chem. Soc.* **1995**, *117*, 3800.
116. Walling, A. E.; Pargas, R. E.; de Dios, A. C. *J. Phys. Chem.* **1997**, *101*, 7299.
117. Tjandra, N.; Bax, A. *J. Am. Chem. Soc.* **1997**, *119*, 9576.
118. Bussemer, B.; Schroder, K. P.; Sauer, J. *Solid State Nucl. Magn. Reson.* **1997**, *9,* 155.
119. Krossner, M.; Sauer, J. *J. Phys. Chem.*, **1996**, *100*, 6199.
120. Pullumbi, P.; Goursot, A.; Valerio, G.; Vetrivel, R.; Salahub, D. R. Poster Abstracts, International Conference on Quantum Chemcial Calculations of NMR and EPR Parameters, BSmolenice, Slovak Republic, 1998, p. P5.
121. Haw, J. F.; Xu, T.; Nicholas, J. B.; Goguen, P. W. *Nature*, **1997**, *389*, 832.
122. Sykes, D.; Kubicki, J. D.; Farrar, T. C.; *J. Phys. Chem.* **1997,** *101*, 2715.
123. Tossell, J. A. *J. Magn. Reson.* **1997,** *127*, 49.
124. Winkler, B.; Blaha, P.; Schwarz, K; *American Mineralogist* **1996,** *81*, 545.
125. Mauri, F; Pfrommer, B.; Louie, S. G. *Phys. Rev. Lett.* **1996,** *77*, 5300.
126. Facelli, J.; Grant, D. M. *Nature* **1993**, *365*, 325.
127. Grant, D. M.; Liu, F.; Iuliucci, R. J.; Phung, C. G.; Facelli, J. C.; Alderman, D. W. *Acta Crystallog. B* **1995**, *51*, 540.

Chapter 2

Modeling NMR Chemical Shifts in Polymers and Amorphous Matter

Isao Ando, Shigeki Kuroki, Hiromichi Kurosu[1], Masahito Uchida[2], and Takeshi Yamanobe[3]

Department of Chemistry and Materials Science, Tokyo Institute of Technology, Ookayama, Meguro-ku, Tokyo 152–8552, Japan

Two theoretical approaches for calculating NMR chemical shift of polymers and its application to structural characterization have been described. One is that model molecules such as dimer, trimer, etc., as a local structure of polymer chains, are in the calculation by combining quantum chemistry and statistical mechanics. This approach has been applied to polymer systems in the solution, amorphous and solid states. Another approach is to employ the tight-binding molecular orbital theory to describe the NMR chemical shift and electronic structure of infinite polymer chains with periodic structure. This approach has been applied to polymer systems in the solid state. These approaches have been successfully applied to structural characterization of polymers

The chemical shift is one of most important NMR parameters(1). The NMR chemical shifts provide detailed information on the structure and electronic structures of polymers in the crystalline state and amorphous state(2).

A polymer chain has an enormous number of chemical bonds. For this, in the solution and amorphous states the NMR chemical shifts of polymers are often the averaged values for all of the possible conformations because of rapid interconversion by rotation about chemical bonds. In solids, however, chemical shifts are often characteristic of specific conformations because of strongly restricted rotation about the bonds. The NMR chemical shift is affected by a change of the electronic structure through the structural change(2). Solid state NMR chemical

[1]Current address: Department of Textile and Apparel Science, Nara Women's University, Kita-Uoya Nishimachi, Nara 630–8263, Japan.
[2]Current address: Tokyo Research Center, Tosoh Company, Hayakawa, Ayase, Kanagawa 252–1123, Japan.
[3]Current address: Department of Materials Engineering, Gunma University, Kiryu, Gunma 376–8515, Japan.

shifts, therefore, give useful information about the electronic structure of a polymer or polymers with a fixed structure. Further, the full chemical shift tensor components can be often determined. The complete chemical shift tensor provides information about the local symmetry of the electron cloud around the nucleus and so provides more detailed knowledge of the electronic structure of the polymer compared with the average chemical shift. In order to establish the relationship between the NMR chemical shift and the electronic structure of polymers, it is necessary to use a sophisticated theoretical method which can take into account the characteristics of polymers.

Some methodologies for obtaining structures and electronic structures of polymers both in the solution, amorphous and solid states by a combination of the observation and calculation of NMR chemical shifts have been developed, and have been applied them to various polymer systems(3-12). Theoretical calculations of NMR chemical shifts for polymer systems have been performed mainly by two approaches. One is that model molecules such as dimer, trimer, etc., as a local structure of polymer chains, are used in the calculation by combining quantum chemistry and statistical mechanics. In particular, this approach has been applied to polymer systems in the solution and amorphous states(2). However, it sometimes has to be recognized that the results of quantum chemical calculations on model molecules are not readily applicable to polymers in the crystalline state because of the existence of long-range intrachain interactions and interchain interactions. Electrons are constrained to a finite region of space in small molecules whereas this is not necessarily the case for polymers, and thus some additional approaches are required. Another approach is to employ the tight-binding molecular orbital(TB MO) theory, which is well known in the field of solid state physics, to describe the electronic structure of linear polymers with periodic structure within the framework of the linear combination of atomic orbitals(LCAO) approximation for the electronic eigenfunctions(4-12). These approaches have led to the determination of the structure and/or electronic structure of polymer systems including polypeptides in the solution, amorphous and solid states. The essence of these two approaches will be described.

APPROACH USING MODEL MOLECULES

Origin of NMR Chemical Shift: The chemical shift of an atom depends upon its electronic and molecular environments(1). The chemical shift for atom A can be calculated by the sum of the following terms.

$$\sigma_A = \sigma^d + \sigma^p + \sigma' \qquad (1)$$

where σ^d is the diamagnetic term, σ^p the paramagnetic term, and σ' the other term which comes from the magnetic anisotropy effect, polar effect and ring current effect. For nuclei with $2p$ electrons such as ^{13}C, ^{15}N,etc., the relative chemical shift is predominantly governed by the paramagnetic term, and for the ^1H nucleus by the first and third terms.

The paramagnetic term is expressed as a function of excitation energy, bond order, and electron density according to the sum-over-state(SOS) method in the simple form as follows.

$$\sigma^p = - C \Sigma <r^{-3}>_{2p} (E_m-E_n)^{-1}Q \qquad (2)$$

where E_m-E_n is the singlet-singlet excitation energy of the *n*th occupied and the *m*th unoccupied orbitals, and Q is a factor including the bond order and electron density. The quantity $<r^{-3}>_{2p}$ is the spatial dimensions for a 2*p* electron while C is the coefficient incorporating the universal constants. This term is calculated by semi-empirical MO or ab initio MO method. The former has some features to give the substantial aspect of the chemical shift behavior associated with the structure and/or the electronic structure. The diamagnetic term is estimated from the calculated electron density. Using these procedures, chemical shifts σ_i of the model molecule with any specified conformation have been calculated.

Medium Effects on NMR Chemical Shift: Most MO calculations of nuclear shielding relate to the case of a molecule in a vacuum. For nuclei forming the molecular skeleton, such as ^{13}C, and nuclei with small shielding ranges, such as ^{1}H, this may not be an unreasonable approximation. This is particularly true if comparison of the theoretical results is to be made with the experimental data taken on a molecule dissolved in an inert solvent.

More reactive atoms especially those with lone pair electrons, such as ^{14}N, ^{15}N, ^{17}O and ^{19}F, are very likely to have their nuclear shieldings influenced by interactions with solvent molecules. Such interactions may be specific, e.g. hydrogen bonding, or nonspecific, e.g. polarisability/polarity, or perhaps a combination of specific and nonspecific solute-solvent interactions. An empirical procedure has been developed for quantitatively unraveling the contributions made to the shielding of solute nuclei by specific and nonspecific interactions.

Nonspecific solute-solvent effects on nuclear shielding may be described by MO calculations which include the influence of solvent polarity/polarisability by means of the solvent dielectric constant, ε. This approach emplays by the use of the solvaton model together with the semi-empirical MO basis set(13-18). This method has successfully accounted for the observed variations of ^{13}C and ^{15}N shielding for a number of solute molecules in solvents with a variety of values of ε. The solvaton model has been demonstrated to correctly predict both the sign and magnitude of the solute shielding variation as the value of ε of the medium is changed. This approach has lead to a good understanding of the various solute-solvent interactions on the ^{13}C chemical shifts of polymers such as poly(vinyl chloride) associated with the stereochemical configurations(15).

Specific solute-solvent interactions, such as hydrogen bonding or protonation, may be included in the calculation of the shielding of solute nuclei by a supermolecule approach. The appropriate structure of the solute-solvent supermolecule may be obtained by the use of molecular mechanics simulations. At the semi-empirical MO level this approach has been successfully used to describe the effects of hydrogen bonding on the nuclear shielding of small molecules.

Applications to Polymers: A polymer chain can assume an enormous number of conformations because of various possibilities of rotation around the chain bonds, which is undergoing molecular motion(19). Thus, the factors governing the appearance of the NMR spectra include the structures, the relative energies of the rotational isomers, the chemical shifts and spin couplings. If molecular motion in the polymer chain is extremely slow on the NMR timescale, the spectrum represents the superposition of the spectra for various conformations. However, if the rotation around the chain bonds is very fast on the NMR timescale, the experimentally observable chemical shift for nucleus A is given as(1,2,20-24)

$$\langle \sigma_A \rangle = \sum_{i=1}^{n} P_i \sigma_i \tag{3}$$

where the numerical indices refer to the preferred conformations, and P_i and σ_i are the probability of occurrence and the chemical shift of the preferred conformation i, respectively. This indicates that the chemical shift of a given nucleus of a combination of a quantum chemical method and a statistical mechanical method as described below. The probability P_i of the preferred conformations in various configurations in vinyl polymers as a function of the racemic units necessary for the calculation of the averaged chemical shift $<\sigma>$ is estimated by statistical mechanics for a polymer chain. For convenience, the planar conformation in any specified configuration of a vinyl polymer is shown in Fig.1. The statistical weight factors are η, 1 and τ for *trans*(T), positive *gauche*(G^+) and negative *gauche* (G^-) conformations, respectively. The statistical weight factor for G^+G^- is ω(the pentane effect)(19). The statistical weight matrices for the characterization of the array of chain conformations of vinyl polymers can be used. The purpose of calculating the probability that the pair of skeletal bonds within the k th dyad, the k th triad, and so on in a chain are in particular rotational states is that such statistical weights are used to construct statistical matrices U = U'U". In such matrices, rows are associated with rotational states about bond i-1 and columns with rotational states about bond i. The statistical weight matrices in the case of pairs of bonds adjoining CHR groups are designated by U". These matrices are expressed as

$$U' = \begin{pmatrix} \eta & 1 & \tau \\ \eta & \omega & \tau \\ \eta & 1 & \tau\omega \end{pmatrix} \tag{4}$$

for the first skeletal bond of a dyad and

$$U''_m = \begin{pmatrix} \eta\omega & 1 & \tau\omega \\ \eta & \omega & \tau\omega \\ \eta\omega & \omega & \tau\omega^2 \end{pmatrix} \tag{5}$$

$$U''_r = \begin{pmatrix} \eta & \omega & \tau\omega \\ \eta\omega & 1 & \tau\omega \\ \eta\omega & \omega & \tau\omega^2 \end{pmatrix} \tag{6}$$

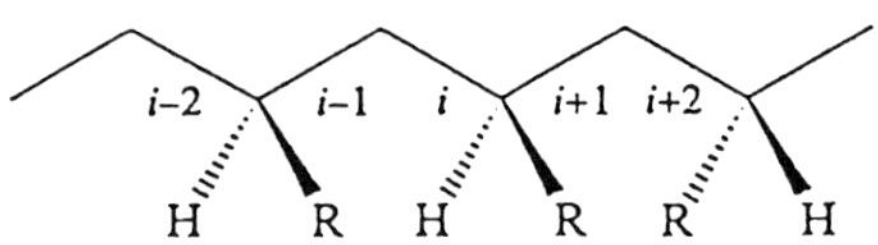

Fig.1 The Planar conformation of any specified configuration of a vinyl polymers

for the second; the former matrix, eq.(5), is for a meso(m) dyad and the latter(eq.(6)) for a racemic(r) one. The configuration partition function Z for the entire chain can be expressed as the sum of the statistical weights for all molecular conformations of the chain consisting of n bonds or x =n/2 repeating units

$$Z = J^*(\prod_{i/2=1}^{n/2-1} U_i' U_i'')J \qquad (7)$$

where J* =(100) and J is the transpose of (111). Let β and γ denote indices from the set T, G^+ and G^-. The probability $P_{\beta\gamma k''}$ that the pair of skeletal bonds within the k th dyad are in the rotational states β and γ, respectively, is the ratio of the sum of the statistical weights for all conformations. It is given by

$$P_{\beta\gamma:k''} = z^{-1} J^*(\prod_{h=1}^{k-1} U_h' U_{h+1}'')(U_k' U_{(\beta\gamma)k}'')(\prod_{i=k+1}^{x-1} U_i' U_{i+1}'')J \qquad (8)$$

in which U" is the matrix representing the second bond of the k th dyad with all of the elements, except for U", replaced by zero. $P_{\beta\gamma k''}$ is estimated by generating Monte-Carlo chains. An averaged chemical shift is then obtained using the values obtained for $P_{\beta\gamma k''}$ and $\sigma_i(1,2)$. Using this developed methodology, the structural behaviors of polyethylene and paraffins, vinyl polymers such as poly(vinyl alcohol) and polypropylene, etc. in the solution have been successfully elucidated on the basis of their observed spectra(1,2,20-24). Also, this can be applied to noncrystalline and crystalline phases in polymers.

In the crystalline state polymer chains assume a fixed conformation. In this case, the structural information obtained from the chemical shift corresponds to the fixed conformation. The calculation of ^{13}C chemical shifts for a dipeptides fragment(N-acetyl-N'-methyl-L-alanineamide)[Ac-L-Ala-NHMe] of poly(L-alanine) and L-alanine residue containing proteins has been attempted using the FPT INDO method in order to understand and predict the ^{13}C chemical shift behavior of polypeptides associated with a secondary structure such as an α-helix, β-sheet, etc., and the determination of secondary structure through the observation of the ^{13}C chemical shift(25). The ^{13}C chemical shifts of the C_β carbon of an alanine residue in various peptides and polypeptides vary significantly depending on the conformation, which may be a right-handed α-helix, β-sheet, or other conformation. Such sizeable displacements of the ^{13}C chemical shifts can be characterized by variations in the electronic structures of the local conformation as defined by the dihedral angles(ϕ, ψ). The calculated contour map for the C_β carbon is shown in Fig.2. From this map, we can estimate the ^{13}C shielding constant for any specified conformation. It has been demonstrated from comparisons of the experimental data and the predicted values given by this chemical shift map that the map successfully predicts the ^{13}C chemical shifts of L-alanine residues in polypeptides and proteins as shown in Table 1(26,27). For example, as seen from this map, the ^{13}C chemical shift of the right-handed α-helix form appears at downfield by 2.5 ppm than that of the β-

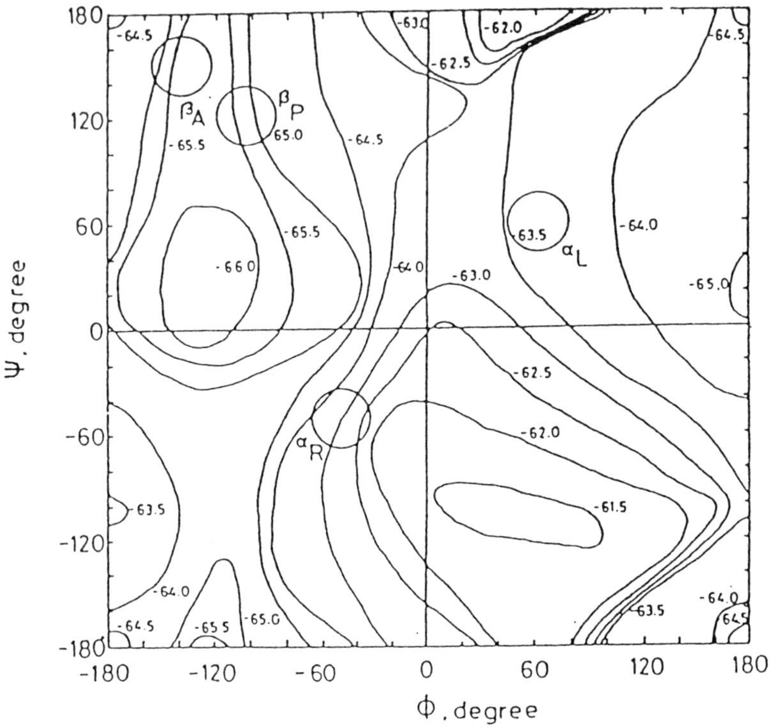

Fig.2 The calculated ¹³C chemical shift map of the C_β carbon in N-acetyl-N'-methyl-L-alanine amide obtaining using the FPT INDO method. The chemical shifts were calculated at 15° intervals for the dihedral angles(φ,ψ).

Table 1 Comparison of the ¹³C chemical shifts of Ala Cβ signals of various polypeptides having different conformations with those of theoretical calculations[a].

Conformation	Dihedral angles[b] (φ and ψ)	Cβ ¹³C chemical shifts(ppm) Observed[c]	Calculated[d]
α_R-helix	-48° and –57°	14.9	-63.0
α_L-helix	48° and 57°	14.9	-63.7
β-sheet	-142 and 145°	19.9	-65.7
3_1-helix	-80° and 150°	17.5	-64.7

[a]Calculated for N-acethyl-L-alanine amide.
[b]Data from G.N.Ramachandran and V.Sasisekharan, Adv.Protein Chem, 1968, 23, 283-438.
[c]Chemical shifts referred to Me_4Si.
[d]Chemical shifts expressed by the nuclear shielding constants.

sheet form. This explains reasonably the experimental result. For chemical shift calculation of ^{15}N nucleus which is also popular in polymers the same method can be used.

Most recently, *ab initio* calculations for the NMR chemical shifts have been available for molecules by remarkable advance of workstation and personal computer in addition to supercomputer(28). This leads to quantitative discussion on the chemical shift behaviors. As an example, *ab initio* MO calculation with the 4-31G basis set using the GIAO-CHF(gauge independent atomic orbital-coupled Hartree-Fock) method on N-acetyl-N'-methyl-L-alanineamide which is the same model molecules as the case of the above FPT INDO calculation was carried out(30). All of the geometrical parameters are energy-optimized. Fig.3a is the calculated isotropic ^{13}C chemical shift map of the C_β carbon as functions of the dihedral angles, where the positive sign means shielding. The whole trend for this map is similar to that obtained by the FPT INDO method as shown in Fig.2. The isotropic shielding constants(σ) for the C_β carbon are 186.4 ppm for the dihedral angles(ϕ, ψ), which correspond to the anti-parallel $\beta(\beta_A)$-sheet conformation, 189.4 ppm for the right-handed α (α_R)-helix,189.6 ppm for the left-handed α (α_L)-helix ;on the other hand, the observed isotropic chemical shifts(δ) are 21.0 ppm for the β_A-sheet, 15.5 ppm for the α_R-helix, and 15.9 ppm for the α_L-helix(31). Such experimental chemical shift behavior is well explained by the calculated behavior. It is found that the change of the dihedral angles dominates the isotropic chemical shift behavior of the L-alanine residue C_β carbon.

In order to understand the isotropic chemical shift behavior, the calculations of principal components for chemical shielding tensor has been made. The (ϕ, ψ) dependences of the σ_{11}, σ_{22}, and σ_{33}(that are defined from the least shielding to the most shielding, respectively.) can be obtained(not shown). The fact that the isotropic C_β chemical shift for the α_R -helix appears in a higher field(15.5 ppm) than that for the β_A -sheet (21.0 ppm) is understood by the explicit differences in the σ_{11} ($\Delta \sigma_{11}$ □9 ppm). The careful investigation of chemical shielding tensor shows that the paramagnetic term for the σ_{11} dominates the total σ value. The 2p-orbital for the C_α -carbon, which corresponds to the angular momentum operator of electron and which contributes to the C_α - C_β σ-bond, is the most effective to the paramagnetic term for the σ_{11},since the magnetic dipole coupling integral of electrons, $<\phi_\nu|\mathbf{L}_\beta/r_B{}^3|\ \phi_\lambda>$($\beta$=1 in this case), is dominantly estimated by the 2p-orbital perpendicular to the σ_{11}.

Fig.3b shows the dependence of isotropic chemical shielding for the Ca-carbon against the main-chain dihedral-angles. From this figure, the isotropic chemical shieldings for the Cα-carbon are 160.4 ppm for the β_A-sheet, 159.6 ppm for the α_R-helix, 159.2 ppm for the α_L-helix, 161.4 ppm for the 31-helix, and 157.9 ppm for silk I and (Ala-Gly)$_n$ form II; for these calculated shielding, the observed isotropic chemical shifts are 48.7 ppm for the β_A-sheet, 53.0 ppm for the α_R-helix, 50.1 ppm for the α_L-helix, 49.7 ppm for the 3_1-helix, and 51.5 ppm for (Ala-Gly)$_n$ form II, respectively. Although it is obvious that there exists the main-chain dihedral-angle dependence on chemical shift for the C_α -carbon, it seems more complicated than that for the C_β -carbon, because the orientation of the chemical shift tensor for the C_α -carbon with respect to the molecular fixed frame is different from one (ϕ,ψ) to

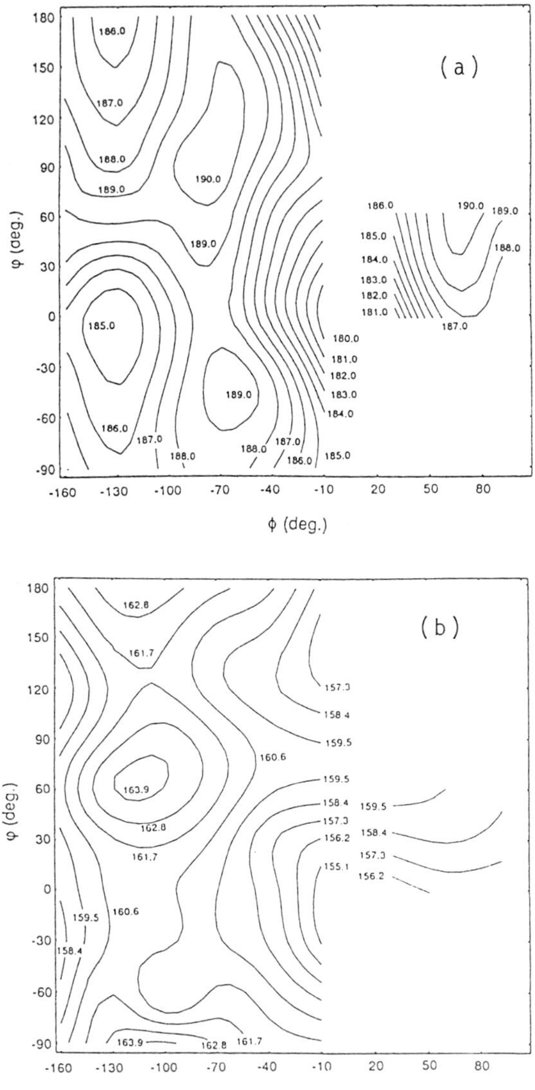

Fig.3 The dependences on the dihedral angles(ϕ,ψ), of the isotropic chemical shielding constant for the L-alanine residue C_β- (a)and C_α-(b) carbons in peptides. Chemical shielding calculations were carried out using the GIAO-CHF method with 4-31G *ab initio* MO basis set. The 4-31G optimized geometries for the model molecules, N-acetyl-N'-methyl-L-alanineamide, were employed.

another, and also because, in a case in which the L-alanine residue carbonyl- or amide-group would form the hydrogen-bond, the hydrogen-bonding structure can also affect on the behavior of chemical shift for the C_α -carbon. As regards the structure of $(Ala\text{-}Gly)_n$ form II and silk I, it needs chemical shift calculation that is taken into consideration of the hydrogen-bonding; the investigation of chemical shift carbonyl-carbon will be needed.

As mentioned above, the principal values of chemical shift tensor give information about three dimensional electronic state of a molecule. However, in order to understand behavior of the principal values, one should obtain information about the orientation of the principal axis system of a chemical shift tensor with respect to the molecular fixed frame. The orientations of the principal axis systems of the chemical shift tensors of L-alanine C_β -carbons in some peptides were calculated, whose L-alanine moieties have different main-chain dihedral-angles,(ϕ,ψ)=(-57.4°,-47.5°)[α_R-helix], (-138.8°,134.7°)[β_A-sheet], (-66.3°,-24.1°)[$3_{10}{}^R$-helix], and (-84.3°, 159.0°) [3_1-helix]. The calculation shows that the σ_{33} component nearly lies along the C_α- C_β bond for all peptides considered here, and also show that the σ_{11} is nearly perpendicular to the plane which is defined by the C_β, the C_α, and the N atoms in L-alanine residue; on the other hand, the σ_{22} is parallel with regard to the plane. These results agree with the experimentally determined direction of σ_{33} of the C_β -carbon in L-alanine amino acid by Naito et al.(31) As seen from Table 2, the σ_{11} component for the dihedral-angles corresponding to the β_A -sheet conformation is 37.06 ppm. This shows downfield shift of about 9 ppm with respect to that for the α_R -helix conformation. This result means that the σ_{11} dominates downfield shift on the isotropic chemical shift of the C_β -carbon for the β_A -sheet conformation. Since the σ_{11} does not orient to a specified chemical bond, it is not easy to comprehend intuitively the chemical shift tensor behavior of the C_β-carbon. However, it is obvious that the through-space interaction between the C_β -methyl group and its surrounding might be important to understand the σ_{11} behavior.

It is well known that hydrogen-bonding plays crucial roles in forming higher-order structure in peptides, polypeptides and proteins. Systematic studies of the hydrogen-binding effects on ^{13}C, ^{15}N and ^{17}O NMR chemical shifts for amide carbonyl-carbon, amide nitrogen and carbonyl-oxygen atoms in solid peptides and polypeptides containing various amino acid residues have been made(33-41). In this article, the hydrogen-bonding effect on the ^{13}C chemical shift will be described. Figure 4 shows plots of the observed isotropic ^{13}C chemical shifts(δ_{iso}) of the glycine(Gly), L-Ala, L-valine(Val), L-leucine(Leu) and L-aspartic acid(Asp) residues against the N...O separation($R_{N...O}$) for hydrogen bonding(>N-H...O=C<) which were determined by X-ray diffraction. It is found that a decrease of $R_{N...O}$ leads to a decrease in shielding shift, and there exists an approximately linear relationship between the 13C chemical shift and $R_{N...O}$ as $\delta_{iso} = a - bR_{N...O}$, where a and b for Gly, L-Ala, L-Val, L-Leu and L-Asp are 206.0 and 12.4, 237.5 and 21.7, 215.4 and 14.2, 202.2 and 10.0, and 199.0 and 9.6, respectively. The a and b values depend on the amino acid residues. These relations indicates that hydrogen bond length can be determined through the observation of the ^{13}C chemical shift of the carbonyl carbon in the amino acid residues in peptide, polypeptides and proteins. It is expected that

Table 2 Calculated ^{13}C chemical shielding for the Cβ carbon of the L-alanine residue in peptides and polypeptides by the 4-31G GIAO-CHF method

sample	^{13}C chemical shielding /ppm Cβ σ_{iso}	σ_{11}	σ_{22}	σ_{33}
α_R-helix	15.72	28.16	22.14	-3.16
β_A-helix	18.74	37.06	21.70	-2.53
3_1-helix	15.94	33.80	19.97	-3.49
$3_{10}{}^R$-helix	15.84	32.47	19.03	-4.00

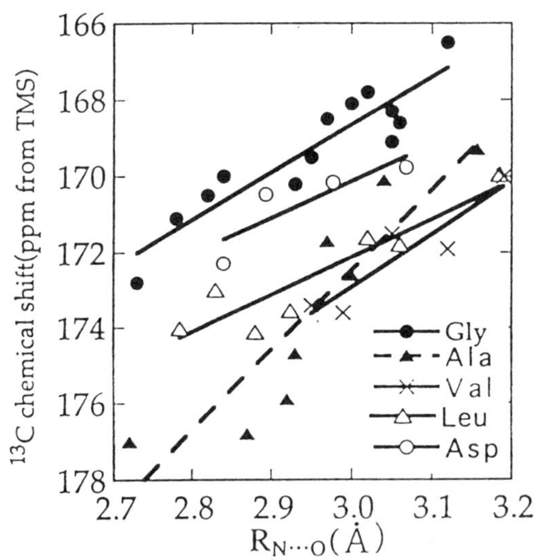

Fig.4 Plots of the observed ^{13}C chemical shifts for the carbonyl carbons in Gly, L-Ala, L-Leu and L-Asp residues in peptides and polypeptides in the solid state against the N…O spearation($R_{N...O}$).

the principal values of the ^{13}C chemical shift tensors(δ_{11}, δ_{22} and δ_{33} in order of increasing shielding) are valuable as parameters for obtaining detailed information on hydrogen-bonding. From the observation and calculation of the principal values, it is found that δ_{11} moves slightly upfield with a decrease in $R_{N...O}$, δ_{22} moves linearly and largely downfield largely, and δ_{33} is independent of $R_{N...O}$, where δ_{11} is in the made sp^2 plane and lies along the direction normal to the C=O bond, δ_{22} lies almost along the C=O bond, and δ_{33} is aligned perpendicular to the amide sp^2 plane. By using these relations, the hydrogen bond lengths and conformations of some polypeptides in the solid state were reasonably determined.

APPROACH OF USING INFINITE POLYMER CHAINS

Sometimes the estimation of the electronic structures of polymer chains necessitates the inclusion of long-range interactions and intermolecular interactions in the chemical shift calculations. To do so, it is necessary to use a sophisticated theoretical method which can take account of the characteristics of polymers. In this context, the tight-binding molecular orbital(TB MO) theory from the field of solid state physics is used, in the same sense in which it is employed in the LCAO approximation in molecular quantum chemistry to describe the electronic structures of infinite polymers with a periodical structure(3-11,36). In a polymer chain with linearly bonded monomer units, the potential energy if an electron varies periodically along the chain. In such a system, the wave function ψ (k) for electrons at a position r can be obtained from Bloch's theorem as follows(36,37):

$$\psi(k) = N^{-1/2} \Sigma \exp(ikj)C_{vn}(k)\,{}_{v}\phi(r\text{-}ja) \qquad (9)$$

where k is the wave number, n is the band index, v is an orbital index in the j th cell, a is the unit vector of translational symmetry. N is the total number of cells, and l is the number of atomic orbitals in the cell. The term ϕ (k) represents the υ th atomic orbital in the j th cell and $C_{vn}(k)$ the expansion coefficient. The formulae needed to calculate the shielding of polymers using the TB MO theory incorporating the SOS method have been derived as a function of k as follows.

$$\sigma^{d}(k) = C\ \Sigma P(k)\langle\phi_{v}|r^{-1}|\phi_{v'}\rangle \qquad (10)$$

$$\sigma^{p}(k) = C'\ \Sigma\langle r^{-3}\rangle_{2p}(E_{m} - E_{n})^{-1}Q \qquad (11)$$

where C and C' are constants and P(k) is the bond order as a function of k. In order to compare the calculated and experimental values of the nuclear shielding, it is necessary to average the calculated data over k, within the first Brillouin zone, as given by

$$\sigma = \Sigma\ D(k)\ \sigma(k) \qquad (12)$$

where D(k) is the density of states, namely the number of states per unit amount of energy. Using this methodology, elucidation of conformation-dependent chemical shift of various polymer chains such as polyethylene, polyacetylene, polypyrrole, polyoxymethylene and polyoxyethylene in the solid state has been studied systematically(3-11). For example, a comparison of the ^{13}C chemical shift tensor of *cis*- and *trans*-polyacetylenes in the solid state is shown in Fig.5. The direction of the principal axes are indicated at the bottom. These results agree well with the experimental data.

This shows that the TB MO calculation correctly predicts the origin of the chemical shift and electronic structure associated with the structure of the polymers. Furthermore, TB INDO/S calculations have been carried out on the seven polyacetylene chains which take an orthorhombic form(9). From these results it has been demonstrated that the chemical shift is very sensitive to intermolecular interactions and the TB MO calculation provides useful information about the band structure.

The TB MO calculation on the ^{15}N chemical shift of polypyrrole in the solid state allows useful information to be extracted from the observed spectra, namely that the two peaks obtained are correctly assigned to the quinoid and aromatic structures.(11,38) (The quinoid structure is closely to the electric conductivity.) A decrease in the band gap leads to a downfield shift. These results on conducting polymers demonstrate that the chemical shift behavior provides information about the band gap which, in turn, is a measure of the electric conductivity. It can be said that TB MO calculations offer useful perspectives in interpreting the results of NMR nuclear shieldings in polymers, both in terms of the structure in the solid state and in understanding the effect of intermolecular interactions on nuclear shieldings. The latter are shown to operate through the electronic structures of the polymers considered.

Most recently, the TB MO theory for chemical shift calculation of polymer crystals with three dimensional structure within the *ab initio* framework has been developed and has been successfully applied to polyethylene crystals(42).

References

1. Ando,I.; Webb,G.A. *Theory of NMR Parameters*, Academic Press, New York, NY, 1983.
2. For example, a) Ando,I. *Encyclopedia of Nuclear Magnetic Resonance*; Grant, D.M.; Harris, R.K., Eds.; John Wiley, New York, NY,1996, pp.176-180. b)Ando,I.; Yamanobe, T.; Kurosu.H.; Webb,G.A. *Ann. Rept. NMR Spectrosc.*, **1990,** 22, 205.
3. Ando,I.; Asakura,T. *Ann.Rept.NMR Spectrosc.*, **1979,** 10A,81.
4. Yamanobe,T.; Chujo, R.; Ando, I. *Mol.Phys.*,**1983,** 50,123.
5. Yamanobe,T.; Ando, I. *J.Chem.Phys.*, **1985,** 83, 3154.
6. Yamanobe, T.; Ando,I.; Saito,H.; Tabeta,R.; Shoji, A,; Ozaki,T. *Chem. Phys.*, **1985,** 99, 259.
7. Yamanobe,T.; Ando,I.; Saito,H.; Tabeta,R.; Shoji, A.; Ozaki,T. *Bull.Chem.Soc.Jpn.*, **1985,** 58, 23.
8. Yamanobe,T.; Sorita,T.; Komoto,T.; Ando, I.; Sato, H. *J.Mol.Structure*, **1987,** 151, 191.

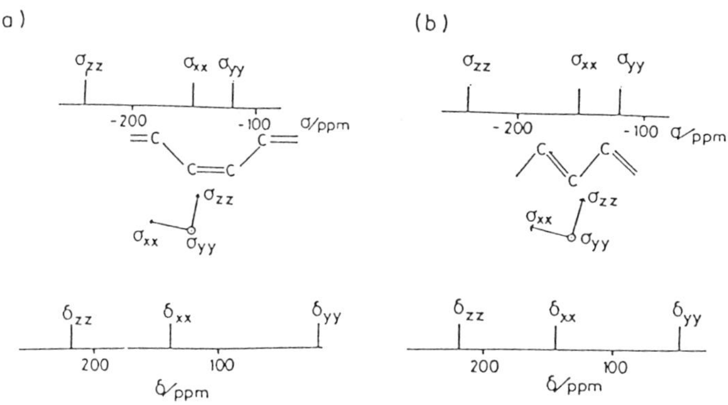

Fig.5 The observed and calculated components of the ^{13}C shielding tensors of *cis*- and *trans*-polyacetylenes. The directions of the principal axes are indicated at the bottom(the angles between the C-H bond and the direction of σ_{zz} for trans- and cis-polyacetylenes are 16°and 17°, respectively).

9. Kurosu,H.; Yamanobe,T.; Komoto, T.; Ando,I. *Chem.Phys.*, **1987,** 116, 391.
10. Ishii,T.; Kurosu,H.; Yamanobe, T.; Ando, I. *J.Chem.Phys.*, **1988,** 89,7315.
11. a)Kurosu,H.; Yamanobe,T.; Ando,I. *J.Chem.Phys.*, **1988,** 89, 5261.(b)Kurosu,H.; Ando,I. *J.Mol.Structure(Theochem),* **1991,** 231, 231.
12. Kikuchi,M.; Kurosu, H.; Ando, I. *J.Mol.Structure,* **1992,** 269,183.
13. Kondo.M.; Ando,I.; Chujo, R.; Nshioka, A. *J.Mag.Reson.*, **1976,** 24,315.
14. Ando,I.; Nishioka,A.; Kondo,M. *J.Mag.Reson.*, **1976,** 21,429.
15. Ando,I.; Kato,Y.; Kondo, M.; Nishioka, A. *Makromol.Chem.*,**1977,** 178, 803.
16. Kondo,M.; Watanabe,S.; Ando,I. *Mol.Phys.*,**1979,** 37,1521.
17. Ando, I.; Webb,G.A.; *Org.Magn.Reson.*, **1981,** 15,111.
18. Ando,I. *Encyclopedia of Nuclear Magnetic Resonance*; Grant, D.M.; Harris, R.K., Eds.; John Wiley, New York, NY,1996, pp.2512-2516.
19. Flory,P.J. *Statistical Mechanics of Chain Molecules*,; Interscience, NewYork, NY,1969.
20. Ando,I.; Nishioka,A.; Asakura,T. *Makromol.Chem.*,**1975,** 176,411.
21. Ando, I.; Nishioka,A. *Makromol.Chem.*,**1975,**176,3089.
22. Ando, I.; Kato,Y.; Nishioka,A. *Makromol.Chem.*,**1976,** 177, 2759.
23. Asakura,T.; Ando,I.; Nishioka, A. *Makromol.Chem.*,**1975,** 176,1151.
24. Asakura,T.; Ando,I.; Nishioka, A. *Makromol.Chem.*,**1977,** 177,1493.
25. Ando,I.; Saito,H.; Tabeta,R.; Shoji,A; Ozaki,A. *Macromolecules,***1984,** 17,457.
26. Saito,H.; Tabeta,R.; Shoji,A.; Ozaki,T.; Ando, I.; Miyata,T. *Biopolymers,* **1984,** 23,2279.
27. Saito,H.; Tabeta,R.; Asakura,T.; Iwanaga,Y.; Shoji,A.; Ozaki,T.; Ando,I *Macromolecules,***1984,** 17,1405.
28. Chesnet, D.B. *Annu.Rept.NMR Spectrosc.,* **1989,**21, 51.
29. Asakawa,N.; Kurosu,H.; and Ando,I. *J.Mol.Structure,* **1994,** 323,279.
30. a)Saito,H.; Tabeta,R.; Shoji,A.; Ozaki,T.; Ando, I. *Macromolecules,***1983,** 16,1050, b)Saito, H.; Tabeta, T.; Shoji, A.; Ozaki, T.; Ando, I.; Asakura, T. *Magnetic Resonance in Biology and Medicine,* Govil, G., Ed.; Tata McGraw Hill, New Dehli, 1985, pp 195-, c)Saito,H.; Ando,I. *Ann.Rept.NMR Spectrosc.,* **1989,** 26,55,d)Ando,I.; Yamanobe, T.; Asakura, T. *Prog.NMR Spectrosc.,* **1990,** 22,349, e)Kurosu, H.; Ando,S.; Yoshimizu, H.; Ando,I. *Ann.Rept.NMR Spectrosc.,* **1994,** 28,189.
31. Naito, A.; Ganapathy, S.; Akasaka, K.; McDowell, C. A. *J.Chem.Phys,* **1981,** 74, 3190.
32. Ladik,J.J. *Quantum Theory of Polymers as Solids*, Plenum Press, NewYork, NY, 1988.
33. Ando, S.; Ando ,I.; Shoji, A.; Ozaki, T. *J.Am.Chem.Soc.*,**1988,** 110, 3380.
34. Asakawa, N.; Kuroki, S.; Kurosu, H.; Ando, I.; Shoji, A.; Ozaki, T. *J.Am.Chem.Soc.,* **1992,** 114, 3261.
35. Tsuchiya,K.; Takahashi,A.; Takeda, N.; Asakawa, N.; Kuroki,S.; Ando, I.; Shoji, A.; Ozaki, T.; *J.Mol.Structure,* **1995,** 350, 233.
36. Kameda,T.; Takeda,T.; Kuroki,S.; Kurosu,H.; Ando,S.; Ando,I.; Shoji,A.; Ozaki, T. *J.Mol.Structure,* **1996,** 384,17.
37. Kameda, T.; Ando,I. *J.Mol.Structure,* **1997,** 412,197.
38. Kuroki, S.; Ando, S.; Ando, I.; Shoji, A.; Ozaki, T. *J.Mol.Structure,* **1990,** 240, 19.; *ibid.*, **1991,** 245, 69.

39. Kuroki, S.; Ando,I.; Shoji, A.; Ozaki,T.; *J.Chem.Soc.,Chem.Commun.*, **1992,** 5,433.
40. Kuroki, S.; Takahashi, A.; Ando, I.; Shoji, A.; Ozaki,T. *J.Mol.Structure,* **1994, 323**,197.
41. Asakawa, N.; Kameda, T.; Kuroki, S.; Kurosu, H.; Ando, S.; Ando, I.; Shoji, A. *Ann.Rept.NMR Spectrosc.*, **1998,** 35,55.
42. Uchida, M.; Kurosu, H.; Toida, Y.; Ando, I. to ACS Series in press.

Chapter 3

NMR and Quantum Chemistry of Proteins and Model Systems

Christina M. Szabo, Lori K. Sanders, William Arnold, Joshua S. Grimley, Nathalie Godbout, Michael T. McMahon, Benjamin Moreno, and Eric Oldfield

Department of Chemistry, University of Illinois at Urbana-Champaign, 600 South Mathews Avenue, Urbana, IL 61801

Abstract. We report the results of experimental and theoretical studies of ^{13}C and ^{15}N shifts in proteins and model systems, together with ^{57}Fe shifts and Mössbauer quadrupole splittings (electric field gradients) in metalloporphyrins and metalloproteins. The ability to relate these spectroscopic observables to structure by using quantum chemical methods opens up new opportunities for predicting and refining protein structure.

Folding a protein into its native conformation generates a very large range of chemical shift non-equivalence for a given type of residue (*1,2*). For protons, the effect is typically ~2-3 ppm (*3*) - although can be considerably more if large ring currents are involved. For ^{13}C, shifts of almost 10 ppm can be observed (*4*), while for ^{15}N, the effects can be extremely large - up to ~33 ppm with valine, for example (*5*), and for the (non-native) ^{19}F nucleus in 4F-tryptophan, folding induces up to a 17 ppm chemical shift range in hen egg white lysozyme (*6*). Explaining these folding-induced shift changes has posed a major challenge for spectroscopists - not least because all of the protein structures involved are only of modest quality, at least as would be judged by "quantum chemical standards". The generally perceived difficulty (or impossibility) of computing chemical shifts in proteins, as well as other spectroscopic observables such as ^{17}O electric field gradients or ^{57}Fe Mössbauer quadrupole splittings, therefore understandably delayed the development of the area. However, several years ago we did begin to investigate chemical shifts in proteins using quantum chemical methods, to see to what extent the experimental results could be reproduced. In initial work with C. Dykstra and J. Augspurger, we found that it was in fact possible to reproduce, to first order, the anticorrelation between ^{13}C and ^{17}O shifts seen in CO-labeled heme proteins (*7*), as well as the correlations between chemical shifts, ^{17}O electric field gradients, and C-O infra-red vibrational stretch frequencies, using a simple model for the electrical polarization of CO (*8*) - the shielding polarizability approach (*9*). This was then successfully applied to predicting ^{19}F shifts in a ^{19}F - Trp labeled protein (*10,11*), in which case both solvation and dynamical averaging effects were included in the calculations (*11*). In both instances, electrostatic fields were shown to dominate the chemical shift trends seen experimentally. It was not clear, though, whether these effects would also dominate the shielding of ^{1}H, ^{13}C, and ^{15}N - the major sites in proteins whose shifts are now routinely recorded using ^{1}H, ^{13}C, ^{15}N multidimensional NMR spectroscopy, since major electronic structural changes due e.g. to "γ-gauche" or torsional effects, as well as significant hydrogen bonding, might reasonably be expected to play a role in shielding.

We therefore set about, with A. C. de Dios and J. G. Pearson, and using the Texas code from P. Pulay (*12*), the process of computing ^{13}C and ^{15}N shieldings in proteins using purely quantum chemical methods. To our delight - it worked (*13*). In particular, we found that we could use quite small amino acid derivatives, *N*-formyl-amino-acid amides, which contain two peptide-like amide groups:

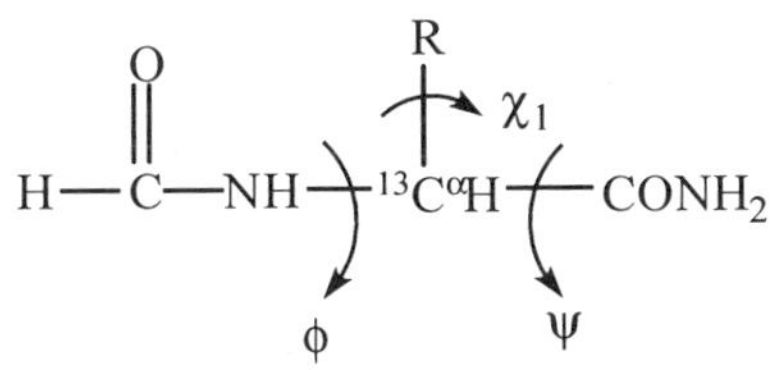

to reproduce protein ^{13}C chemical shifts with quite good accuracy (*13*), while the slightly larger *N*-formyl-ala-valine amide:

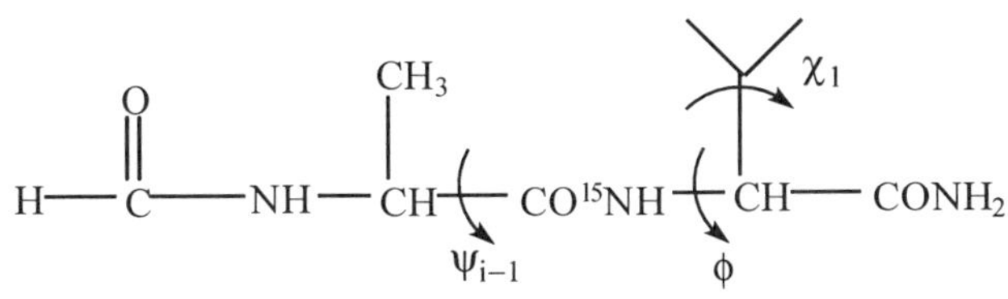

provided a good description of the ^{15}N shifts for valine residues, especially when hydrogen bond partners were introduced (*13*). This study (*13*) showed the feasibility of predicting ^{13}C, ^{15}N as well as ^{19}F NMR chemical shifts from known structures, albeit with the caveat that we found it essential to use a standard set of bond lengths and three-atom bond angles (based on small peptide crystal structures) in order to reproduce the experimental results. In later work, we showed that complete shielding surfaces - how e.g. the C^{α} shift varies as a function of the backbone torsion angles (ϕ and ψ)-could be evaluated, and that more importantly, these relationships could be used to refine (*14*) and predict (*15*) some aspects of peptide and protein local structure. For a review of these early developments see e.g. Refs. *16*, *17*.

In more recent work, we have begun to investigate a number of other topics. In particular, we have extended our shielding calculations to include metal ions (*18*,*19*), here using the density functional theory B3LYP approach described by Bühl (*20*), in addition to successfully predicting both ligand shifts (*21-23*) and metal and ligand electric field gradients (*22,24*). When combined with the results of infra-red spectroscopy and IR vibrational frequency calculations (*23,25*), we have now been able to develop a detailed geometric model of how CO binds to heme proteins, which is in accord with all of the NMR, IR and Mössbauer spectroscopic results (*23*), and which helps clarify how CO binding to heme proteins in blood and muscle is discriminated against, in favor of O_2.

Since it is not possible to discuss each of these topics in detail here, we shall focus first on our more recent studies on ^{13}C shielding in amino-acids and peptides, comparing theory with experiment, and exploring the shielding calculations in more detail. Our aim is to begin to better understand the various structural correlations with shielding seen experimentally. We also briefly discuss the electrostatic potential surfaces in the amide-group containing amino-acid, asparagine, using both x-ray diffraction and quantum chemical approaches. In the long term, it seems quite likely that obtaining charge density, electrostatic potential and perhaps other electrostatic properties from "NMR" (i.e. the SCF) calculations will become routinely useful when investigating protein structure, and protein-ligand interactions.

We consider first the localized molecular orbital (LMO) contributions to shielding (*26*) in helical and sheet geometries, calculated by using the deMon program (*27*), and we show in Figure 1 the LMO contributions to C^α shielding for the alanine model. The major LMO contributions can be broken down into five categories: C^α-C^β, C^α-C^0, C^α-H^α, C^α-N and the carbonyl oxygen lone pair contributions, as shown in Table I for glycine, alanine, valine (χ_1 = 180°) and phenylalanine (χ_1 = 180°, χ_2 = 90°). As may be seen from Figure 1 and Table I there is good agreement with the expected ≈6 ppm typical C^α isotropic shift separation (5.8, 8.1, 7.1, 7.2) between helical and sheet residues, Δ, with C^α in the sheets being more shielded. The largest LMO contribution to Δ comes from the C^α -C^0 contribution, which correlates with a change in the ψ torsion angle, as shown above. This is not surprising since ψ undergoes the largest change on transition from a helix (ψ ≈ -60°) to a sheet (ψ ≈ 120°), but more surprising is the large contribution to the helix-sheet separation from C^α -H^α, Table I. Here, the LMO analysis gives a less clear correlation with the structural change, and a more detailed analysis using canonical MO's appears necessary, and will be reported elsewhere (*28*). We find that the LMO contributions to C^α shielding in valine (χ_1 = 180°) and phenylalanine are very similar to those found for C^α in alanine, as shown in Figure 2 and Table I. That is, the C^α-C^0 contribution dominates the overall helix-sheet separation.

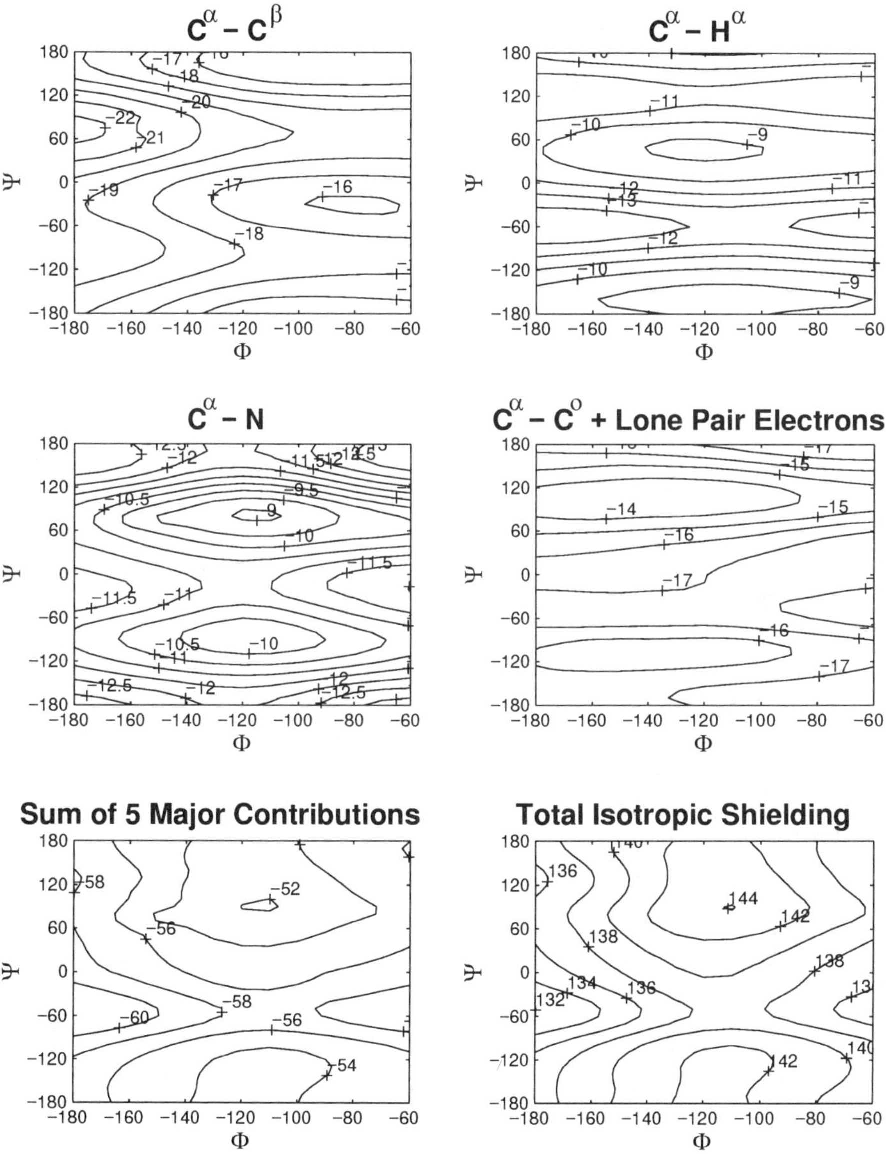

Figure 1. LMO contributions to shielding for C^{α} in *N*-formyl-alanine amide as a function of ϕ,ψ computed using the deMon program. The individual, sum of major, and overall contributions to shielding are shown.

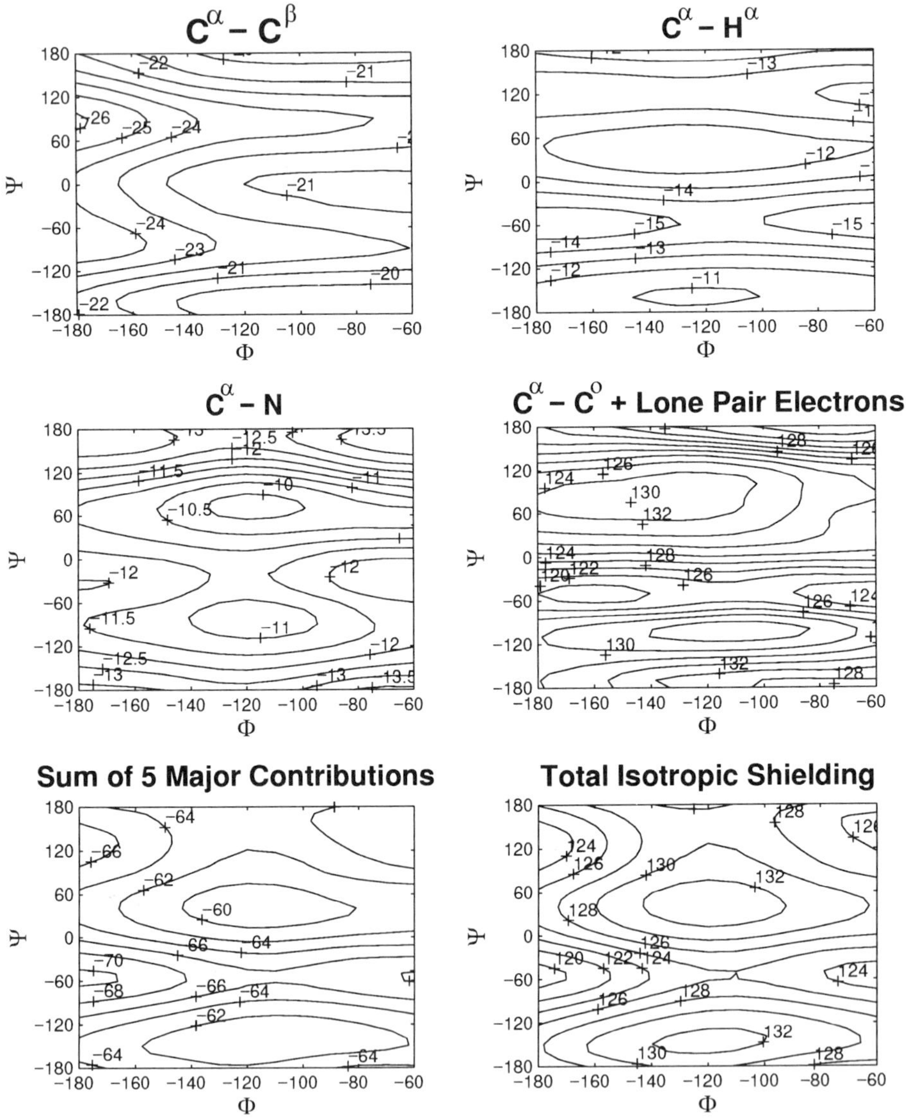

Figure 2. LMO contributions to shielding for C^{α} in *N*-formyl-valine amide ($\chi_1 = 180°$) as a function of ϕ,ψ.

Table I. Major LMO contributions to $^{13}C^{\alpha}$ isotropic shielding in *N*-formyl-glycine amide, *N*-formyl-alanine amide, *N*-formyl-valine amide, and *N*-formyl-phenylalanine amide (deMon program, PW91, uniform IGLO II basis set)

	Glycine			Alanine			Valine			Phenylalanine		
Contribution	Helix	Sheet	Δ	Helix	Sheet	Δ	Helix	Sheet	Δ	Helix	Sheet	Δ
C^{α}-H^{α}/C^{α}-C^{β}	-8.8	-9.6	.8	-16.5	-17.5	1.0	-21.5	-22.4	.9	-17.9	-18.5	.6
C^{α}-C^{o}	-14.7	-12.9	-1.8	-15.3	-12.1	-3.2	-15.6	-13.1	-2.5	-17.0	-14.4	-2.6
C^{α}-H^{α}	-13.5	-11.5	-2.0	-13.5	-11.2	-2.3	-15.7	-13.4	-2.3	-16.2	-14.3	-1.9
C^{α}-N	-11.3	-10.0	-1.3	-11.3	-10.2	-1.1	-12.0	-11.1	-.9	-12.9	-12.4	-.5
Lone Pair (Σ)	-3.0	-2.1	-.9	-3.3	-1.4	-1.9	-3.3	-2.0	-1.3	-3.5	-1.9	-1.6
Σ 5 Major LMO	-51.3	-46.1	-5.2	-59.9	-52.4	-7.5	-68.1	-62.0	-6.1	-67.5	-61.5	-6.0
Total Isotropic	143.5	149.3	-5.8	135.2	143.3	-8.1	123.1	130.2	-7.1	124.8	132.0	-7.2

Table II. Major LMO contributions to $^{13}C^{\alpha}$ shielding tensors in *N*-formyl-valine amide (deMon program, PW91, uniform IGLO II basis set)

	σ_{11}			σ_{22}			σ_{33}			$\overline{\sigma}_{11\text{-}22\text{-}33}$		
Contribution	Helix	Sheet	$\Delta\sigma_{11}$	Helix	Sheet	$\Delta\sigma_{22}$	Helix	Sheet	$\Delta\sigma_{33}$	Helix	Sheet	$\Delta\overline{\sigma}$
C^{α}-C^{β}	-4.2	-40.6	36.4	-25.3	-31.7	6.4	-34.9	5.2	-40.1	-21.5	-22.4	.9
C^{α}-C^{o}	-35.0	-25.4	-9.6	-24.5	-6.9	-17.6	12.7	-7.0	19.7	-15.6	-13.1	-2.5
C^{α}-H^{α}	-34.4	14.0	-48.4	5.1	-22.2	27.3	-17.9	-32.0	14.1	-15.7	-13.4	-2.3
C^{α}-N	-3.9	-22.8	18.9	-17.7	4.4	-22.1	-14.4	-14.9	.5	-12.0	-11.1	-.9
Lone Pair (Σ)	-8.8	-4.2	-4.6	-1.4	-.9	-.5	0.4	-.8	1.2	-3.3	-2.0	-1.3
Σ 5 Major LMO	-86.3	-79	-7.3	-63.8	-57.3	-6.5	-54.1	-49.5	-4.6	-68.1	-62.0	-6.1
Tensor Element	113.4	101.9	11.5	120.1	134.2	-14.1	135.9	154.3	-18.4	123.1	130.2	-7.1

For C^{β} in alanine, there are four main LMO contributions to shielding, from C^{α} and from the three H^{β} atoms. The three H atoms contribute differently to C^{β} shielding, as shown in Figure 3, but when these contributions to C^{β} shielding are combined with those of C^{β}-C^{α}, we find good accord with the total overall isotropic shielding surface, shown in Figure 4.

From the perspective of one interested in obtaining new or improved protein structures, it should be noted that chemical shielding surface calculations are of considerable interest because they permit the determination not only of chemical shieldings (or chemical shifts) as a function of ϕ, ψ, χ - including experimentally inaccessible areas - but they also provide the elements of the chemical shielding tensors, Table II, as well as their orientations, Figure 5. Table II shows how the individual LMO contributions to the individual shielding tensor elements vary for typical helical (ϕ = -60, ψ = -60) and sheet (ϕ = -120°, ψ = 120°) valine χ_1 = 180° geometries. These results indicate that the span (or breadth) of the C^{α} shielding tensor for valine in sheet residues is much larger than that observed for helical sites, 52.4 versus 22.5 ppm, Table II, in general accord with the results of Hartree-Fock calculations presented previously, in which large differences in tensor span between helical and sheet residues were reported (*29*). These calculations also showed a major difference in the orientation of the C^{α} shielding tensor on transition from a helical to a sheet geometry, as shown in Figure 5. In neither the helical nor the sheet geometries is the C^{α} tensor oriented along any obvious molecular axis. The question next arises: How accurate are the results of such shielding tensor calculations? If they are quite accurate, then use can be made of them in structure refinement and determination, which could then lead to more detailed insights into the relationships between structure and function, using quantum chemical methods.

In early work, we showed (*30*) that pure amino-acid single crystal shielding tensor magnitudes and orientations could be well described using Hartree-Fock methods, which then led us to use chemical shifts to help predict (*31*) and refine (*14*) elements of protein solution structure. We then investigated C^{α} shieldings in peptides in the solid-state (*15*), finding good agreement with crystallographically-determined ϕ, ψ torsion angles by using the Bayesian probability approach (32). In this method, we simply compute the likelihood, Z, that a computed property, $P^{calc}(\phi, \psi)$, is equal to the experimental value, P^{expt}, from:

$$Z(\phi,\psi) = \exp^{-\left(\frac{P^{expt} - P^{calc}(\phi,\psi)}{W}\right)^2}.$$

So, if the three tensor elements, σ_{ii}, are known experimentally for a given C^{α} site, the likelihood that a given ϕ,ψ pair can account for the experimental results is given by:

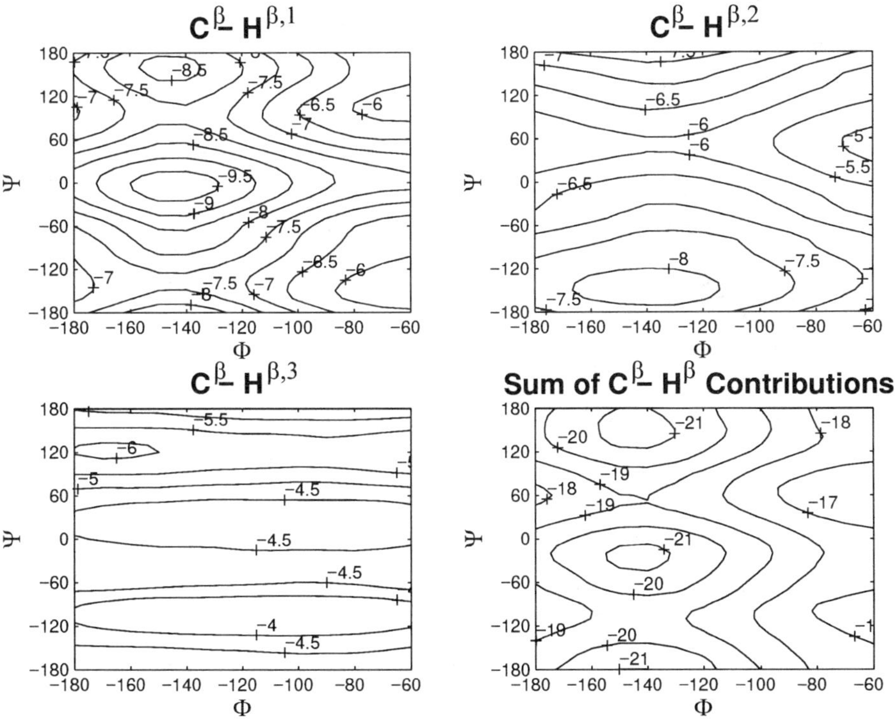

Figure 3. C^{β}-H LMO contributions to C^{β} shielding in *N*-formyl-alanine amide as a function of ϕ,ψ.

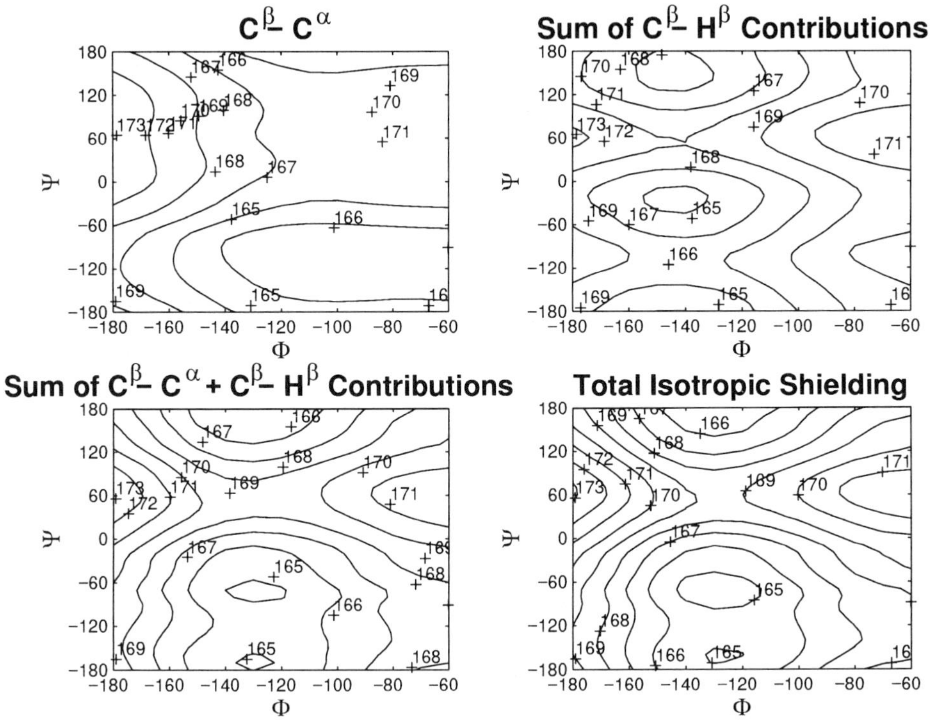

Figure 4. C^{β}-C^{α}, C^{β}-H^{β} and total LMO contributions to C^{β} shielding in *N*-formyl-alanine amide as a function of ϕ,ψ.

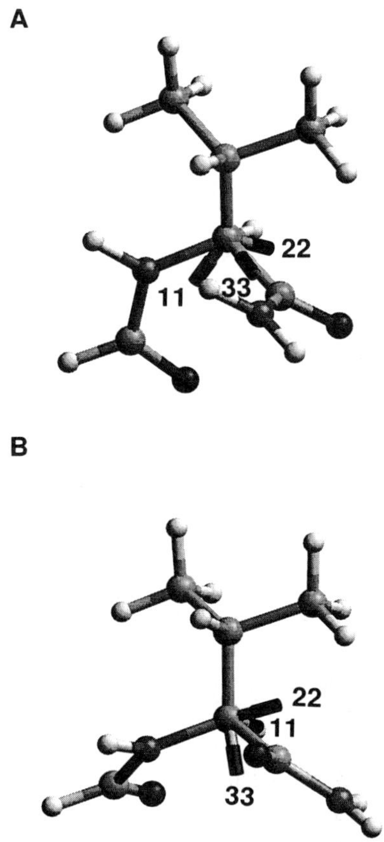

Figure 5. Orientation of the principal components of the C^{α} shielding tensor for *N*-formyl-valine amide (χ_1 = 180°). **A**, helical (ϕ = -60°, ψ = -60°). **B**, sheet (ϕ = -120°, ψ = 120°).

$$^{3}Z(\phi,\psi) = \exp -\left(\frac{(\sigma_{11}^{\,expt} - \sigma_{11}^{\,calc}(\phi,\psi))}{w_1}\right)^2 \exp -\left(\frac{(\sigma_{22}^{\,expt} - \sigma_{22}^{\,calc}\phi,\psi)}{w_2}\right)^2 .....$$

where w_i are search-width parameters, related to the standard deviations between theory and experiment (*15*). The method was found to work quite well for alanine peptides, with an rms error in (ϕ, ψ) from crystallographic values of ~10° being found using just a single labeled site. Also of interest is the fact that once the orientation of the shielding tensor is known, then the solution "chemical shift anisotropy" can be determined. The solution CSA or "$\Delta\sigma$" is determined experimentally from dipole-dipole/chemical shift anisotropy interference in relaxation and has the definition: $\Delta\sigma = \sigma_{par} - \sigma_{perp}$, where σ_{par} is the shielding along the $C^{\alpha}H^{\alpha}$ (or NH^{α}) bond vector and σ_{perp} is the average shielding perpendicular to this axis. Experimental results for calmodulin and ubiquitin, two small proteins, have been reported by Tjandra and Bax (*33*) and Case has reported a comparison between DFT computed $\Delta\sigma$ values and experiment (*34*) which show quite good agreement. We obtain similar results with our deMon-computed Ala C^{α} surface, Figure 6A, with a slight improvement when using HF results (*29*), Figure 6B. Even better agreement is obtained when our specific (Gly, Ala, Val, Ser, Thr, Ile, Phe) surfaces are used, with the best correlations being found with ubiquitin, Figure 6C. These results suggest that $\Delta\sigma$ values, as well as isotropic shifts, may play a role in protein structure refinement.

More recently, we have begun to investigate other shielding tensors in the solid-state. For example, we show in Figure 7A the ^{13}C "magic-angle" sample spinning (MAS) NMR spectrum of a selectively ^{13}C methyl-labeled tripeptide, alanyl-[^{13}C]alanyl-alanine hemi-hydrate (*35*), whose structure is known (*36*) and was confirmed in this laboratory on the sample used for NMR spectroscopy. By recording spectra at several spinning speeds, it is possible to recover the principal elements of the methyl ^{13}C shielding tensor (*37*). In this case, we find $\delta_{11} = 39.9 \pm 0.3$ ppm, $\delta_{22} = 20.1 \pm 0.4$ ppm and $\delta_{33} = 2.4 \pm 0.2$ ppm. These values are in generally good agreement with those obtained from previously published results (Ref *29*; the shielding surfaces can be found at http://feh.scs.uiuc.edu) by using the known ϕ, ψ values, as shown in Figure 7B, where a slope of -1.07 and an R^2 value of 0.99 are obtained. This is a particularly interesting result since at first sight it appears to suggest that the methyl group might be rigid. However, this is typically not the case for methyl groups, as found for example by ^2H NMR (*38*), and in fact as noted with crystalline alanine by R. Griffin (private communication), these results actually suggest that the methyl group spends most of its time located in potential minima, even though it frequently jumps between these minima, on the NMR timescale. If this idea is correct, it should be possible to detect methyl protons

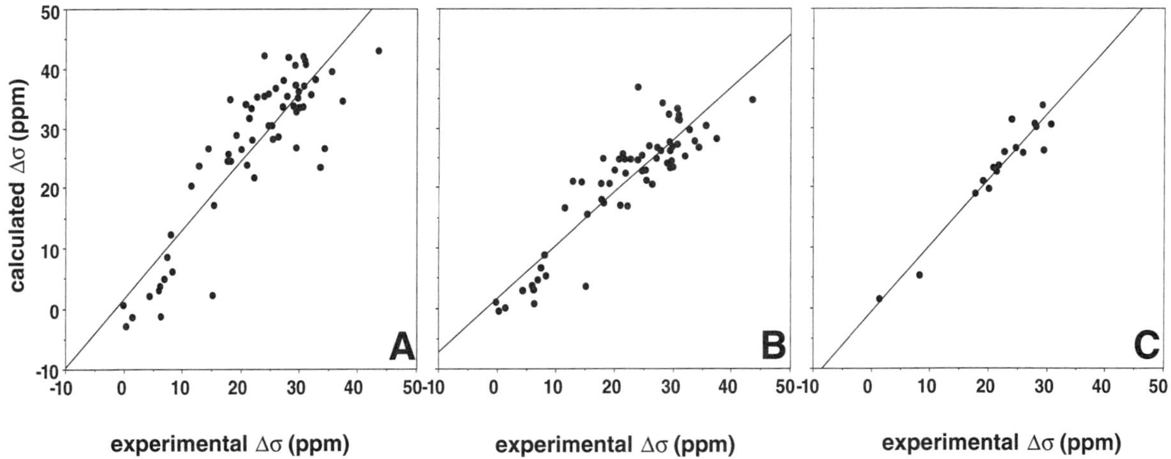

Figure 6. Theory versus experiment comparison of the C^{α} solution CSA ($\Delta\sigma = \sigma_{par} - \sigma_{perp}$) for ubiquitin and calmodulin. **A**, ubiquitin plus calmodulin, DFT alanine C^{α} surface used for all data points, $\chi^2/\nu = 10.7$. **B**, ubiquitin plus calmodulin, Hartree-Fock alanine C^{α} surface used for all data points, $\chi^2/\nu = 5.7$. **C**, ubiquitin alone, individual gly, ala, val, ile, ser, thr, phe Hartree-Fock C^{α} surfaces (http://feh.scs.uiuc.edu) used, $\chi^2/\nu = 1.7$.

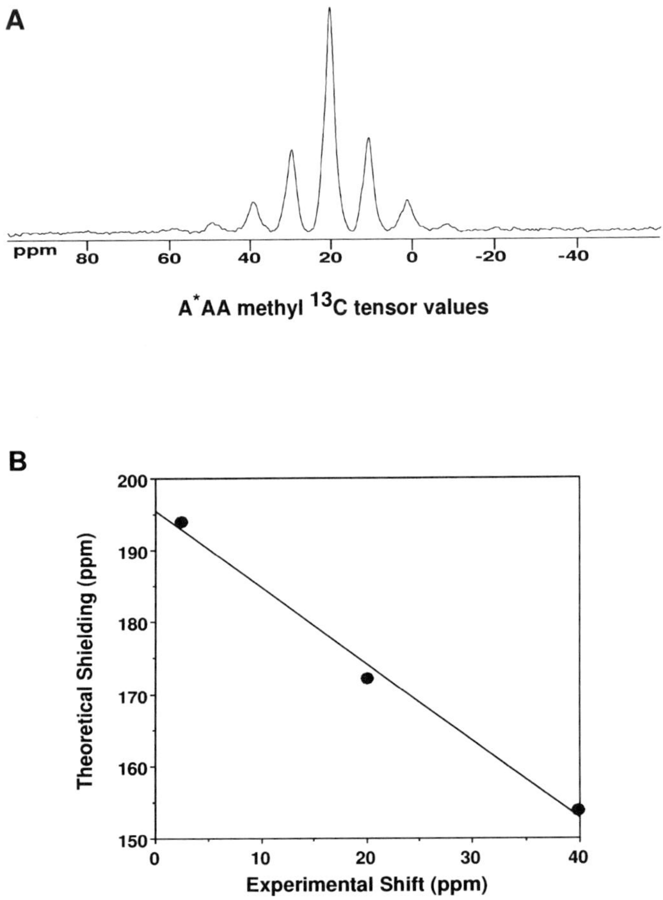

Figure 7. $^{13}C^{\beta}$ alanine shielding tensor results. **A**, 11 Tesla ^{13}C CP-MAS NMR spectrum of crystalline ala-[^{13}C]ala-ala, ν_r = 1.2 kHz. **B**, graph showing experimental shift tensor elements versus computed shieldings: slope = 1.07, R^2 = 0.99.

crystallographically in peptides, and it might also be possible to use quantum chemistry to accurately predict their locations.

To explore these points further we have obtained high-resolution X-ray diffraction data for a series of amino-acids and peptides. We show in Figure 8 a high-resolution electron density map, ρ(**r**), for the system *N*-acetyl-L-leucine, obtained by using Mo K^{α} radiation and an array detector. There is clearly a large electron density associated with each of the methyl hydrogens, Figure 8A. Also of interest is the observation that the methyl H-C-C-H torsion angles determined crystallographically, which define the methyl group orientation, are almost identical to those we obtain from a quantum chemical geometry optimization, Figure 8B, in which there is an rms error of only 8°.

These results raise another interesting question: How accurate are the charge densities, ρ(**r**), and the electrostatic potentials, Φ(**r**), which can be deduced from X-ray data, and how well do they compare with the results of quantum chemical calculations? This question has been the topic of debate by crystallographers for some time, and in recent work using synchrotron sources and area detectors, Flaig and co-workers (*39,40*) have found good agreement using proline and aspartic acid. We have extended this approach to cover other amino-acids and peptides of interest in our NMR studies, and we show in Figure 9 a comparison between the Φ(**r**) of L-asparagine obtained from diffraction data (Figure 9A) and from HF calculations (Figure 9B), together with a comparison between ρ(**r**) values determined at the bond critical points (Figure 9C) as described by Bader (*41*). There is clearly good agreement, although crystal lattice effects cause small differences, as noted by Flaig et al. (*39*) and will need to be incorporated into the quantum chemical calculations (*30*).

Why are these results of interest from an NMR spectroscopist's viewpoint? The principal reason is that obtaining NMR spectra and interpreting them in conventional structural terms (such as a peptide or protein 3D structure), or carrying out e.g. an MO analysis of the different contributions to shielding or the EFG, are really just initial steps in making structure/function correlations, such as better understanding enzyme mechanisms or drug interactions. NMR is exceptionally useful for obtaining structures, and chemical shifts and shift anisotropies can play a role in both initial structure determination and in refinement. But when good structures are obtained, which in the future could well involve quantum chemical local geometry optimizations (*42*), the next steps will be to probe more functional questions, such as: what is the charge density and the electrostatic potential in an active site, or a drug binding site? Here, it is worth pointing out that whenever chemical shifts or electric field gradients (in our case, very often Mössbauer quadrupole splittings) are computed, the SCF part of the calculation already contains all that is needed to evaluate ρ(**r**), Φ(**r**), and $\nabla^2\rho(\mathbf{r})$, the parameters typically used to describe the electrostatic structure of a molecule. So for very little added expense, considerable new and potentially interesting information about a

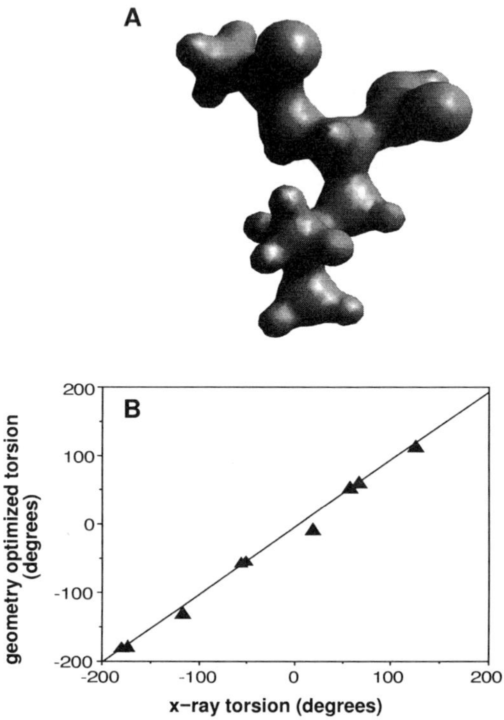

Figure 8. Methyl-orientation results for the three methyl groups in *N*-acetyl-L-leucine. **A**, x-ray determined $\rho(\mathbf{r})$ charge density map. **B**, graph showing relation between the H-C-C-H methyl torsion angles determined from the x-ray results and those determined by using a quantum chemical geometry optimization. The rms error is 8°.

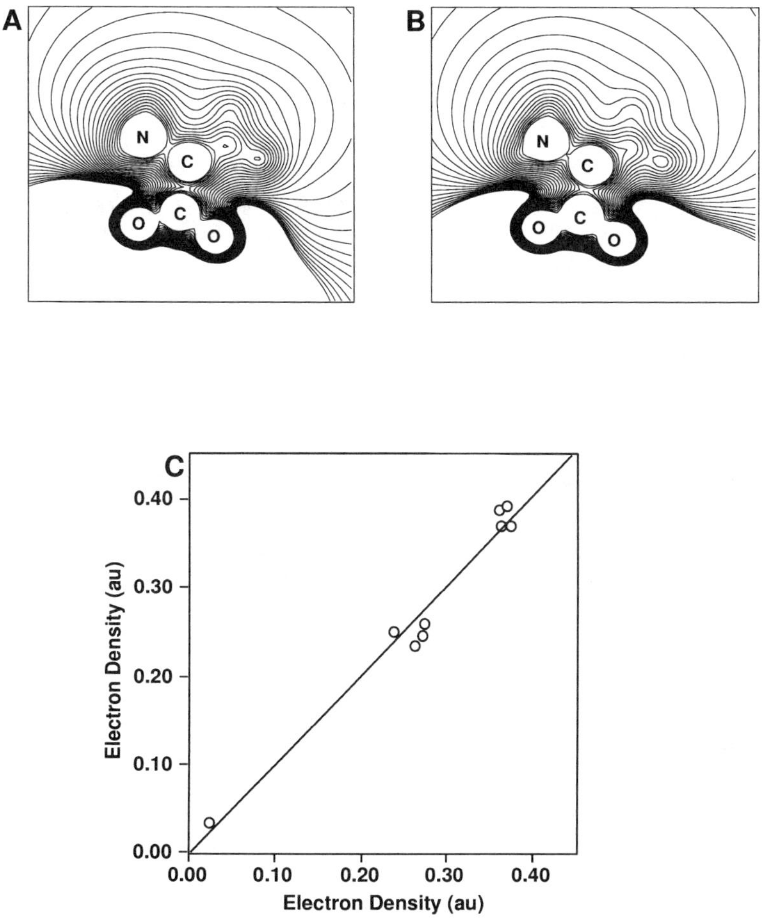

Figure 9. Electrostatic potential results for L-asparagine$\cdot H_2O$ in the OCN plane: **A**, X-Ray, **B**, Hartree-Fock/6-31G**, contours from -0.02 to 0.85 $e(4\pi\varepsilon_o a_o)^{-1}$. **C**, graph showing correlation between experimental $\rho(\mathbf{r})$ and Hartree-Fock/ 6-31G** results: slope = 1.02, $R^2 = 0.973$.

molecule can be obtained. The question of course is - is this information reliable? In early calculations and diffraction experiments, accurate results may only have been possible for very small molecules, but now, much larger molecules can be investigated both crystallographically (using synchrotron or other bright x-ray sources and area detectors) and computationally, and the results obtained appear quite encouraging. This should not be particularly surprising, since obtaining accurate EFG and chemical shift results implies that both ground and excited states are well described, and the electrostatic properties are just ground state properties. Nevertheless, it is comforting to actually see good agreement between theory and (X-ray) experiment. Of course, while such investigations may be of less interest when investigating, say, a simple metal carbonyl like $Fe(CO)_5$ (*43*), they could become very important when trying to solve problems such as how O_2 vs. CO discrimination is effected in metalloproteins, or how a drug binds to a target protein molecule.

This leads us finally to a brief discussion of our recent work on metalloproteins, and their model systems, metalloporphyrins. Here, the basic long-term objectives are to obtain a better understanding of how CO and O_2, and their isoelectronic counterparts the isocyanides (RNC) and nitrosoalkanes/nitrosoarenes (RNO), bind to Fe, as well as to probe the structure of cytochrome *c*, a small protein involved not only in electron-transfer, but in apoptosis, or programmed cell death (*44*).

In our initial studies, we have been interested in computing metal-ion NMR chemical shifts in metalloproteins and model systems. Here, the work of Bühl (*20*) was particularly helpful since it showed, for small Fe and Rh-containing organometallics, that use of the B3LYP hybrid exchange-correlation functional (*45*) gave good predictions of the metal shifts (*20*). We therefore used this approach to investigate ^{57}Fe NMR chemical shifts in metalloproteins, using several model systems made and characterized crystallographically in this laboratory (*46*). We found quite reasonable agreement with experiment for CO myoglobin, RNC myoglobin and cytochrome *c* systems, which extend the shift range discussed by Bühl by ~10,000 ppm (*19,20*). The agreement with experiment was less than that found with the small molecules, but this is hardly surprising given the complexity of these systems. This introduces not only uncertainties in the model crystallographic structure determination, but there could also be differences between the model system and protein local structure. Nevertheless, the results obtained were exciting since they did show good ^{57}Fe shift agreement between theory and experiment for both myoglobin and cytochrome *c*, but in CO-myoglobin good agreement was only found when using the coordinates for our synthetic Fe-CO metalloporphyrin, which contains a linear and untilted Fe-C-O subunit (*23*). The nature of the possible "tilts" and "bends" of Fe-C-O as a mechanism for CO/O_2 discrimination (see e.g. Ref. *47*) is the topic of "lively debate" (see e.g. Ref. *48*), and our results support the undistorted model. For example, the ^{57}Fe shifts computed at the crystallographic geometries are in error by ~6000 ppm (*19*), as shown in Figure 10.

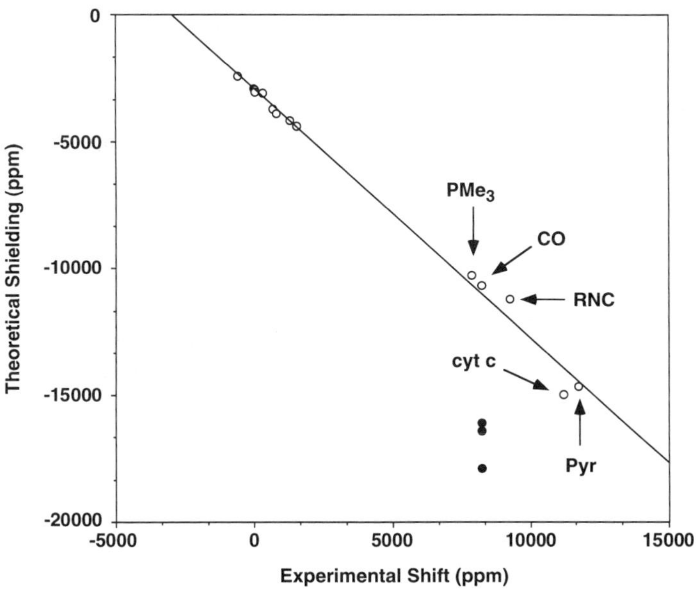

Figure 10. Graph showing the correlation between experimental ^{57}Fe NMR chemical shifts (points at top left are organometallics, from Bühl, Ref. 20, at bottom right, they are for metalloporphyrins, see Ref. 19) and theoretical G94/B3LYP results. The solid circles represent shieldings computed directly from protein crystal structures containing tilted/bent Fe-C-O subunits, and are inconsistent with experiment. See Refs. 19,23 for more details.

Other properties are also available to help determine the local structure of the Fe-C-O subunit in metalloproteins - including ^{13}C and ^{17}O NMR isotropic chemical shifts, ^{17}O electric field gradients, as well as Mössbauer quadrupole splittings. We have employed quantum chemistry to make structure-spectroscopic correlations with each of these parameters, and then used the Bayesian approach to predict ligand tilt (τ) and bend (β), in much the same way as we outlined previously for peptide ϕ, ψ determinations (*15,23*). We investigated first the ^{57}Fe Mössbauer quadrupole splittings in a large number of organometallics and metalloporphyrins, finding an rms error between theory and experiment of only ~0.2 mm sec^{-1} (*24*). We then computed probability surfaces, $P(\tau,\beta)$ for each spectroscopic observable, and deduced the likely τ,β values for the A_0 substate of CO-myoglobin (*23*). The results indicated $\tau \approx \beta \lesssim 5°$, or very close to linear. This view is supported by the results of additional DFT calculations by de Dios and Earle (*25*), who investigated the effects of electrostatic field perturbations of the vibrational frequency, the ^{13}C shift, the ^{17}O shift, as well as the ^{17}O EFG in a simple FeCO cluster (*23,25*). Overall, our results indicate a close-to-linear Fe-C-O subunit in all CO-heme proteins. While an analysis of just the energetics of distortion indicate a fairly soft-potential for τ,β distortions, the ^{13}C, ^{17}O and ^{57}Fe NMR, ^{17}O and ^{57}Fe EFGs, and the IR results, all support the linear model, with electrostatic fields controlling the small differences seen between different proteins and protein substates (*23*).

In metalloproteins, it should also be possible to probe the electronic structure of the heme metallocycle itself, using ^{13}C and ^{15}N shifts, as reported by Grant et al. for some simple covalent metal ion complexes of tetraphenylporphyrins (*49*). We have synthesized two somewhat more complex [$^{15}N_4$]-tetraphenylporphyrins, containing CO and PhNO groups, as metalloprotein models. A typical ^{15}N MAS NMR spectrum is shown in Figure 11A, together with a theory-versus-experiment correlation for the shielding tensor elements for the CO and PhNO metalloporphyrins, Figure 11B. As with simpler model compounds, there is good agreement between theory and experiment. In the future, we plan to extend such DFT calculations to the ^{13}C sites in cytochromes and myoglobins, where some 25 years ago we observed a large chemical shift dispersion for the non-protonated carbons in ferrocytochrome *c* (*50*). While likely a challenging task, using ^{13}C, ^{15}N and ^{57}Fe NMR chemical shifts (and ^{57}Fe EFGs) to help probe both electronic and electrostatic structure in this electron-transfer protein should be of great interest.

In summary then, over the past five years, since the last NMR chemical shift meeting in Maryland (*51*), progress with protein NMR shift (and EFG) calculations has been very exciting. At the time of the last US meeting, our view was that electric fields played a major role in the shielding differences seen due to folding for most nuclei. This now only seems likely for ^{19}F in fluoroaromatic amino-acids and for the CO-heme conformational substates. This is particularly fortunate, however, since chemical shifts are now seen to be primarily probes of ϕ, ψ and χ, the backbone and

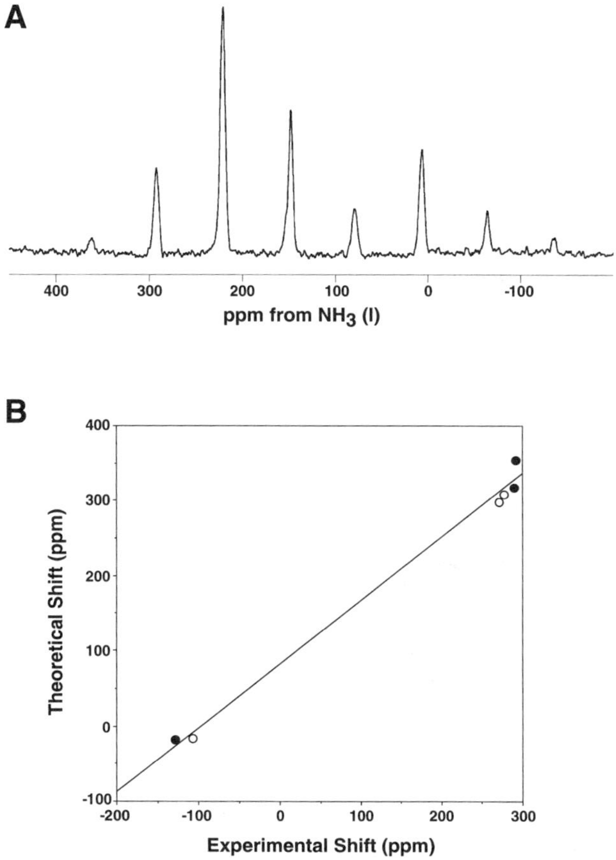

Figure 11. ^{15}N NMR MAS NMR spectra and tensor results for $[^{15}N_4]$-ring labeled metalloporphyrins. **A**, 8.45 Tesla ^{15}N MAS NMR spectrum of Fe(tetraphenylporphyrinate)(PhNO)(1-methylimidazole), ν_r = 2.6 kHz. **B**, graph showing correlation between experimental and DFT computed ^{15}N tensor elements for Fe(TPP)(PhNO)(1-MeIm) (●) and Fe(TPP)(CO)(1-MeIm) (○). The mean experimental and theoretical shieldings over the four non-equivalent sites in each molecule are shown since the solid state shifts are not specifically assigned. Slope = 0.847, R^2 = 0.995.

side-chain torsion angles, which means they can be used in structure refinement. Then, when better structures are obtained with proteins (or already, with smaller model systems), the same calculations we use to evaluate shieldings and EFGs can be used to probe...electrostatics. Since diffraction-derived electrostatic properties are already in quite good agreement with quantum chemical predictions, it seems likely that new insights into protein function (involving e.g. hydrogen bonding, enzymology, and drug design) will be obtained early on in the next millennium, by using shifts, shift anisotropies and quantum chemistry (and of course dipolar-couplings (*52*)) to help refine structure, and quantum chemistry to help probe function.

Acknowledgments. This work was supported by the United States Public Health Service (National Institutes of Health grants HL-19481 and GM-50694), and utilized computational facilities made available by the National Center for Supercomputing Applications (supported by NSF grant CHE97-0020N). We are grateful to Professor P. Coppens for assistance with multipole refinement of our x-ray data.

Literature Cited

1. McDonald, C. C.; Phillips, W. D. *J. Am. Chem. Soc.* **1967**, *89*, 6332-6341.
2. Allerhand, A.; Childers, R. F.; Oldfield, E. *Biochemistry* **1973**, *12*, 1335-1341.
3. Ösapay, K.; Case, D. A. *J. Am. Chem. Soc.* **1991**, *113*, 9436-9444.
4. Wang, J.; Hinck, A. P.; Loh, S. N.; LeMaster, D. M.; Markley, J. L. *Biochemistry* **1992**, *31*, 921-936.
5. Torchia, D. A.; Sparks, S. W.; Bax, A. *Biochemistry* **1989**, *28*, 5509-5524.
6. Lian, C.; Le, H.; Montez, B.; Patterson, J.; Harrell, S.; Laws, D.; Matsumura, I.; Pearson, J.; Oldfield, E. *Biochemistry* **1994**, *33*, 5238-5245.
7. Park, K. D.; Guo, K.; Adebodun, F.; Chiu, M. L.; Sligar, S. G.; Oldfield, E. *Biochemistry* **1991**, *30*, 2333-2347.
8. Augspurger, J. D.; Dykstra, C. E.; Oldfield, E. *J. Am. Chem. Soc.* **1991**, *113*, 2447-2451.
9. Augspurger, J.; Pearson, J.; Oldfield, E.; Dykstra, C. E.; Park, K. D.; Schwartz, D. *J. Magn. Res.* **1992**, *100*, 342-357.
10. Luck, L. A.; Falke, J. J. *Biochemistry*, **1991**, *30*, 4248-4256; Luck, L. A.; Falke, J. J. *Biochemistry* **1991**, *30*, 4257-4261.
11. Pearson, J. G.; Oldfield, E.; Lee, F. S.; Warshel, A.; *J. Am. Chem. Soc.* **1993**, *115*, 6851-6862.
12. Pulay, P.; Wolinski, K.; Hinton, J. F. *The Texas Program*; University of Arkansas, Fayetteville, AR, 1991; Wolinski, K.; Hinton, J. F.; Pulay, P. *J. Am. Chem. Soc.* **1990**, *112*, 8251-8260.
13. de Dios, A. C.; Pearson, J. G.; Oldfield, E. *Science* **1993**, *260*, 1491-1496.

14. Pearson, J. G.; Wang, J.-F.; Markley, J. L.; Le, H.; Oldfield, E. *J. Am. Chem. Soc.* **1995**, *117*, 8823-8829.
15. Heller, J.; Laws, D. D.; King, D. S.; Wemmer, D. E.; Pines, A.; Havlin, R. H.; Oldfield, E. *J. Am. Chem. Soc.* **1997**, *119*, 7827-7831.
16. de Dios, A. C. *Prog. NMR Spectroscopy* **1996**, *29*, 229-278.
17. Oldfield, E. *J. Biomol. NMR* **1995**, *5*, 217-225.
18. Godbout, N.; Oldfield, E. *J. Am. Chem. Soc.* **1997**, *119*, 8065-8069.
19. Godbout, N.; Havlin, R.; Salzmann, R.; Wojdelski, M.; Debrunner, P. G.; Oldfield, E. *J. Phys. Chem.* **1998**, *102*, 2342-2350.
20. Bühl, M. *Chem. Phys. Lett.* **1997**, *267*, 251-257.
21. Havlin, R. H.; McMahon, M.; Srinivasan, R.; Le, H.; Oldfield, E. *J. Phys. Chem.* **1997**, *101*, 8908-8913.
22. Salzmann, R.; Kaupp, M.; McMahon, M.; Oldfield, E. *J. Am. Chem. Soc.* **1998**, *120*, 4771-4783.
23. McMahon, M.; de Dios, A. C.; Godbout, N.; Salzmann, R.; Laws, D. D.; Le, H.; Havlin, R. H.; Oldfield, E. *J. Am. Chem. Soc.* **1998**, *120*, 4784-4797.
24. Havlin, R. H.; Godbout, N.; Salzmann, R.; Wojdelski, M.; Arnold, W.; Schulz, C. E.; Oldfield, E. *J. Am. Chem. Soc.* **1998**, *120*, 3144-3151.
25. de Dios, A. C.; Earle, E. M. *J. Phys. Chem. A* **1997**, *101*, 8132-8134.
26. Kutzelnigg, W.; Fleischer, U.; Schindler, M. in *Deuterium and Shift Calculation*; Diehl, P.; Fluck, E.; Günther, H.; Kosfeld, R.; Seelig, J., Eds.; *NMR Basic Principles and Progress*; Springer-Verlag: Berlin, 1990; Vol. 23, pp 166-262.
27. Malkin, V. G.; Malkina, O. L.; Casida, M. E.; Salahub, D. R. *J. Am. Chem. Soc.* **1994**, *116*, 5898-5908; Malkin, V. G.; Malkina, O. L.; Eriksson, L. A.; Salahub, D. R. *Modern Density Functional Theory: A Tool for Chemistry; Theoretical and Computational Chemistry*; Seminario, J. M., Politzer, P., Eds.; Elsevier: Amsterdam, 1995; Vol. 2.
28. Ruiz-Morales, Y.; Schreckenbach, G.; Ziegler, T.; Oldfield, E., to be published.
29. Havlin, R. H.; Le, H.; Laws, D. D.; deDios, A. C.; Oldfield, E. *J. Am. Chem. Soc.* **1997**, *119*, 11951-11958.
30. Janes, N.; Ganapathy, S.; Oldfield, E. *J. Magn. Res.* **1983**, *54*, 111-121.
31. Le, H.; Pearson, J. G.; de Dios, A. C.; Oldfield, E. *J. Am. Chem. Soc.* **1995**, *117*, 3800-3807.
32. Box, G. E. P.; Tiao, G. C. *Bayesian Inference in Statistical Analysis*; John Wiley and Sons, Inc.: New York, NY, 1992.
33. Tjandra, N.; Bax, A. *J. Am. Chem. Soc.* **1997**, *119*, 9576-9577.
34. Sitkoff, D.; Case, D. A. *Prog. NMR Spectroscopy* **1998**, *32*, 165-190.
35. Moreno, B.; Arnold, W.; Szabo, C. M., to be published.
36. Fawcett, J. K.; Camerman, N.; Camerman, A. *Acta Cryst., Sect. B* **1975**, *B31*, 658-665.
37. Herzfeld, J.; Berger, A. E. *J. Chem. Phys.* **1980**, *73*, 6021-6030.

38. Keniry, M. A.; Kintanar, A.; Smith, R. L.; Gutowsky, H. S.; Oldfield, E. *Biochemistry* **1984**, *23*, 288-298.
39. Koritsánszky, Flaig, R.; Zobel, D.; Krane, H.-G.; Morgenroth, W.; Luger, P. *Science* **1998**, *279*, 356-358.
40. Flaig, R.; Koritsanszky, T.; Zobel, D.; Luger, P. *J. Am. Chem. Soc.* **1998**, *120*, 2227-2238.
41. Bader, R. F. W. *Atoms in Molecules. A Quantum Theory*; Clarendon Press: Oxford, 1990.
42. Pearson, J. G.; Le, H.; Sanders, L. K.; Godbout, N.; Havlin, R. H.; Oldfield, E. *J. Am. Chem. Soc.* **1997**, *119*, 11941-11950.
43. Salzmann, R.; Kaupp, M.; McMahon, M.; Oldfield, E. *J. Am. Chem. Soc.* **1998**, *120*, 4771-4783.
44. Hengartner, M. O. *Nature*, **1998**, *391*, 441-442.
45. Becke, A. D. *J. Chem. Phys.* **1993**, *98*, 5648-5652; Lee, C.; Yang, W.; Parr, R. G. *Phys. Rev. B* **1988**, *37*, 785-789.
46. Salzmann, R.; McMahon, M. T.; Godbout, N.; Sanders, L. K.; Wodjelski, M.; Oldfield, E. submitted to *J. Am. Chem. Soc.*
47. Stryer, L. *Biochemistry*, Fourth Edition; W. H. Freeman and Company: New York, NY, 1995.
48. Service, R. F. *Science* **1995**, *269*, 921-922.
49. Solum, M. S.; Altmann, K. L.; Strohmeier, M.; Berges, D. A.; Zhang, Y.; Facelli, J. C.; Pugmire, R. J.; Grant, D. M. *J. Am. Chem. Soc.* **1997**, *119*, 9804-9809.
50. Oldfield, E.; Allerhand, A. *Proc. Natl. Acad. Sci. U.S.A.* **1973**, *70*, 3531-3535.
51. Augspurger, J. D.; Dykstra, C. E.; Oldfield, E.; Pearson, J. G. in *Nuclear Magnetic Shieldings and Molecular Structure*; Tossell, J. A., Ed.; NATO ASI Series C; Kluwer Academic Publishers: Boston, MA, 1993, Vol. 386; pp 75-94.
52. Tjandra, N.; Bax, A. *Science* **1997**, *278*, 1111-1114.

Chapter 4

NMR in Catalysis: Theoretical and Experimental Approaches

John B. Nicholas[1] and James F. Haw[2]

[1]**Environmental Molecular Sciences Laboratory, Pacific Northwest National Laboratory, P.O. Box 999, Richland, WA 99352**
[2]**Loker Hydrocarbon Research Institute and Department of Chemistry, University of Southern California, University Park, Los Angeles, CA 90089–1661**

Theoretical and Experimental NMR techniques provide powerful tools for the investigation of heterogeneous catalysis. Recent advances in *in situ* NMR techniques are summarized, as are advances in theoretical methods. The utility of our combined theoretical/experimental approach is illustrated by studies of the pentamethylbenzenium cation and the 1,3-dimethylcyclopentenyl cation in zeolite HZSM-5, acetylene adsorption on MgO, and the isopropyl cation on frozen SbF_5. We also discuss the role of the basicity of adsorbates in the formation of stable carbenium ions on zeolites.

Heterogeneous catalysis (*1-4*) is essential to many of the world's key industries, including petroleum refining and the synthesis of fine chemicals, solvents, monomers and polymers. A greater basic understanding of heterogeneous catalysis, resulting from the collective efforts of many researchers, can reasonably be expected to lead to important new processes and better economics for existing processes. Furthermore, heterogeneous catalysis will also play an important role in the redesign of existing manufacturing processes to minimize their environmental impact. Although these investigations are easily justified based on national priorities, the characterization of catalytic reaction mechanisms provides significant basic scientific challenges and opportunities for curiosity-driven research.

For the past decade, the major thrust of one of the authors has been the elucidation of reaction mechanisms in heterogeneous catalysis on zeolites, metal oxides and other materials. The primary experimental tool has been *in situ* NMR spectroscopy, and this is increasingly coupled with theoretical calculations carried out by the other author.

NMR is firmly established as a research tool in catalysis, and the interested reader is referred to *NMR Techniques in Catalysis* (*5*), edited by Bell and Pines and published in 1994, for an overview of the diverse research objectives and approaches in NMR studies of catalytic materials.

Carbon is present in a majority of the reactants of interest in heterogeneous catalysis, and ^{13}C is a very attractive nucleus for chemical investigations by NMR, especially with isotopic enrichment. Whereas the expense of stable-isotope chemicals would appear to be a disadvantage, labeling has advantages that go far beyond improved analytical sensitivity. Selective isotopic labeling is central to most of the experimental methods used to study mechanism in solution and gas phase chemistry, and our objective is to elucidate reaction mechanisms in heterogeneous catalysis. The fate of a ^{13}C label—be it completely scrambled, selectively scrambled or unaltered—has been a central observation of a number of our published *in situ* studies, and, on more than one occasion such a finding was an unplanned benefit of labeling for the sake of sensitivity. Of course other nuclei, including ^{1}H, ^{27}Al, ^{29}Si and ^{15}N, can also be very informative in NMR studies of catalysis, but ^{13}C has figured in a great many of our investigations and here we will focus on that nucleus to a similar extent.

How are NMR measurements of adsorbates and reactive species on catalysts made? This has been the subject of several recent reviews (*6-8*). In brief, there are a number of approaches that differ in the extent to which they are intended to reproduce the conditions of an actual catalysis experiment. Heterogeneous catalysis in a microreactor is typically carried out in a carrier gas flow stream at temperatures of 673 K or even higher. Products emerging from the microreactor are most commonly analyzed with an on-line gas chromatograph (GC). The contact time of the reactant on the catalyst typically ranges between several tenths of a second to several tens of seconds. In contrast, the acquisition of a ^{13}C spectrum of an adsorbate on a typical catalyst at 298 K takes several minutes or tens of minutes, even with enriched compounds. As the sample temperature increases to the range commonly used in catalysis, NMR sensitivity rapidly decreases, and the discrepancy in time scale becomes greater.

Fortunately, many interesting questions are really about the surface chemistry of a particular adsorbate on a catalytic material. It is often sufficient to adsorb that species onto an active catalyst sample, using good vacuum line techniques, and measure the NMR spectrum at room temperature or below, as needed, to freeze out dynamics or prevent reaction. ^{13}C experiments are commonly done using magic-angle spinning; slow spinning speeds give several orders of side bands and these can be analyzed using the method of Herzfeld and Berger (*9*) to obtain principal components of the chemical shift tensor.

As far as actual catalysis goes, different kinds of *in situ* NMR experiments have been developed that are intended to reproduce some of the conditions in a catalytic reactor. Most simply, the sample is sealed in a glass ampoule and heated outside of the probe between NMR measurements at room temperature (*10*). In other cases variable temperature MAS NMR probes are used to combine the steps of heating and measuring spectra (*11*). These experiments are still a long way removed from the conditions of a flow reactor. Attempts to make *in situ* NMR measurements more closely correspond to the conditions of actual catalysis have included transient heating with a laser (*12*) and high temperature MAS probes with some degree of reagent flow (*13*).

A different approach to *in situ* NMR studies makes use of the pulse-quench catalytic reactor (*14-15*). This has all of the characteristics of a standard research microreactor including provision for continuous and/or pulse introduction of reactants and gas chromatography for analysis of volatile products that exit the reactor. The most significant feature of the pulse-quench reactor is provision for rapidly switching the gas stream flowed over the catalyst bed to cryogenically cooled nitrogen so as to very rapidly reduce the catalyst temperature. Instrumentally, this is achieved through the use of high speed, pneumatically actuated valves and computer control. The temperature of the catalyst bed is measured by a high speed thermocouple in contact with zeolite pellets on the downstream end of the catalyst bed. Measurements show that the catalyst temperature decreases 50 K in the first 80 ms of a quench, a further 100 K decrease occurs in the following 90 ms, and the catalyst temperature is ambient or lower in less than one second. Chemical reaction rates are strongly temperature dependent; once the catalyst temperature drops the first 150 K, the extent of further conversion of reactants to products in the time required for the quench to reach room temperature will usually be negligible.

Chemical equilibria are frequently temperature dependent, and we should not expect that the organic phase trapped in the quenched catalyst will always be identical to that existing immediately prior to the quench. However, the organic material in the quenched catalyst should be sufficiently related to its antecedents that inferences about structures present at higher temperatures will be possible. Another advantage of the pulse-quench reactor that proved useful is the fact that products are removed from the catalyst bed in this experiment. This has proven useful for the observation of hydrolytically unstable species which form in reactions that also generate water (*16*). In sealed ampoules or rotors, the water can not escape the catalyst bed; in the pulse-quench reactor, the water is swept out in the gas stream. Most of the experiments that motivated the calculations in this contribution were performed using "conventional" sealed-rotor methods, but the pentamethylbenzenium study (*vide infra*) would not have been possible without the pulse-quench reactor.

However the sample is prepared, we measure ^{13}C spectra of one or more adsorbates on the catalyst, and then need to interpret the spectra to deduce the structure of adsorption complexes or reactive intermediates formed on the catalyst. In many cases the complexes and intermediates formed are unusual and exotic species for which the interpretation of the spectra may be far less than routine. This is where *ab initio* chemical shift calculations are essential. In diffraction methods, such as x-ray or neutron diffraction, one can more-or-less easily invert the experimental data to yield molecular structure. There is no straightforward relationship between chemical shift data and structure; theoretical calculations provide the bridge between experiment and theory. In a typical study, we model the adsorbates on clusters that represent catalyst active sites, using experience and chemical intuition to create our initial structures.

These adsorbate-catalyst complexes are optimized, their NMR spectra are calculated, and the results are compared to experiment. If significant disagreement is found, the structures are modified and the procedure is repeated until a clear interpretation of the chemistry emerges.

From a purely theoretical point of view we must consider how the NMR calculations are done, and the factors that influence their accuracy. Early attempts at the theoretical calculation of NMR data were hampered by computational expense and the problem of gauge dependence, common to all magnetic phenomena. The idea of using gauge-including atomic orbitals (GIAOs, also known as London orbitals), which have a formal dependence on magnetic field strength and thereby circumvent the gauge origin problem, was first implemented at the *ab initio* Hartree-Fock level by Ditchfield in 1974 (*17*). Theoretical NMR calculations are now possible using a variety of methods (*18*). In 1990, Wolinski, Hinton, and Pulay reformulated the GIAO approach in terms of analytic derivative techniques (*19*) making NMR calculations at the restricted Hartree-Fock (RHF) level much more tractable.

As might be expected, NMR calculations that ignore electron correlation often give poor results, especially for molecules which typically require a correlated treatment in order to predict other properties accurately. For example, a good description of multiple bonds and lone pairs generally requires a correlated method. Thus, RHF NMR predictions for molecules such as CO and acetonitrile are poor (*20*). Furthermore, it has recently been shown that isotropic chemical shift calculations at the RHF level are unreliable for benzenium (*21*) and related carbenium ions which we often encounter in catalysis.

Thus, another significant advance in theoretical NMR was the recent development of methods such as the GIAO-MP2 technique of Gauss (*22*), which include the effect of electron correlation on magnetic properties. Chemical shift calculations at the MP2 level significantly improve the calculated isotropic chemical shifts of neutral molecules with triple bonds and the benzenium ion discussed above. Of course, the treatment of these difficult cases also requires an adequate basis set. In many cases a triple-zeta basis set is needed in order to provide enough flexibility for accurate NMR predictions. Diffuse functions and multiple polarization functions may also be required. Gauss has since extended the GIAO formalism to include higher order correlation at the MP3, MP4, and coupled cluster levels (*23-24*).

Implicit in the above discussion was the assumption that the geometry of the molecules under study are also determined to sufficient accuracy. Thus, we generally also need correlated methods for the geometry optimizations prior to NMR calculations. Because many of the systems we study are large, we have found that density functional theory (DFT) methods are often very useful, providing the effects of electron correlation in a computationally expedient fashion. In the following we provide several examples that demonstrate the utility of NMR calculations in the interpretation of experimental studies in catalysis.

The Pentamethylbenzenium Ion in HZSM-5

There is considerable controversy with regard to the formation of persistent (long-lived) carbenium ions in zeolite solid acids (*8*). The conventional view has been that stable carbenium and even carbonium ions are readily formed from even weakly basic molecules such a propene. However, there is increasing evidence that the formation of persistent species requires a strongly basic precursor, and is thus only possible in select cases (*25*). Consequently, we feel that any positive claims of persistent carbenium species in zeolites warrant exceptional proof.

Using the pulse flow catalytic reactor mentioned earlier, we were able to create a pentamethylbenzenium (Figure 1) in zeolite HZSM-5 (*16*). The cation was synthesized in the channel intersections or pore mouths by alkylating benzene or toluene with methanol at 523 K. This cation cannot be prepared in detectable concentration when the reaction is carried out in a sealed rotor, as is commonly done in *in situ* NMR studies. In contrast, when alkylation is carried out in a flow reactor, the co-product—water—is removed, and the cation accumulates as a significant product in the zeolite.

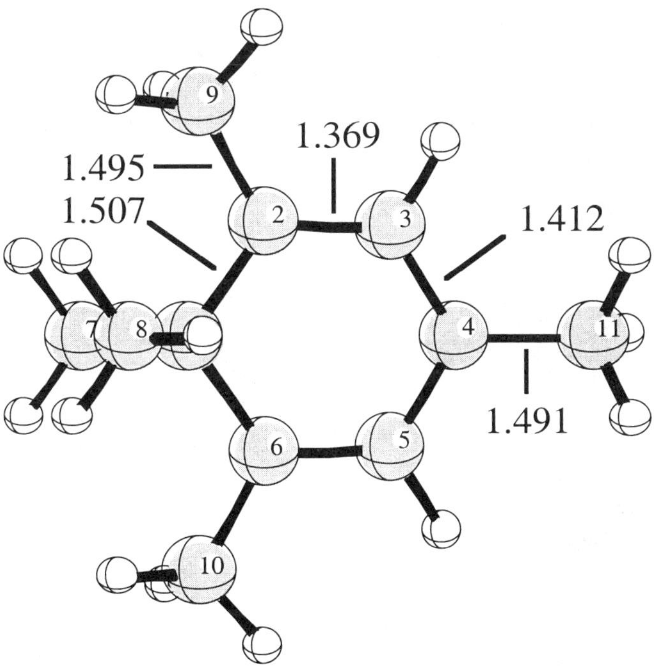

Figure 1. The C_S geometry of the pentamethylbenzenium cation with carbons labeled as referred to in the text. Selected bond lengths in Å are also shown. (Reproduced from reference 13. Copyright 1995 American Chemical Society.)

The experimental ^{13}C NMR spectra of this cation exhibits clear signals at 206, 190, 139, and 58 ppm, and incompletely resolved signals between 23 and 26 ppm. To verify the identity of this species, we first determined the geometry with DFT methods. The optimizations were done at the B3LYP level with the 6-311G** basis set. The lowest energy minimum geometry has C_1 symmetry. A rotation of the C7 methyl group gives C_s symmetry, and only raises the energy by less than 0.1 kcal/mol. In order to make the NMR calculation tractable, we used the C_s geometry.

In C_s symmetry, the two methyl groups bonded to C1 are not equivalent; the orientation of the two methyls with regards to the plane of the ring is different. However, experimentally, dynamical averaging will make these two carbons equivalent in the NMR spectra. The predicted ^{13}C isotropic chemical shifts were calculated at the GIAO-MP2 level with a tzp basis set on C and dz on H. The isotropic chemical shifts are referenced to the value for carbon in TMS, obtained at the same level of theory. The theoretical isotropic chemical shifts are: C1, 65 ppm; C2 and C6, 209 ppm; C3 and C5, 139 ppm; C4, 191 ppm; C7, 23 ppm; C8, 35 ppm; C9 and C10, 28 ppm; C11, 29 ppm. For carbons 2-7 the theoretical isotropic shifts agree to within a few ppm. For C1 the agreement is less good, with the theoretical prediction being 7 ppm downfield of the experimental value. This discrepancy may be due to the effects of the dynamical interconversion of the positions of the C7 and C8 methyl groups. We could also consider that for the same reason the observed chemical shifts for C7 and C8 are better represented by the average of the two theoretical values; 29 ppm. Overall, the agreement between and theoretical and experimental chemical shifts is sufficiently close to verify the experimental interpretation of a stable carbenium species in the zeolite.

The 1,3-dimethylcyclopentenyl Carbenium Ion in HZSM-5

A very similar study to the one described above regards the formation of cyclopentyl carbenium ions in HZSM-5 (Haw, J.F.; Nicholas, J.B.; Deng, F.; Song, W.; Wang, Z. ; Xu, T. to be submitted) The 1,3-dimethylcyclopentenyl carbenium ion (Figure 2) is an apparent intermediate in the hydrocarbon synthesis in methanol-to-olefin (MTO) chemistry. This carbenium ion was also created using the catalytic flow reactor. Under flow conditions, the half-life of the carbenium ion in the catalyst bed is measured isotropic resonances are at 250, 148, 33, and 24 ppm. To verify the identity of the 1,3-dimethylcyclopentyl carbenium ion we first optimized the geometry at the MP2/6-311+G* level of theory. The core electrons were not included in the correlation treatment. Chemical shieldings were then calculated with the GIAO method at the MP2 level using a tzp basis set on carbon and dz on the hydrogens.

We show the MP2/6-311+G* optimized geometry of the cation in Figure 2. Selected bond distances are shown. The molecule has C_{2v} symmetry. The C1-C2 bond lengths (1.395 Å) are very similar to the C-C bond lengths in benzene (1.400 Å at MP2/6-311+G*), consistent with the delocalized nature of the electron density and sp^2

hybridization. The C2-C4 (1.477 Å) and C2-C3 (1.490 Å) bond lengths also reflect partial double-bond character. In contrast, the C3-C3 bond length (1.530 Å) is consistent with a C-C single bond (1.541 Å in ethane at MP2/6-311+G*).

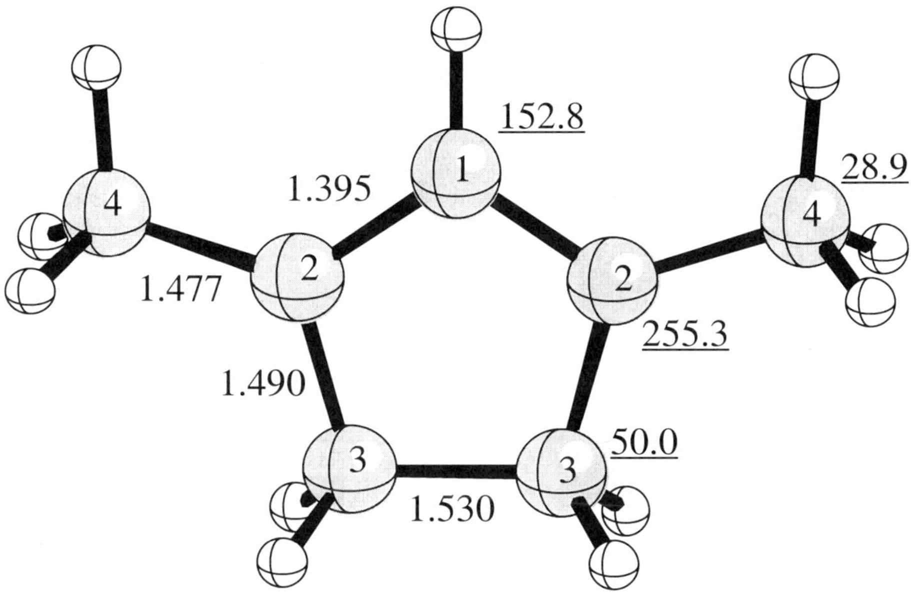

Figure 2. The MP2/6-311+G* optimized geometry of the 1-3-dimethylcyclopentyl carbenium ion. Selected bond distances in Å are shown. The GIAO-MP2/tzp/dz values of the ^{13}C isotropic chemical shifts are underlined.

Also shown in Figure 2 are the GIAO-MP2/tzp/dz values of the isotropic ^{13}C chemical shifts. The predicted chemical shift for C2 and C3 is 255.3 ppm, which compares to the experimental value of 250 ppm. For C2, theory predicts a chemical shift of 152.8 ppm, whereas the experimental value is 148 ppm. The CH_2 carbons are predicted to have isotropic chemical shifts of 50.0 ppm which compare to the measured values of 33 ppm. Finally, the methyl carbons have theoretical values of 28.9 ppm, whereas the experimental chemical shifts are 24 ppm. In all cases the theoretical values are downfield of the experimental chemical shifts. The differences are generally ≈5 ppm, with the exception of C3. The source of this larger difference is not clear. Still, the agreement is sufficient to verify the presence of the 1,3-dimethylcyclopentyl carbenium ion within the zeolite.

Acetylene Adsorption of Nano-sized MgO

The first two examples both involved the creation of cationic species on an acidic zeolite. In both cases we did not need to model the interaction of the cation with the zeolite framework; good agreement was obtained with just calculation of the isolated cation. Apparently, the cation is not strongly perturbed by the presence of the zeolite. Such fortunate circumstances are rare. Here we show an example of how theoretical NMR calculations can help elucidate chemistry on a basic metal oxide surface, in particular, the adsorption of acetylene on MgO (*26*). For this study we needed to model the active sites of the catalyst, for which there are many possibilities. It is assumed the reactive sites are those in which Mg and O are substantially less coordinated than in the bulk. "Corner" sites are those in which Mg or O are three-coordinate, whereas "Edge" sites have four-fold coordination. These sites are where the strongest binding of the adsorbates are obtained.

Experimentally, we used in situ ^{13}C NMR with magic angle spinning (MAS) to measure spectra of the species formed by adsorption of acetylene on MgO powders. Three signals were observed following the initial adsorption of acetylene, 77, 96, and 121 ppm. Two additional peaks at 27 and 58 ppm became apparent after the catalyst was stored at 298 K for 12 hours.

We again used DFT methods for our theoretical determination of the geometries. Optimizations were done using the B3LYP exchange-correlation functional and the tzvp basis set. Models of the surface ranging in size from Mg_4O_4 to Mg_8O_8 were explored. The MgO models were held fixed in crystallographic positions. Our initial theoretical calculations indicated that acetylene was strongly bound to corner sites. The molecule would lose a proton to a surface oxygen and form an acetylide with carbon bound to surface Mg.

We considered several such acetylides bound to corner and edge sites on the various size models of the surface. The optimized geometry of one of these acetylides is shown in Figure 3a. Further reaction of the acetylide would lead to the formation of a vinylidene, shown in Figure 3b. We also optimized geometries for several vinylidenes on the different surface models.

To compare with the experimental NMR data, the ^{13}C isotropic chemical shifts of all of the acetylide and vinylidene complexes were calculated. These systems were too large for us to be able to do chemical shift calculations at the MP2 level. We thus used the GIAO method at the RHF level. Fortunately, the RHF level of theory appears to be adequate for these complexes. We used a mixed basis set for the GIAO, with tzplarge+ on C, tzplarge on H, tzp on the Mg and O associated with the adsorbed species, and dzp on the remaining Mg and O (*26*).

For the acetylide model shown in Figure 3a, the calculated chemical shifts are 101.0 and 120.9 ppm, which correspond to the experimental values of 96 and 121 ppm mentioned above. In contrast, the calculated chemical shifts for the vinylidene are

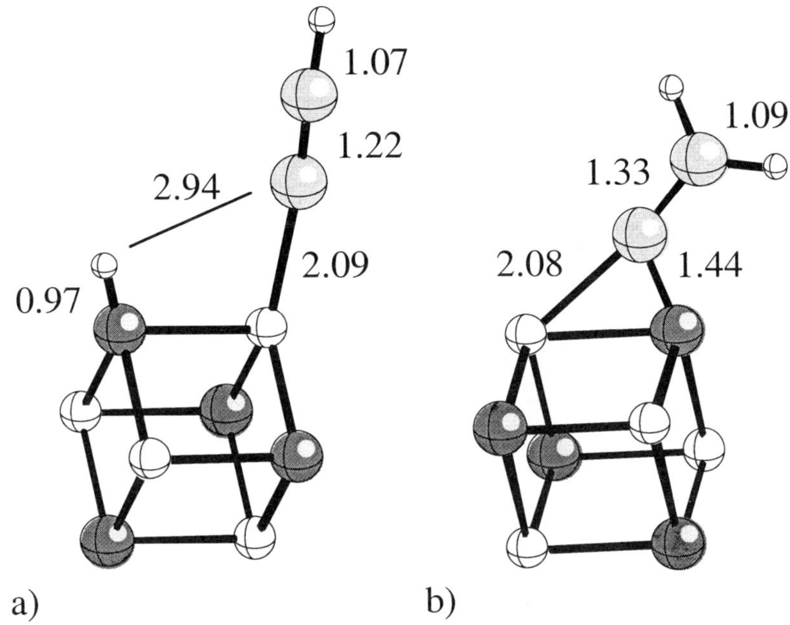

Figure 3. a) The B3LYP/tzvp optimized geometry of the acetylide on Mg_4O_4. b) The B3LYP/tzvp optimized geometry of the vinylidene on Mg_4O_4. Selected bond lengths in Å are also shown. (Reproduced from reference 25. Copyright 1998 American Chemical Society.)

103.7 and 233.1 ppm. While one carbon has an isotropic chemical shift in reasonable agreement with experiment, the shift of the other carbon is clearly not observed in the experimental spectra. We were thus able to verify the presence of the acetylide, and rule out the existence of the vinylidene. Similar conclusions could be drawn from the calculations using the larger surface models.

As noted earlier, additional signals are observed on the aged catalyst. A variety of schemes can be proposed by which acetylene undergoes further reaction to create new species. One possibility for the source of the two signals is the formation of some type of ethoxide. The ethoxide can involve either an oxygen that is formally part of the surface, or an oxygen that comes from a hydroxyl group, which are also known to be present on the surface. Figure 4 shows optimized geometries for these two cases. The GIAO-RHF isotropic chemical shifts for the first "surface" ethoxide are 19.3 and 49.1 ppm, in reasonable but not excellent agreement with the experimental values of 27 and 58 ppm. The chemical shifts of the other surface ethoxy are 23.8 and 55.1 ppm, giving

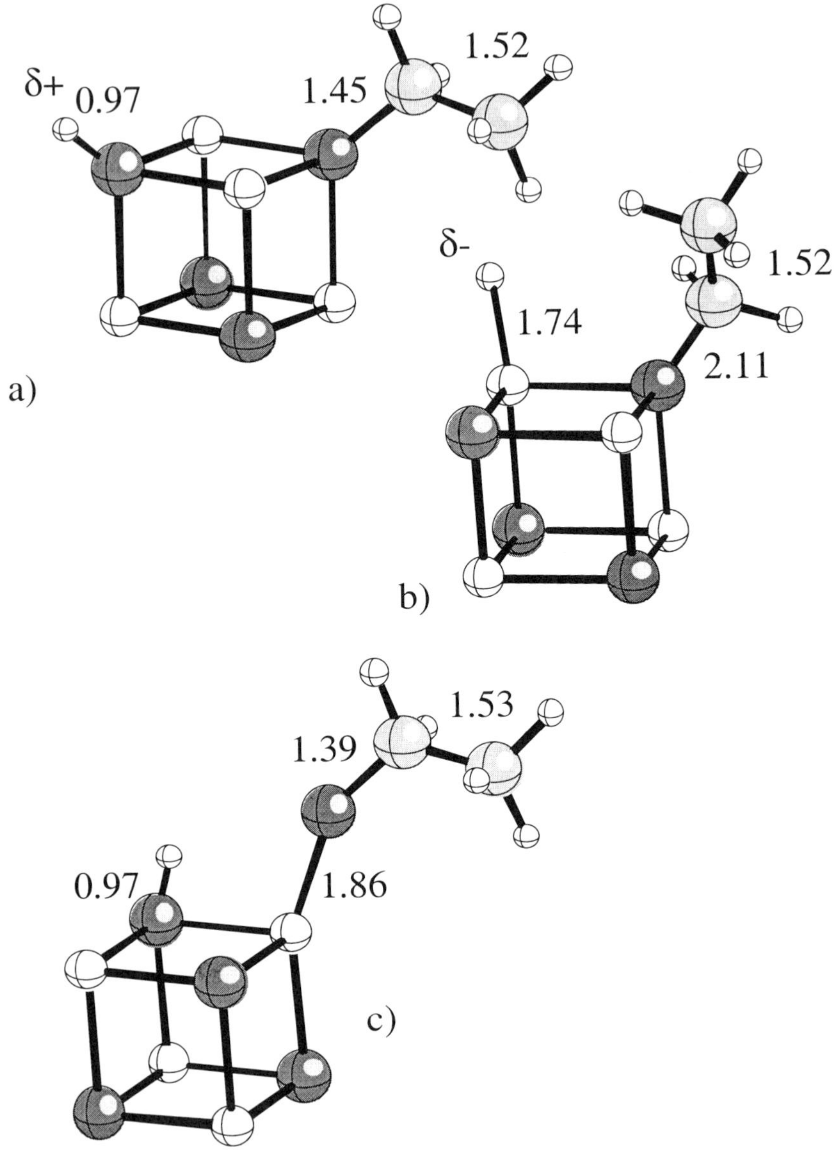

Figure 4. a, b) Models of the ethoxide bound to a surface oxygen. c) A Model of the ethoxide that might be formed with a surface hydroxyl group. All geometries were optimized at the B3LYP/tzvp level. Selected bond lengths in Å are also shown. (Reproduced from reference 25. Copyright 1998 American Chemical Society.)

a better agreement. The chemical shifts for the dangling ethoxy are 24.7 and 57.1 ppm, in excellent agreement with the experimental chemical shifts. While the theoretical results so far obtained cannot unambiguously differentiate between surface and dangling ethoxy, it is clear that some form of ethoxy was present and was detected by the NMR experiment. Other theoretical results were obtained and where possible, compared to experimental data. In particular, the IR frequencies for all the acetylide, vinylidene, and ethoxy species were calculated and compared to experimental spectra. The interpretation of the theoretical and experimental IR was also consistent with the formation of only the acetylide. In addition, the binding energies of all species were compared. The acetylide is thermodynamically favored over the vinylidene. The one resonance which we will not address in this discussion is that which is observed at 77 ppm. This signal is attributed to physiadsorbed acetylene.

The Isopropyl Cation on Solid Lewis Acids

For our final example of theoretical NMR in catalysis, we again turn to carbenium ion chemistry. Here we study the formation of the isopropyl cation on frozen SbF_5, a strong Lewis acid (*27*). In contrast to the studies presented earlier, with this system we were able to experimentally measure the chemical shift tensor. Because the full tensor is naturally obtained from NMR calculations, a comparison can readily be made. In addition, for the isopropyl cation we also studied the effect the medium (in this case, the charge balancing anion) had on the chemical shift tensor.

Experimentally, the isopropyl cation was prepared by the low temperature reaction of 2-bromopropane-2-^{13}C with frozen SbF_5. The ^{13}C spectrum was measured at 83 K using slow speed magic angle spinning. Analysis of the spectrum using the method of Herzfeld and Berger yielded tensor values of δ_{11} = 497 ppm, δ_{22} = 385 ppm and δ_{33} = 77 ppm for the central carbon, which results in an isotropic chemical shift of 320 ppm.

Theoretically, we optimized the geometry using the MP2/6-311++G** level of theory. We calculated the chemical shift tensor using the GIAO-MP2 method with the same tzp/dz basis set scheme described earlier. We also calculated NMR data using the GIAO method at the DFT (B3LYP) level using the same tzp/dz basis set and the MP2/6-311++G** geometries. The lowest energy geometry of the isopropyl cation has C_2 symmetry. The calculated isotropic shift for the C_2 structure is 8 ppm larger than that from experiment. The moderate discrepancy in the isotropic shift actually belies larger and partially offsetting discrepancies in the principal components. For example, δ_{11} was calculated to be 545 ppm, but the experimentally measured value was only 497 ppm.

As expected for C_2 symmetry, the cation is formally chiral, and can racemize by methyl group rotation, passing through a C_s transition state. Our calculated barrier for this process is ca. 0.7 kcal/mol, so this process is rapid on the NMR time scale, even at temperatures well below 77 K.

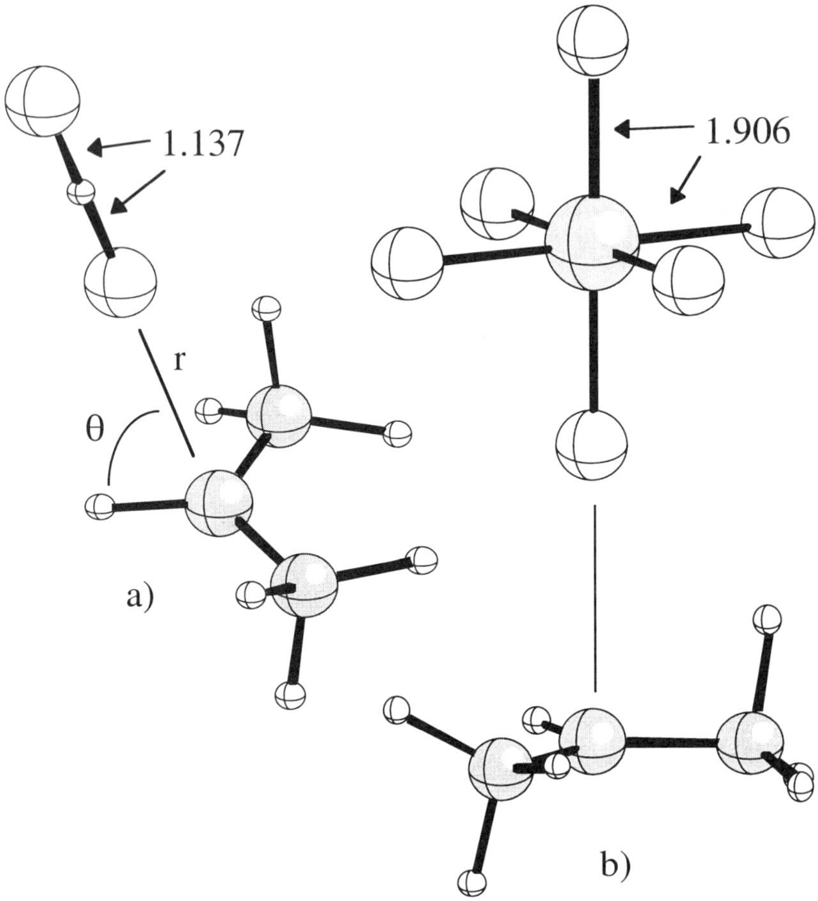

Figure 5. a) The isopropyl cation coordinated to FHF^- counterion. b) The isopropyl cation coordinated to SbF_6^-. (Reproduced from reference 26. Copyright 1998 American Chemical Society.)

Although the *isotropic* ^{13}C shift of the central carbon is unaffected by racemization, this dynamic process causes changes in sign of several of the off-diagonal elements of the chemical shift tensor. We accounted for racemization by generating and symmetrizing the tensors for the two enantiomers in the same reference frame, then averaging the respective components. We then obtained the principal components by diagonalizing the averaged tensor. As shown in Table I for both the MP2 and DFT methods, racemic averaging (RA) changes each principal component by no more than 4 ppm, and this effect is considered no further.

Table I. Theoretical and Experimental ^{13}C NMR data for the Isopropyl Cation.

System	Method	**r**, θ (Å, °)	δ_{iso} ppm	δ_{11} ppm	δ_{22} ppm	δ_{33} ppm
C_2	MP2		328	545	409	30
C_2	MP2/RA		328	545	406	34
C_2	DFT		330	553	411	25
C_2	DFT/RA		330	553	410	26
C_2/FHF^-	MP2	2.8, 90	316	527	392	30
C_2/FHF^-	MP2	2.8, 70	316	516	403	28
C_2/FHF^-	MP2	3.0, 90	320	532	397	30
C_2/FHF^-	MP2	3.0, 70	320	524	408	28
C_2/SbF_6^-	DFT	2.8, 90	317	530	396	24
C_2/SbF_6^-	DFT	2.8, 70	316	523	404	23
	Exptl.		320	497	385	77

As reasonable approximations to the effect of medium, we chose the models defined in Figure 5a-b. The geometry of the individual cation and anion were frozen, and their orientation was fixed by values of **r** and θ suggested by the $(CH_3)_3C^+Sb_2F_{11}^-$ crystal structure. Due to program limitations, GIAO-MP2 calculations were not possible using SbF_6^- as a counter ion. Thus, a single FHF^- was chosen as a substitute. A few representative MP2 results are shown in Table I to demonstrate that treatment of the anion can change the calculated isotropic shift and principal components. GIAO calculations for the much larger SbF_6^- ion were only possible at the DFT level. Again, we obtained upfield corrections in the isotropic shift, appreciable corrections to δ_{11} and modest improvements in δ_{22}. δ_{33} was little improved by inclusion of the anion.

Further Work on Carbenium Ions in Zeolites

Theory helps the experimentalists in many ways; this volume is on chemical shift calculations, but the other ways in which theoretical chemistry guides NMR studies of catalysis should not be overlooked. Indeed, further theoretical work on two of the cations discussed above has helped us understand why some carbenium ions persist indefinitely in zeolite solid acids as stable species at 298 K, and others do not (*25*). The three classes of carbenium ions we were most concerned with, the indanyl cation, the dimethylcyclopentenyl cation, and the pentamethylbenzenium cation (Scheme 1), could all be formally generated by protonation of an olefin. We actually synthesized them in the zeolites by other routes, but we suspected that the simplest "parent olefins" of these cations must be very basic hydrocarbons, otherwise the carbenium ions might just transfer protons back to the conjugate base site on the zeolite. Experimental values were not available for any of the parent olefins shown below, so we calculated the proton affinities (enthalpies) by first determining the

B3LYP/TZVP geometries of all species shown in Scheme 1, followed by MP4(sdtq)/6-311+G* single-point energy calculations. The theoretical proton affinities are shown; for comparison, the experimental PA for pyridine is 222 kcal/mol. It would appear that olefins that protonate on a zeolite to yield persistent carbenium ions are very basic indeed.

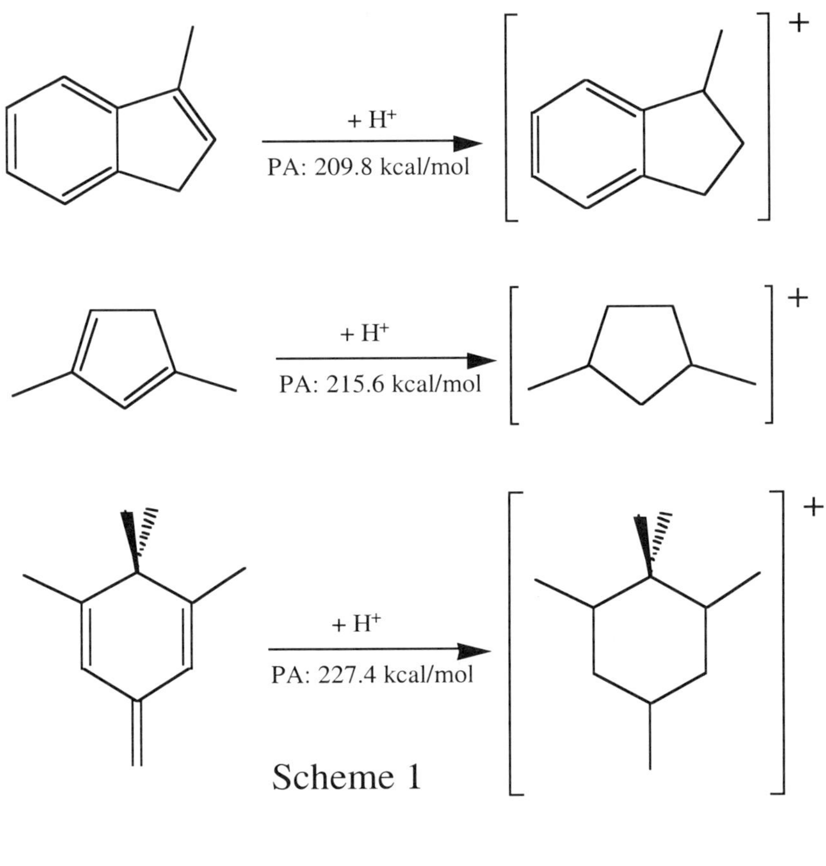

Scheme 1

Summary and Conclusion

We have shown several examples of how theoretical NMR calculations can be used in conjunction with experimental studies to further our understanding of catalysis. The agreement between the results of the two approaches is often excellent, and in all cases shown the agreement is sufficient to aid the interpretation of the chemistry. The synergy between the theoretical and experimental approaches provides benefits beyond what either approach could provide separately. We expect theoretical predictions of NMR data will become more widely used as both theorists and experimentalists gain a better appreciation of the methods. These techniques will become even more useful as increases in computer speed and the development of better algorithms allow use to treat larger systems and attain better accuracy.

Acknowledgments

JBN is funded by the U.S. Department of Energy (DOE) Office of Basic Energy Sciences (BES). JFH is supported by the National Science Foundation (CHE-9528959) and DOE-BES (Grant No. DE-FG03-93ER14354). Computer resources were provided by the Scientific Computing Staff, Office of Energy Research, at the National Energy Research Supercomputer Center (NERSC), Berkeley, CA. Part of this work was performed at the W. R. Wiley Environmental Molecular Sciences Laboratory, a national scientific user facility sponsored by the DOE Office of Biological and Environmental Research and located at Pacific Northwest National Laboratory (PNNL). PNNL is operated for DOE by Battelle.

References

1. van Santen , R. A.; Kramer, G. J. *Chem. Rev.* **1995**, *95*, 637.
2. Corma, A. *Chem. Rev.*, **1995**, 95, 559.
3. Hattori, H. *Chem. Rev.*, **1995**, 95, 537.
4. Barteau, M.A. *Chem. Rev.*, **1996**, 96,1413.
5. *NMR Techniques in Catalysis*; Bell, A; Pines, A., Eds.; Marcel Dekker: New York, NY; 1994.
6. Haw, J.F; Xu, T., In *Advances in Catalysis*, Gates , B. C., Ed.; 1998, Vol. 42; 115.
7. Xu, T.; Haw , J. F., *Topics in Catalysis*, **1997**, 4, 109.
8. Haw, J.F.; Nicholas, J.B; Xu, T.; Beck, L.W.; Ferguson, D.B. *Acc. Chem. Res.*, **1996**, 29, 259.
9. Herzfeld, J.; Berger, A. E., *J. Chem. Phys.* **1980**, 73, 6021.
10. Anderson, M.W.; Klinowski, J. , *Nature*, **1989**, 339, 200.
11. Haw, J.F.; Richardson, B.R.; Oshiro, I.S.; Lazo, N.D.; Speed, J.A., *J. Am. Chem. Soc.*, **1989**, 111, 2052.
12. Ernst, H.; Freude, T.; Mildner, T.,*Chem. Phys. Lett.*, **1994**, 229, 291.
13. Hunger, M.; Horvath, T.; *J. Chem Soc. Chem. Comm.* **1995**, 1423.
14. Haw, J. F. ; Goguen, P. W. ; Xu, T. ; Skloss, T. W.; Song, W.; Wang, Z., *Angewandte Chemie*, **1998**, 37, 948.
15. Goguen, P. W. ; Xu, T.; Barich, D. H.; Skloss, T. W.; Song, W.; Wang, Z.; Nicholas, J. B.; Haw, J. F., *J. Am. Chem. Soc.*, **1998**, 120, 2650.
16. Xu, T.; Barich, D. H.; Goguen, P. W.; Song, W.; Wang, Z.; Nicholas, J. B.; Haw, J. F., *J. Am. Chem. Soc.*, **1998**, 120, 4025.
17. Ditchfield,R., *Mol. Phys.* **1974**, 27, 789.
18. *Nuclear Magnetic Shieldings and Molecular Structure*; Tossell, J. A. Ed.; Kluwer Academic: Boston, MA, 1993.
19. Wolinski, K.; Hinton, J.F.; Pulay, P., *J. Am. Chem. Soc.* **1990**, 112, 8251.
20. Gauss, J., *J. Chem. Phys.* **1993**, 99, 3629.
21. Sieber, S.; Schleyer, P. v. R.; Gauss, J., *J. Am. Chem. Soc.* **1993**, 115, 6987.

22. Gauss, J., *Chem. Phys. Lett.* **1992**, 191, 614.
23. Gauss, J., *Chem. Phys. Lett.* **1994**, 229, 198.
24. Gauss, J.,; Stanton, J. F.; *J. Chem. Phys.* **1995**, 103, 3561.
25. Nicholas, J.B.; Haw, J.F.; *J. Am. Chem. Soc.* **1998**, 120, 11804.
26. Nicholas, J.B.; Khier, A,; Xu, T.; Krawietz, T. R.; Haw, J.F., *J. Am. Chem. Soc.* **1998**, 120, 10471.
27. Nicholas, J.B.; Xu, T.; Barich, D.H.; Torres, P.D.; Haw, J.F., *J. Am. Chem. Soc.*, **1996**, 118, 4202.

Chapter 5

Effects of a Static Electric Field on Molecular Magnetic Properties Employing the *CTOCD* Method: Shielding Polarizabilities of CO, H_2O, and CH_4 Compounds

M. B. Ferraro[1,2], M. C. Caputo[1], and C. Ridruejo[1]

[1]Departamento de Física, Facultad de Ciencias Exactas y Naturales, Universidad de Buenos Aires, 1428, Ciudad Universitaria, Buenos Aires, Argentina

[2]Consejo Nacional de Investigaciones Científicas y Técnicas, 1917 Rivadavia, Buenos Aires, Argentina

A procedure based on formal annihilation of the diamagnetic contribution to the quantum mechanical electron current density, via a continuos transformation of its origin all over the molecular domain, *CTOCD-DZ* method, is applied for determining shielding polarizabilities to first order in a perturbing electric field. Analytical expressions for the third-rank tensors have been implemented in the *SYSMO* suite of programs employing the coupled Hartree-Fock (*CHF*) approach. In the limit of exact eigenfunctions to a model Hamiltonian, the *CTOCD-DZ* expressions reduce to conventional terms. In any calculation relying on the algebraic approximation, irrespective of size and quality of the (gaugeless) basis set employed, all the components of the magnetic shielding polarizabilities evaluated within these methods are origin independent. Test calculations have been carried out in CO, H_2O and CH_4 compounds.

The presence of a time-independent electric field produces changes in the molecular second-rank tensors, i. e., the magnetic susceptibility $\chi_{\alpha\beta}$ and the magnetic shielding $\sigma^I_{\alpha\beta}$, of the I*th* nucleus, which are the usual response properties to the presence of an external magnetic field. The contributions of Buckingham to the understanding of the behavior of molecular properties in the presence of electromagnetic fields are fundamental. The dependence of those properties on the electric fields is one of the topics in which his influence has been very important (*1,2*).

The application of a static electric field polarizes the electronic charge distribution and leads to changes in molecular magnetic susceptibility and nuclear magnetic shielding, which can be rationalized in terms of response tensors of higher rank.

In the presence of two external perturbations, i. e., the spatially uniform time-independent electric and magnetic fields **E** and **B**, and of a permanent dipole moment μ_I, on nucleus **I**, the energy of the molecule, evaluated in the singlet electronic state $|\psi_a\rangle$ is, employing Buckingham notation (*1, 2*) to denote molecular tensors,

$$W_a = W_a^{(0)} - \mu_\alpha E_\alpha - \frac{1}{2}\alpha_{\alpha\beta}E_\alpha E_\beta + \ldots - \frac{1}{2}\chi_{\alpha\beta}B_\alpha B_\beta + \ldots + \sigma^I_{\alpha\beta}\mu_{I\alpha}B_\beta + \ldots$$
$$-\frac{1}{2}\chi_{\alpha\beta\gamma}B_\alpha B_\beta E_\gamma + \ldots + \sigma^I_{\alpha\beta\gamma}\mu_{I\alpha}B_\beta E_\gamma + \ldots \quad (1)$$

Magnetic susceptibility and nuclear magnetic shielding of a chosen nucleus I, in the presence of an external weak, homogeneous electric field **E** may be expanded as (*3*)

$$\chi_{\alpha\beta}(E) = \chi_{\alpha\beta} + \chi_{\alpha\beta\gamma}E_\gamma + \frac{1}{2}\chi_{\alpha\beta\gamma\delta}E_\gamma E_\delta + \ldots, \quad (2)$$

$$\sigma^I_{\alpha\beta}(E) = \sigma^I_{\alpha\beta} + \sigma^I_{\alpha\beta\gamma}E_\gamma + \frac{1}{2}\sigma^I_{\alpha\beta\gamma\delta}E_\gamma E_\delta + \ldots \quad (3)$$

The third-rank tensors $\chi_{\alpha\beta\gamma}$ and $\sigma^I_{\alpha\beta\gamma}$ describe non linear response of the electron cloud to first order in **E**. These quantities are sometimes referred to as hypermagnetizabilities, which are related to Cotton-Motton effect (*4-6*), (CME- the birefringence of light in gases in a constant magnetic field); and as shielding polarizabilities (*7-14*).

Many groups have been active in this field recently. Augspurger and co-workers have carried out calculations for a wide range of nuclei in extended series of molecules (*10-13*). Cybulski and Bishop (*14*) employed *MP3* and linearized coupled cluster to doubles (LCCD) plus finite field numerical differentiation calculations on N_2, HF, CO and H_2, and, very recently, Ågren *et al.* have calculated the *MSCF* cubic response (*15*). Bishop and Cybulski also adopted *SCF* and *MP2* methods for calculating the electric field dependence of magnetic nuclear shielding (*9*) and susceptibility (*7*). The review by Raynes (*16*) is an excellent introduction to the subject. He pointed out that the effect of an electric field on the chemical shielding is invoked to explain experimental observations such as intermolecular interactions in gases, effects of solvent in liquids, intramolecular electric fields in solids.

SCF and *MSCF* electric field dependence of the magnetizability and nuclear magnetic shielding have been studied by Rizzo *et al.* (*17-19*) within *GIAO* basis sets. The use of London orbitals guarantees invariance of theoretical estimates in a change of coordinate system, which is a basic requirement in the computation of magnetic response properties.

Although faster convergence of *GIAO* calculations might then be preferable in numerical studies (*20*), *CTOCD* schemes for $\chi_{\alpha\beta}$ and $\sigma^I{}_{\alpha\beta}$ are easier to implement at any level of accuracy (*21-25*) and become competitive provided that proper basis sets are employed (*26*), and are nicely suited to satisfy the constraints of charge and current conservation by annihilation of either diamagnetic, *CTOCD-DZ* method, or paramagnetic, *CTOCD-PZ* method, contributions to electronic current density via continuos transformation of origin (Keith and Bader (*27*) have presented the idea of continuos transformation of origin by the first time.).

The present paper makes a brief revision of the *CTOCD-DZ* method developed by Lazzeretti and Zanasi (*28*), via formal annihilation of diamagnetic contributions to the electronic current density induced in the presence of static homogeneous electric and magnetic fields. We have implemented the equations reported in ref. (*28*) transforming the non-Hermitian operators into Hermitian ones and employing the coupled Hartree-Fock approach (*CHF*) to calculate origin independent $\sigma^I{}_{\alpha\beta\gamma}$, nuclear shielding polarizabilities, for the set of molecules: CO, H_2O and CH_4. Indeed, shielding polarizability evaluated within this method is origin independent and the constraints for charge and current conservation are exactly fulfilled, no matter the size of the basis set employed to perform the calculations. The main emphasis is placed in understanding physical facts and testing the accuracy of the results. Those results do not depend on the choice of the origin of coordinates but their accuracy does depend on the quality of the basis set. We compare critically our results with those of other authors and employ different basis sets described in the **Results** section.

Molecular Magnetic Properties in the Presence of a Static Electric Field.

We shall briefly review some definitions to compute molecular magnetic properties in the presence of a static electric field, i. e., hypermagnetizabilities and shielding polarizabilities, $\chi_{\alpha\beta\gamma}$ and $\sigma^I{}_{\alpha\beta\gamma}$.

Rayleigh-Schrödinger perturbation theory provides the third-order energy.

$$W_a^{(3)} = (1/\hbar^2) \sum_{j\neq a} \sum_{k\neq a} \omega_{ja}^{-1}\, \omega_{ka}^{-1}\, \langle a|H^{(1)}|j\rangle \langle j|H^{(1)}|k\rangle \langle k|H^{(1)}|a\rangle$$

$$- (1/\hbar^2) \langle a|H^{(1)}|a\rangle \sum_{j\neq a} \omega_{ja}^{-2} \langle a|H^{(1)}|j\rangle \langle j|H^{(1)}|a\rangle$$

$$- (2/\hbar) \sum_{j\neq a} \omega_{ja}^{-1} Re\, \{\langle a|H^{(1)}|j\rangle \langle j|H^{(2)}|a\rangle\} \tag{4}$$

where $|a\rangle \equiv |\psi_a{}^{(0)}\rangle$ is the unperturbed electronic reference (singlet) state, and $|j\rangle$ is the j-*th.* excited state. $H^{(1)}$ and $H^{(2)}$ denote the first-order and second-order interactions Hamiltonians. In the presence of static external electric field $\boldsymbol{E}$ and magnetic field $\boldsymbol{B}$, and of an intrinsic magnetic moment μ_I, on nucleus **I**, the electronic first-order Hamiltonian contains three first-order terms

$$H^E = eE_\alpha R_\alpha\,, \tag{5}$$

$$H^B = (e/m_e c) \sum_{i=1,n} \boldsymbol{A}_i^{\boldsymbol{B}}.\boldsymbol{p}_i = (e/2mc)\, B_\alpha L_\alpha\,, \tag{6}$$

$$H^{\mu I} = (e/m_e c) \sum_{i=1,n} A_i^{\mu I}.p_i = (e/mc)\, M_I^{n}{}_{\alpha}\, \mu_{I\alpha} \tag{7}$$

using the notation of refs. (*29,30*). The vector potentials $A_i^B = A^B(r_i)$ and $A_i^{\mu I} = A^{\mu I}(r_i)$, are defined as

$$A^B(r) = \frac{1}{2} \mathbf{B} \times (\mathbf{r} - \mathbf{r_0}), \qquad \mathbf{A}^{\mu I}(\mathbf{r}) = \frac{\mu_I \times (\mathbf{r} - \mathbf{R_I})}{|r - R_I|^3}, \tag{8}$$

(r_0, the origin of the vector potential may be put equal to **0**, without loss of generality).

The third-rank diamagnetic and paramagnetic contributions to electric field dependent magnetic susceptibility and nuclear magnetic shielding, to first order in E are

$$\chi_{\alpha\beta\gamma} = -\frac{\partial^3 W_a^{(3)}}{\partial P_\alpha \partial B_\beta \partial E_\gamma} = \chi^d{}_{\alpha\beta\gamma} + \chi^p{}_{\alpha\beta\gamma}, \tag{9}$$

$$\chi^d{}_{\alpha\beta\gamma} = (e^3/4m_e c^2 \hbar) \sum_{j\neq a} 2\,\omega_{ja}^{-1} Re\{\langle a|\sum_{i=1,n}(r^2\delta_{\alpha\beta} - r_\alpha r_\beta)_i|j\rangle \langle j|R_\gamma|\, a\rangle\}, \tag{10}$$

$$\chi^p{}_{\alpha\beta\gamma} = -(e^3/4m_e^2c^2)\{L_\alpha, L_\beta\, R_\gamma\}_{-2}, \tag{11}$$

$$\sigma^I{}_{\alpha\beta\gamma} = \frac{\partial^3 W_a^{(3)}}{\partial \mu_{I\alpha} \partial B_\beta \partial E_\gamma} = \sigma^{Id}{}_{\alpha\beta\gamma} + \sigma^{Ip}{}_{\alpha\beta\gamma}, , \tag{12}$$

$$\sigma^{dI}{}_{\alpha\beta\gamma} = -(e^2/2m_e c^2\hbar) \sum_{j\neq a} 2\,\omega_{ja}^{-1} Re\{\langle a|\sum_{i=1,n}(r_i E^i{}_{I\gamma}\,\delta_{\alpha\beta} - r_{i\alpha} E^i{}_{I\beta})\,|j\rangle\langle j|R_\gamma|a\rangle\}, \tag{13}$$

$$\sigma^{pI}{}_{\alpha\beta\gamma} = (e^3/2m_e^2c^2)\{M_I^n{}_\alpha, L_\beta\, R_\gamma\}_{-2} \tag{14}$$

In equations 11 and 14 we have employed the definition *(30)*

$$\{A, B, C\}_{-2} \equiv (-4\, Tr\, [F^{(A)}(X^{(B)}\Delta X^{(C)} - X^{(C)}\Delta X^{(B)})] + [B, C, A] + [C, A, B] \tag{15}$$

where *[B, C, A]* and *[C, A, B]* are permutations of the perturbators involved in the expression between square brackets. Equation 15 is a third-rank tensor written employing the McWeeney procedure *(31)* for the CHF approach. $F^{(A)}$ represents the first-order perturbated Fock matrix, $X^{(A)}$ matrices are computed only once to solve the first-order *CHF* problem for each perturbator, and Δ_{pq} is the overlap matrix between the atomic orbitals, χ_p and χ_q, of an orthonormal basis set χ. In the actual calculations, employing non orthogonal basis sets of Gaussian functions, it is customary to orthogonalize them according to the Löwdin procedure.

Electronic Current Density in the Presence of Static Electric and Magnetic Fields.

The third-order interaction energy, equation 4, contains contributions, which can be expressed in terms of the second-order electron current density vector $\mathbf{J}^{\mathbf{BE}}$ induced by the fields,

$$W^{BBE} = -(1/2c) \int J^{BE} \bullet A^B \, dr, \qquad W^{\mu I\, BE} = -(1/c) \int A^{\mu I} \bullet J^{BE} \, dr \tag{16}$$

The expression for J^{BE} is derived via the general quantum mechanical definition *(32)*, introducing the perturbation expansion for the current density and the a-state molecular wave-function (depending on n-electron space–spin coordinates χ_i), $\psi_a(\chi_1,\chi_2,\ldots\chi_n)$, i. e.

$$J = J^{(0)} + J^B + J^{\mu I} + J^{BE} + \ldots \tag{17}$$

$$\psi_a = \psi_a^{(0)} + \Psi_a \bullet B + \psi_a^{E} \bullet E + \psi_a^{\mu I} \bullet \mu_I + E \bullet \psi_a^{EB} \bullet E + B \bullet \psi_a^{BE} \bullet B + \ldots \tag{18}$$

The $J^{(0)}$ in equation 17 vanish identically for any diamagnetic system, e. g. , the singlet state considered here.

The first- and second-order electronic wavefunctions necessary for further development are obtained from Rayleigh-Schrödinger perturbation theory (see ref. (*28*) for details).

Gauge invariance of magnetic properties is associated with the continuity equation *(33,34)*. In a gauge transformation of the vector potential

$$A^{B'} \rightarrow A^{B''} = A^{B'} + \nabla f, \qquad A^{B''} = \frac{1}{2} B \times (r - r') \tag{19}$$

where $f = f(r)$ is an arbitrary function well-behaved for $r \rightarrow \infty$, the third-rank interaction energy, equation 16, and the response properties, equations 9 and 12, are left-unchanged provided that the integral

$$\int J^{BE} \bullet \nabla f \, dr = \int \nabla \bullet (J^{BE} f) \, dr - \int f \, \nabla \bullet J^{BE} \, dr \tag{20}$$

vanishes. By applying the Gauss theorem, the first volume integral on the right-hand side is converted into a surface integral, and vanishes owing to the boundary conditions usually assumed for ψ_a and J^{BE} , i. e., ψ_a, $J^{BE} \rightarrow 0$ for $r \rightarrow \infty$. Thus the integral on the left-hand side vanishes if the continuity equation, $\nabla \bullet J^{BE} = 0$, is satisfied.

Transformation Laws for the Current Density in a Change of Coordinate System. In the coordinate transformation

$$r' \rightarrow r'' = r' + d \tag{21}$$

which can be described as a gauge transformation 19 with $f = (r'' - r') \bullet A^{B'}$, the transformation law for the diamagnetic and paramagnetic contributions to the current density is *(28)*

$$J_d^{BE}(r - r'') = J_d^{BE}(r - r') + J_d^{(r'' - r') \times BE}(r), \tag{22}$$

$$J_p^{BE}(r - r'') = J_p^{BE}(r - r') + J_p^{(r'' - r') \times BE}(r), \tag{23}$$

Using the hypervirial relationship *(33)*

$$\langle a|P_\alpha|j\rangle = -i\, m_e\, \omega_{ja}\, \langle a|R_\alpha|j\rangle, \tag{24}$$

Lazzeretti and Zanasi proved *(28)* that

$$J_p^{(r''-r')\times BE}(r) = -J_d^{(r''-r')\times BE}(r) \tag{25}$$

so that

$$J^{BE}(r) = J_d^{BE}(r-r'') + J_p^{BE}(r-r'') = J_d^{BE}(r-r') + J_p^{BE}(r-r') \tag{26}$$

is origin independent for *exact* eigenfunctions to any model Hamiltonian. Within the exact coupled Hartree-Fock *(CHF)* method, the current density $J^{BE}(r)$ is invariant in a coordinate transformation. In actual calculations, employing the algebraic approximation, this condition is only partially met, depending on the quality of the basis set.

Advantage of the *CTOCD* Method: Continuous Transformation of the Origin of the Current Density. The *CTOCD* method for theoretical determination of hypermagnetizabilities and shielding polarizabilities is reported in detail in ref. *(28)*. In this section we make only a very brief description of the theory involved in the formulation of *CTOCD-DZ* $\chi_{\alpha\beta\gamma}$ and $\sigma^I{}_{\alpha\beta\gamma}$.

The scheme named 'continuos transformation of origin of the current density', see *(27,21,22,28)* proves that the transformed diamagnetic current density tensor,
$J_d^{BE}(r - r'')$, can be formally annihilated in every point r, all over the molecular domain, by considering the d shift in equation 21 as a function of r, and choosing $d(r) = r$ in equation 22, that is setting $r'' = r'$. So that

$$J_d^{BE}(r-r') = -J_d^{(r-r')\times BE}(r) \tag{27}$$

As the diamagnetic term is set to zero, the procedure is named *CTOCD-DZ*. Total current becomes completely paramagnetic in form, and contains two terms that are expressed within the original coordinates system as a function of r, i. e.

$$J^{BE}(r) = J_p^{BE}(r-r') + J_p^{(r-r')\times BE}(r) \tag{28}$$

As the total current density is an invariant, comparison between equations 26 and 28 necessarily implies,

$$J_p^{(r-r')\times BE}(r) = J_d^{BE}(r-r') \tag{29}$$

for every **r**, (provided that the hypervirial condition 24 holds). The formally annihilated diamagnetic contribution reappears as a new 'paramagnetic' term. (See ref (*28*) for a detailed discussion about the new paramagnetic term along the direction of the external magnetic field)

By employing equation 28 for the current density within expressions 16 and right-hand side of equations 9 and 12, new definitions for total magnetic properties are arrived at in the form

$$\chi_{\alpha\beta\gamma} = \chi^{p}{}_{\alpha\beta\gamma} + \chi^{\Delta}{}_{\alpha\beta\gamma} \tag{30}$$

$$\sigma^{I}{}_{\alpha\beta\gamma} = \sigma^{pI}{}_{\alpha\beta\gamma} + \sigma^{\Delta I}{}_{\alpha\beta\gamma} \tag{31}$$

where

$$\chi^{\Delta}{}_{\alpha\beta\gamma} = -\frac{e^3}{4m_e^2c^2}\varepsilon_{\beta\lambda\mu}\{P_\lambda, R_\gamma, U_{\mu\alpha}\}_{-2} \tag{32}$$

$$\sigma^{\Delta I}{}_{\alpha\beta\gamma} = \frac{e^3}{2m_e^2c^2}\varepsilon_{\beta\lambda\mu}\{P_\lambda, R_\gamma, T_{I\mu\alpha}{}^{n}\}_{-2} \tag{33}$$

and the definitions of the Hermitian operators $\boldsymbol{U}$ and $\boldsymbol{T}_I$ are

$$U_{\alpha\beta} = \frac{1}{2}\sum_{i=1,n}[(r_{i\alpha} - r'_\alpha)\, l^{i}{}_{\beta}(r') + l^{i}{}_{\beta}(r')[(r_{i\alpha} - r'_\alpha)] \tag{34}$$

$$T^{n}{}_{I\alpha} = \frac{1}{2}\sum_{i=1,n}[(r_{i\alpha} - r'_\alpha)M^{i}{}_{I\beta}(r') + M^{i}{}_{I\beta}(r')[(r_{i\alpha} - r'_\alpha)] \tag{35}$$

The *CTOCD-DZ* expressions 32 and 33 reduce to the conventional $\chi^{d}{}_{\alpha\beta\gamma}$, equation 10, and $\sigma^{dI}{}_{\alpha\beta\gamma}$, equation 13, if the hypervirial constraint 24 is satisfied, as it can be proven by direct substitution. It has also been demonstrated *(21)* that the *CTOCD-DZ* approach is equivalent to the Geertsen approach *(35-37)*, as far as average properties are concerned.

In a change of coordinate system 21, the contributions to the shielding polarizabilities transform

$$\sigma^{pI}{}_{\alpha\beta\gamma}(\boldsymbol{r}'') = \sigma^{pI}{}_{\alpha\beta\gamma}(\boldsymbol{r}') - \frac{e^3}{2m_e^2c^2}\varepsilon_{\beta\lambda\mu}\, d_\lambda\{M^{n}{}_{I\alpha}, P_\mu, R_\gamma\}_{-2} \tag{36}$$

$$\sigma^{\Delta I}{}_{\alpha\beta\gamma}(\boldsymbol{r}'') = \sigma^{\Delta I}{}_{\alpha\beta\gamma}(\boldsymbol{r}') + \frac{e^3}{2m_e^2c^2}\varepsilon_{\beta\lambda\mu}\, d_\lambda\{M^{n}{}_{I\alpha}, P_\mu, R_\gamma\}_{-2} \tag{37}$$

By comparing equations 36 and 37 one can notice that total *CTOCD-DZ* nuclear magnetic shieldings are independent of the coordinate system, as there is exact cancellation between terms arising from variation of Δ and p components in any calculation employing the algebraic approximation, e. g., adopting gaugeless basis sets of arbitrary quality.

CTOCD-DZ $\chi_{\alpha\beta\gamma}$ depends only linearly on the shift of origin, as it is demonstrated in ref. *(28)*.

Results

A set of small molecules CO, H_2O and CH_4 has been considered in the present study. Zero-order molecular orbitals are expanded over atomic Gaussians functions, three different basis sets have been employed. The first one, hereafter referred to as ***I***, (10s6p4d/6s4p)-[5s3p2d/3s2p] set, is an *ad hoc* basis set, developed by Sadlej *(38)*. In this work we are dealing with two simultaneous perturbations on the molecule, electric and magnetic fields. The basis set must be suitable to produce good values for both, nuclear magnetic shieldings and their polarizabilities. Thus, we have chosen basis set ***II***, reported by Sadlej (*38,39*) and added to the set the electric field derivatives of a STO-3G basis set to get a (13s12p7d/6s7p)-[6s5p3d/3s3p] set. The general theory, which justifies this procedure, is discussed by Sadlej *(38,39)*, and by Lazzeretti and Zanasi (*40*). The coefficients and exponents of the added derivative functions are given in ref. (*41*). Basis set ***III*** is taken from a compilation by Huzinaga (*42*) and is described in ref. (*43*) as basis set ***IV***, (11s7p3d1f/6s3p1d)-[8s7p3d1f/5s3p1d] set.

The calculations reported in this work have been carried out by the SYSMO computer programs (*44*), modified by us to implement a new coupled Hartree-Fock (CHF) section to describe the *CTOCD-DZ* shielding polarizabilities.

A description of the effect of a uniform electric field on the nuclear shielding was given by Buckingham (*3*). The change in the average magnetic shielding, assuming an electric field fixed relative to the molecule, is

$$\sigma^{I}_{Av} = -A_\gamma E_\gamma - B_{\gamma\delta} E_\gamma E_\delta \tag{38}$$

where (Einstein summation is implied)

$$A_\gamma = -(1/3)\, \sigma^{I}_{\alpha\alpha\gamma} \tag{39}$$

$$B_{\gamma\delta} = -\frac{1}{3(1+\delta_{\gamma\gamma})}\, \sigma^{I}_{\alpha\alpha\gamma\delta} \tag{40}$$

The number of non-vanishing elements $\sigma^{I}_{\alpha\beta}$, $\sigma^{I}_{\alpha\beta\gamma}$ and $\sigma^{I}_{\alpha\alpha\gamma\delta}$, for a given nucleus depends on the local "on-site" symmetry at the position of the nucleus (*45*).

In tables **I-V** we report σ^{I}_{Av} and all non-vanishing elements $\sigma^{I}_{\alpha\beta\gamma}$ for the heavy nucleus. For the hydrogen atoms, we report only those elements of $\sigma^{I}_{\alpha\beta\gamma}$ that contribute to the average shielding in equation 38. The A_γ values, equation 39, have been obtained via the *CTOCD-DZ* approach and compared with other authors calculations.

CO Molecule. Table **I** summarizes the results of the nuclear magnetic shielding and its polarizabilities for the carbon and oxygen atoms in CO. The molecule was placed along the *z* axis, with the positive direction from O to C, assuming the C_1 symmetry group.

The $\sigma^{\Delta I}_{\alpha\beta\gamma}$ values, equation 33, are compared with the $\sigma^{dI}_{\alpha\beta\gamma}$, equation 13, for each basis set. The *CTOCD-DZ* contributions depend on the quality of the basis set.

.Table I: CO nuclear shielding constants (in ppm) and *CTOCD-DZ* shielding polarizabilities[a] (in ppm au) for basis sets *I-III*.

	Carbon			Oxygen		
	I	*II*	*III*	*I*	*II*	*III*
σ_{Av}	3.77	-18.25	-23.96	-56.50	-79.08	-86.68
σ^{d}_{xxz}	-3.02	-2.87	-2.85	12.03	12.05	12.41
σ^{Δ}_{xxz}	-9.15	-9.41	-2.04	16.76	10.95	12.61
σ^{p}_{xxz}	-464.18	-541.07	-541.37	2098.66	2233.94	2285.19
σ_{xxz}	-473.33	-550.48	-543.41	2115.42	2244.89	2297.80
σ^{d}_{zzz}	-27.33	-27.14	-27.12	44.26	44.38	44.38
σ^{Δ}_{zzz}	-38.92	-34.05	-25.77	52.81	38.82	44.12
σ^{p}_{zzz}	0.00	0.00	0.00	0.0	0.0	0.0
σ_{zzz}	-38.92	-34.05	-25.77	52.81	38.82	44.12
A_z	328.52	378.33	370.86	-1427.88	-1509.5	-1546.57

	Ref. (*18*)	Refs. (*7*, *9*)	Ref. (*18*)	Refs. (*7*,*9*)
$\sigma_{Av.}$	-23.80	-25.00	-84.40	-86.80
A_z	374.7	376.90	-1553.0	-1561.5

[a] $\sigma_{\alpha\beta\gamma} = \sigma^{\Delta}_{\alpha\beta\gamma} + \sigma^{p}_{\alpha\beta\gamma}$

Only for basis set ***III***, $\sigma^{\Delta I}_{\alpha\beta\gamma}$ shielding polarizabilities are very close to the $\sigma^{dI}_{\alpha\beta\gamma}$ values for both nuclei. The paramagnetic $\sigma^{pI}_{\alpha\beta\gamma}$ contributions, are approximately of the same quality for basis set ***II*** and ***III***. Our results are compared with those of Bishop and Cybulski (*7,9*), and with those of Coriani *et al* (*18*), both for the SCF level, and the agreement is very good, even if, those authors have used larger basis sets.

H_2O Molecule. The molecule is placed in the *xz* plane assuming the C_1 group. Table **II** reports the hydrogen magnetic shielding and shielding polarizabilities. The corresponding results for the oxygen atom are depicted in Table **III**.

The A_γ values for the proton, are in very good agreement with the literature's values (*17, 46*). We have also made the comparison taking into account that Grayson and Raynes (*47*) reported A^{H1}_z = 91.0 ppm au, but employing the C_s symmetry group and placing the *z*-axis in the O-H direction. Our A_γ values transformed to that geometry are A^{H1}_z = 88.16, 85.36 and 84.9 ppm au, for basis sets ***I***, ***II*** and ***III***, respectively.

For the oxygen atom, $\sigma^{\Delta I}_{\alpha\beta\gamma}$ shielding polarizabilities are in very good agreement with the $\sigma^{dI}_{\alpha\beta\gamma}$ values for basis set ***III*** and the A^{O1}_z values are also in very good agreement with those of refs. (*17, 46*).

Table II H_2O: Proton nuclear shielding (in ppm) and *CTOCD-DZ* shielding polarizabilities[a] (in ppm. au) for basis sets *I-III*.

	I	*II*	*III*		*I*	*II*	*III*
σ_{Av}	51.3	32.14	31.17				
σ^{d}_{xxy}	-15.91	-15.92	-15.89	σ^{d}_{xxz}	7.04	7.01	6.71
σ^{Δ}_{xxy}	-16.08	-13.18	-15.02	σ^{Δ}_{xxz}	3.63	3.28	4.83
σ^{p}_{xxy}	-71.40	-73.09	-70.27	σ^{p}_{xxz}	51.86	51.05	46.97
σ_{xxy}	-87.48	-86.27	-85.29	σ_{xxz}	55.49	54.33	51.80
σ^{d}_{yyy}	-35.29	-35.25	-35.23	σ^{d}_{yyz}	8.42	8.40	8.01
σ^{Δ}_{yyy}	-34.67	-31.26	-33.18	σ^{Δ}_{yyz}	4.86	1.89	4.00
σ^{p}_{yyy}	-18.02	-19.36	-18.58	σ^{p}_{yyz}	44.27	42.29	42.30
σ_{yyy}	-52.69	-50.62	-51.76	σ_{yyz}	49.13	44.18	46.30
σ^{d}_{zzy}	-15.36	-15.32	15.32	σ^{d}_{zzz}	18.91	18.85	18.53
σ^{Δ}_{zzy}	-18.07	-14.25	-13.83	σ^{Δ}_{zzz}	11.12	10.60	16.70
σ^{p}_{zzy}	-74.81	-77.15	-76.42	σ^{p}_{zzz}	14.39	14.42	8.04
σ_{zzy}	-92.88	-91.40	-90.25	σ_{zzz}	25.51	25.02	24.74
A_y	77.68	76.10	75.77	A_z	-43.38	-41.18	-40.95
	Ref. *(17)*	Ref. *(46)*				Ref.*(17)*	Ref. *(46)*
σ_{av}	30.80	32.34		A_z		-42.70	-47.20
A_y	79.60	78.50					

[a] $\sigma_{\alpha\beta\gamma} = \sigma^{\Delta}_{\alpha\beta\gamma} + \sigma^{p}_{\alpha\beta\gamma}$

Table III: H_2O: Oxygen nuclear shielding constant (in ppm) and *CTOCD-DZ* shielding polarizabilities[a] (in ppm. au) for basis sets *I-III*.

	I	*II*	*III*
σ_{Av}	326.42	328.45	327.97
σ^{d}_{xxz}	-10.90	-10.87	-11.05
σ^{Δ}_{xxz}	-15.10	-4.22	-10.17
σ^{p}_{xxz}	-257.30	-276.45	-283.57
σ_{xxz}	-272.40	-280.67	-293.74
σ^{d}_{yyz}	-4.94	-4.96	-5.43
σ^{Δ}_{yyz}	-20.45	-11.32	-7.83
σ^{p}_{yyz}	-530.59	-520.98	-536.93
σ_{yyz}	-551.04	-532.31	-544.76
σ^{d}_{zzz}	-5.29	-5.40	-6.03
σ^{Δ}_{zzz}	-10.98	-0.33	-6.30
σ^{p}_{zzz}	-327.98	-340.62	-349.11
σ_{zzz}	-338.96	-340.95	-355.41
A_z	387.46	384.64	397.97
	Ref(*17*)	Ref(*46*)	
σ_{Av}	328.40	329.06	
A_z	378.20	381.60	

[a] $\sigma_{\alpha\beta\gamma} = \sigma^{\Delta}_{\alpha\beta\gamma} + \sigma^{p}_{\alpha\beta\gamma}$

CH_4 Molecule. We have made all the calculations employing the C_1 symmetry group for this compound. The carbon atom is placed in the origin of coordinates. We inform the hydrogen magnetic shieldings and their polarizabilities in Table **IV**. The agreement between $\sigma^{\Delta I}{}_{\alpha\beta\gamma}$ and $\sigma^{dI}{}_{\alpha\beta\gamma}$ values is excellent for the three employed basis sets.

To make the comparison between our A_γ values and those taken from the literature *(18)*, we have taken into account that Coriani *et al (18)* adopted C_{3v} geometry and placed H_1 on the $+z$ axis. After transforming our A_γ values to that geometry we have found $A^{H1}{}_z$ = 80.61, 78.74, and 76.46 ppm au for basis sets ***I***, ***II*** and ***III***, respectively, to be compared with 78.7 ppm. au. *(18)*.

For the carbon atom, the A_γ values reported in Table **V**, are identically zero by symmetry. The only non vanishing $\sigma_{\alpha\beta\gamma}$ component, 47.26 and 50.73 ppm au, for basis sets ***II*** and ***III***, respectively, is successfully compared with the 51.6 ppm au. value obtained by Coriani *et al (18)*. Basis set convergence is quite well achieved for both nuclei.

Table IV: CH_4 Proton nuclear shielding (in ppm) and their *CTOCD-DZ* shielding polarizabilities[a] (in ppm au) for basis sets *I-III*.

	I	*II*	*III*		*I*	*II*	*III*
σ_{Av}	44.55	31.81	31.58				
$\sigma^d{}_{xxy}$	-18.07	-17.98	-17.81	$\sigma^d{}_{xxz}$	-12.64	-12.50	-12.42
$\sigma^\Delta{}_{xxy}$	-18.30	-15.32	-16.88	$\sigma^\Delta{}_{xxz}$	-15.75	-14.04	-12.25
$\sigma^p{}_{xxy}$	-64.16	-66.53	-63.35	$\sigma^p{}_{xxz}$	-40.45	-40.82	-40.67
σ_{xxy}	-82.46	-81.85	-80.23	σ_{xxz}	-56.20	-54.86	-52.92
$\sigma^d{}_{yyy}$	-37.73	-37.68	-37.70	$\sigma^d{}_{zzz}$	-27.98	-27.95	-27.90
$\sigma^\Delta{}_{yyy}$	-36.84	-35.24	-36.45	$\sigma^\Delta{}_{zzz}$	-25.62	-23.93	-26.75
$\sigma^p{}_{yyy}$	-0.76	-0.61	0.46	$\sigma^p{}_{zzz}$	-4.54	-5.69	-3.44
σ_{yyy}	-37.60	-35.85	-35.99	σ_{zzz}	-30.16	-29.62	-30.19
$\sigma^d{}_{zzy}$	-16.24	-16.13	-16.05	$\sigma^d{}_{yyz}$	-10.32	-10.31	-10.27
$\sigma^\Delta{}_{zzy}$	-18.91	-16.71	-15.47	$\sigma^\Delta{}_{yyz}$	-10.99	-8.59	-9.66
$\sigma^p{}_{zzy}$	-58.50	-59.09	-58.03	$\sigma^p{}_{yyz}$	-42.28	-42.75	-41.40
σ_{zzy}	-77.41	-75.80	-73.50	σ_{yyz}	-53.27	-51.34	-51.06
A_y	65.82	64.50	63.24	A_z	46.54	45.27	44.73
(Ref.*18*)					*I*	*II*	*III*
σ_{Av} : 31.4 A_z : 78.70				A_z (geometry Ref. *18*)	80.61	78.74	76.46

[a] $\sigma_{\alpha\beta\gamma} = \sigma^\Delta{}_{\alpha\beta\gamma} + \sigma^p{}_{\alpha\beta\gamma}$

Table V: CH_4 Carbon nuclear shielding (in ppm) and their *CTOCD-DZ* shielding polarizabilities[a] (in ppm au) fro basis sets *I-III*.

	I	*II*	*III*	Ref. (*18*)
σ_{Av}	199.4	195.17	193.95	193.40
σ^{d}_{xzx}	8.74	8.97	8.48	
σ^{Δ}_{xzx}	4.90	5.49	8.44	
σ^{p}_{xzx}	42.36	45.24	42.29	
σ_{xzx}	47.26	50.73	50.73	51.60

[a] $\sigma_{\alpha\beta\gamma} = \sigma^{\Delta}{}_{\alpha\beta\gamma} + \sigma^{p}{}_{\alpha\beta\gamma}$

Before starting with the **Conclusions** section we must point out that we have made all the calculations using the C1 symmetry, even if the local symmetry is higher than that for H_2O and CH_4 molecules, and it may be exploited in this kind of calculation. We made that choice, not symmetry at all, because we had to check the computational implementation of the $\sigma^{\Delta I}{}_{\alpha\beta\gamma}$ quantities. We are preparing a next manuscript for a large set of molecules employing the higher-local symmetry for each of them.

Conclusions. For all the nuclei in the three molecules treated here, our and other authors' SCF calculations yield similar results.

Rizzo *et al* reported in refs. (*17,18*) results of nuclear shieldings and their polarizabilities for the same compounds but including correlation effects. By inspection of those values, we conclude that correlation affects dramatically both, nuclear magnetic shieldings and their polarizabilities for the CO compound but it is not so much important for H_2O and CH_4 molecules. In the case of the H_2O molecule, the basis sets employed here must be improved since it is evident that they are not suitable enough to provide results of the same accuracy as those taken from the literature.

Our calculations do not include vibrational corrections. Bishop and Cybulski have computed them for nuclear magnetic shielding and shielding polarizabilities of CO (*9*).

We have reported the $\sigma^{\Delta I}{}_{\alpha\beta\gamma}$ and the $\sigma^{dI}{}_{\alpha\beta\gamma}$ values because they must be identical in the *Hartree-Fock limit*. The $\sigma^{\Delta I}{}_{\alpha\beta\gamma}$ values depend very much on the quality of the basis set while the $\sigma^{dI}{}_{\alpha\beta\gamma}$ values are almost independent of that quality. The addressed reason is that the $\sigma^{dI}{}_{\alpha\beta\gamma}$ evaluation requires the calculation of the first-order electric field perturbed density matrix, but it is necessary to compute three first-order CHF perturbations to get $\sigma^{\Delta I}{}_{\alpha\beta\gamma}$ values.

The goal of the *CTOCD-DZ* method is that it provides nuclear magnetic shieldings and shielding polarizabilities which are origin independent, that is total shielding polarizabilities which are origin independent in calculations employing any finite basis set. In the limit of exact eigenfunctions of a model Hamiltonian, e. g., within the *exact Hartree-Fock limit,* $\sigma^{dI}{}_{\alpha\beta\gamma} = \sigma^{\Delta I}{}_{\alpha\beta\gamma}$. Both contributions to the shielding polarizabilities, $\sigma^{\Delta I}{}_{\alpha\beta\gamma}$ and $\sigma^{pI}{}_{\alpha\beta\gamma}$ are calculated within the same approximation. The accuracy of the shielding polarizabilities computed within the *CTOCD-DZ* approach is affected by the quality of the basis set, but they are origin independent for any (gaugeless) basis set, because the constraints for charge and current conservation are exactly satisfied.

Acknowledgments.
We thank P. Lazzeretti and R. Zanasi for their helpful assistance. Financial support of the University of Buenos Aires (UBACYT-TX063), Argentinian CONICET (Consejo Nacional de Investigaciones Científicas y Técnicas), and Agencia Nacional de Promoción de Ciencia y Técnica, are gratefully acknowledged.

Literature Cited

1. Buckingham, A. D., *Adv. Chem. Phys.* **1967**, *12*, 107.
2. Buckingham, A. D., Orr, B. J., Quart. Rev. **1967**, *21*, 195.
3. Buckingham, A. D., *Canad. J. Chem.* **1960**, *38*, 300.
4. Cotton, A., Mouton, H., *Compt. Rend*. **1907**, *141*, 317.
5. Cotton, A.,Mouton, H., *Compt. Rend*. **1907**, *141*, 349.
6. Cotton, A., Rapp. Cons. Phys. Solvay **1932**, 418.
7. Cybulski, S. M., Bishop, D. M., *J. Chem. Phys.*, **1994**, *101*, 424.
8. Rizzo, A., Helgaker, T., Ruud, K., Barszczewicz, A., Jaszunski, M., *J. Chem. Phys.*, **1994**, *102*, 8953.
9. Bishop, D. M., Cybulski, S. M., Molec. Phys., **1993**, *80*, 199.
10. Augspurger, J. D., *et al.*, *J. Magn. Res.*, **1992**, *100*, 342.
11. Augspurger, J. D., Dykstra, C. E., *J. Phys. Chem.*, **1991**, *95*, 9230.
12. Augspurger, J. D., Dykstra, C. E., *Molec. Phys.*, **1992**, *76*, 229.
13. Augspurger, J. D., Dykstra, C. E., Oldfield, E., *J. Am. Chem. Soc.*, **1991**, *113*, 2447.
14. Cybulski, S. M., Bishop, D. M., *Chem. Phys. Lett.*, **1996**, *250*, 471.
15. Jonsson, D., Norman, P., Vahtras, O., Ågreen, H., J. *Chem. Phys.*, **1996**, *105*, 6401.
16. Raynes, W. T., *Encyclopedia of NMR*, eds. Grant, D. M., Harris, R. K., Willey. London, **1996**, pp.1846.
17. Rizzo, A., Helgaker, T., Ruud, K., Barszczewicz, A., Jaszunski, M.,Jørgensen, P., *J. Chem. Phys.*, **1995**, *102*, 8953.
18. Coriani, S., Rizzo, A., Ruud, K., Helgaker, T., *Molec. Phys.*, **1995**, *88*, 931.
19. Coriani, S., Rizzo, A., Ruud, K., Helgaker, T., *Chem. Phys.*, **1997**, *216*, 53.
20. Enevoldsen, T., Oddershede, J., *Molec. Phys.*, **1995**, *86*, 235.
21. Lazzeretti, P., Malagoli, M., Zanasi, R., *Chem. Phys. Lett.*, **1994**, *220*, 299.
22. Coriani, S., Lazzeretti, P., Malagoli, M., Zanasi, R., *Theoret. Chim. Acta*, **1994**, *89*, 181.
23. Zanasi, R., Lazzeretti, P., Malagoli, M., Piccinini, F., *J. Chem. Phys.*, **1995**, *102*, 7150.
24. Lazzeretti, P., Malagoli, M., Zanasi, R., *J. Chem. Phys.*, **1995**, *102*, 9619.
25. Lazzeretti, P., Zanasi, R., *Int. J. Quantum, Chem.* **1996**, *60*, 249.
26. Zanasi, R., *J. Chem. Phys.*, **1996**, *105*, 1460.
27. Keith, T. A., Bader, R. F. W., *Chem. Phys. Lett.*, **1993**, *210*, 223.
28. Lazzeretti, P., Zanasi, R., *Molec. Phys.*, **1996**, *89* , 157.
29. Lazzeretti, P., *Adv. Chem. Phys.*, **1987**, *75*, 507.
30. Lazzeretti, P., Malagoli, M., Zanasi, R., *Strategies and Applications in Quantum Chemistry*, eds. Ellinger, Y., Defranceschi M., Klewer Academic Publishers. Netherlands, **1996**, pp 279.
31. Dodds, J. L., Mc Weeney, R, Raynes, W. T., Riley, J. P., *Mol. Phys.*, **1977**, *33*, 611, and references therein.

32. Landau, L. D., Lifshitz, E. M., *Quantum Mechanics*, Pergamon Press. Oxford, **1981**.
33. Epstein, S. T., *The Variation Method in Quantum Chemistry*, Academic Press. New York, **1974**.
34. Landau, L. D., Lifshitz, E. M., *The Classical Theory of Fields*, Pergamon Press. Oxford, **1979**.
35. Geertsen, J., *J. Chem. Phys.*, **1989**, *90*, 4892.
36. Geertsen, J., *Chem. Phys. Lett.*, **1991**, *179*, 479.
37. Geertsen, J., *Chem. Phys. Lett.*, **1992**, *188*, 326.
38. Sadlej, A. J., *Coll. Czech. Chem. Commun.*, **1986**, *53*, 1995.
39. Sadlej, A. J., *Theoret. Chim. Acta*, **1991**, *79*, 123.
40. Lazzeretti, P., Zanasi, R., *J. Chem. Phys.*, **1986**, *84*, 3916.
41. Grayson, M., Raynes, W. T., *Mol. Phys.*, **1994**, *81*, 533.
42. Huzinaga, S., *Approximate Atomic Functions*, University of Alberta, Edmonton, **1971**.
43. Ruud, K., Helgaker, T., Kobayashi, R., Jørgensen, P., Bak, K. L., Jensen, H. J. Aa., J. Chem. Phys., **1994**, *100*, 8178.
44. Lazzeretti, P., Zanasi, R., *Phys. Rev. A*, **1985**, *32*, 2607.
45. Raynes, W. T., Ratcliffe, R., *Mol. Phys.,* **1979**, *37*, 971.
46. Grayson, M., Raynes, W. T., *Chem. Phys. Lett.*, **1994**, *218*, 270.
47. Grayson, M., Raynes, W. T., *Chem. Phys. Lett.*, **1993**, *214*, 473.

Chapter 6

Extremely Fast Calculation of ^{13}C Chemical Shift Tensors Using the Bond Polarization Theory

Ulrich Sternberg and Wolfram Prieß

Friedrich-Schiller-Universität Jena, Institut für Optik und Quantenelektronik, Max-Wien-Platz 1, D-07743 Jena, Germany

The semi-empirical bond polarization theory provides a fast method for the calculation of ^{13}C chemical shift tensors with an accuracy level comparable to *ab initio* methods for most molecules. There are no restrictions in the size of the molecular system, so that the method can be applied to large molecules or crystal lattices. Chemical shift calculations were carried out for several small molecules and compared to *ab initio* results. Additionally the calculated chemical shift tensor for the fullerence C_{60} is presented.

The calculation of magnetic shielding parameters are of great importance for the interpretation of the chemical shift in NMR spectra. However, the calculation of chemical shift tensors is still an adventurous task. Currently isotropic chemical shifts can be estimated from empirical increments and correlations and these estimates are in most cases precise enough to reveal the topology or constitution of a molecule. However, if tensorial chemical shifts should be estimated from the structure of a molecule, in most cases *ab initio* quantum chemical calculations on the IGLO (1), LORG (2) or GIAO (3) level must be carried out. In general this approach is limited to medium size molecules and strictly speaking it reflects the situation of an isolated molecule in the gas phase. De Dios *et al.* proposed a simulation scheme for the calculation of magnetic shielding parameters in amino acids and peptides which includes the polarizing effects of surrounding molecules (4).

The authors of this work proposed a semi-empirical scheme for the calculation of ^{13}C chemical shift tensors based on the bond polarization theory (5). This method can reproduce ^{13}C chemical shift tensors with deviations from experiment comparable to the errors of the *ab initio* methods. One major advantage is that the calculations can be performed for large molecular systems with hundreds of atoms even on a PC computer. In contrast to the *ab initio* method a set of empirical parameters is needed for the calculations. In the case of the bond polarization theory these parameters can be estimated directly from experimental chemical shifts solving a set of linear equations.

Once the semi-empirical parameters are determined, the bond polarization method can calculate the chemical shift values very quickly. In this work we compare our calculations with *ab initio* results taken from literature to establish the capabilities and limits of this method. Due to the restrictions of *ab initio* methods this comparison is limited to smaller isolated molecules. In this case a comparison with experimental values neglects the influences of solvents or crystal lattices.

The Bond Polarization Model

The bond polarization model gives the chemical shift of an atom a as the sum over the N_a bonds of this bond. The bond contributions are formed of a component for the unpolarized bond (which also includes the inner shell contributions to the magnetic shielding) and a polarization term. The bond contributions are represented by a tensor with its principal axes along the basis vectors of the bond coordinate system. The transformation from the bond coordinate system into a common cartesian system is given by the transformation matrix $\boldsymbol{D}^i$.

$$\delta^a_{\alpha\beta} = \sum_i^{N_a} D^i_{\alpha\alpha'} D^i_{\beta\beta'} [\, \delta^{ai}_{\alpha'\beta'} + A^{ai}_{\alpha'\beta'} V_{ai}] \qquad (1)$$

Within the framework of the bond polarization the shift components of the unpolarized bond δ^{ai} and the parameter $\boldsymbol{A}^{ai}$ giving the polarization influence on the chemical shift are determined empirically from solving a set of linear equations 1 for a number of substances where both the chemical shift tensor and the molecular structure are known. The bond polarization energies V_{ai} are calculated as effect of surrounding net atomic charges q_x on atomic hybrids χ. With the bond polarity parameter d_i the polarization energy can be calculated.

$$V_{ab} = \frac{\sqrt{1-d_i^2}}{2} \sum_{x \in B}^{N_B} q_x \left\{ <\chi_a| \frac{1}{|\vec{R}_{ax} - \vec{r}|} |\chi_a> - <\chi_b| \frac{1}{|\vec{R}_{bx} - \vec{r}|} |\chi_b> \right\} \qquad (2)$$

The origin of equations 1 and 2 was explained in more detail in (6). The calibration of the bond polarization parameters for ^{13}C chemical shift tensors for $C(sp^3/sp^2)$-C, $C(sp^3/sp^2)$-H, $C(sp^3/sp^2)$-O, and $C(sp^3)$-N σ-bonds as well as $C(sp^2)$-C and $C(sp^2)$-O π-bonds was described in detail in (5). Structural data and tensorial single crystal ^{13}C NMR data of 20 crystalline substances including sugars, polycyclic aromatic compounds, and amino acids were used in this calculation.

Calculations

The molecular structures used in the calculations were optimized using a molecular force field program. The force field parameters were derived from various sources: The bond length was taken into accound according to a model of Dewar and Llano (7). The VALBOND method of Root et al. (8) was used for the calculation of bond angels. The dihedral angles were parametrized according to the PIMM method (9).

A special term was introduced to force the planarity of sp^2-carbon environments. Van der Waals interactions were described using the intramolecular MM2X parameters as given from Halgren (10). Atomic charges used for electrostatic interaction as well as the bond polarization calculations were calculated using the bond polarization method (11). A detailed description of the force field program will be given in a future publication.

At first the calculations were performed on eleven simple molecules: ethane, propane, isobutane, neopenthane, ethene, 2-butene, methanol, dimethyl ether, dimethoxymethane, trimethoxymethane, and tetramethoxymethane. The correlation between calculated and experimental principal values is shown in Figure 1. The correlation coefficient is $R^2 = 0.97$ and the root mean square deviation of the calculated from the experimental values is 13 ppm.

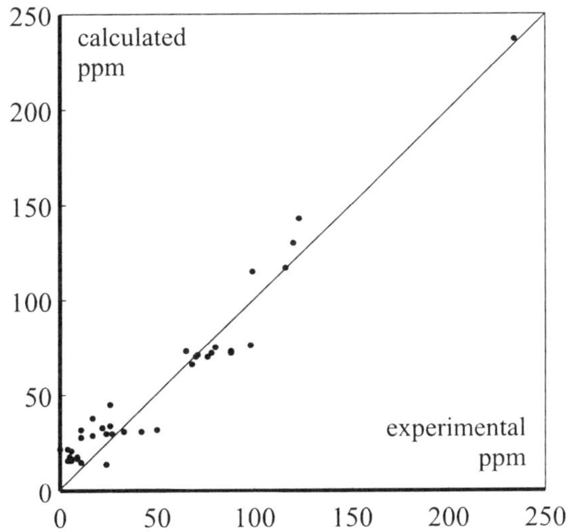

Figure 1. Correlation between calculated and experimental principal values of ^{13}C chemical shift tensors

The principal values of the calculated chemical shift tensors are compared to experimental values and IGLO *ab initio* calculations taken from literature in Table I. The experimental and most of the calculated values are given in the order δ_{11} / δ_{22} / δ_{33}. However in cases where the orientation of the principal axes was known from symmetry considerations or *ab initio* calculations, and this order was not adequately reproduced in our calculations the components were sorted according to their corresponding tensor orientation in the local coordinate system. In many of these small molecules the tensor axes are defined by the symmetry of the molecules. In other cases the axes are at least loosely defined by nearly symmetric local bond surroundings. For example the δ_{33} component of the CH_3 chemical shift tensor is close to the direction of the non CH bond or the δ_{33} component of a sp^2 carbon is perpendicular to the plane of the π system.

Table I. Calculated principal values of ^{13}C chemical shift tensors compared to experimental values and *ab initio* results (the reference U.F. denotes a private communication from Ulrich Fleischer, Ruhr University Bochum)

molecule	group	BPOL	experiment	*ab initio*	ref.
ethane	CH_3	28/28/16	11/11/4	16/16/5	(12)
propane	CH_3	30/29/22	30/15/4	24/17/-3	(12)
	CH_2	38/32/22	not given	17/11/4	(13)
isobutane	CH_3	31/31/17	42/33/6	36/29/-1	(12)
	CH	33/33/43	26/26/22	11/11/19	(14)
neopentane	CH_3	32/32/18	50/50/5	45/45/1	(12)
	C	43/43/43	not given	28/28/28	U.F.
ethene	CH_2	237/130/14	234/120/24	273/113/13	(15)
2-butene	CH_3	30/30/16	27/27/5	32/29/0	(12)
	CH	229/110/22	232/113/37	not given	(16)
methanol	CH_3	70/66/21	76/68/6	76/71/9	(12)
dimethyl ether	CH_3	73/72/16	88/88/6	85/81/8	(12)
dimethoxymethane	CH_3	76/70/18	98/70/9	74/64/4	(12)
	CH_2	127/91/51	not given	not given	
trimethoxymethane	CH_3	75/73/15	80/65/11	75/67/7	(12)
	CH	143/117/115	123/116/99	115/103/80	(14)
tetramethoxymethane	CH_3	72/71/17	78/71/9	65/61/3	(12)
	C	161/161/161	not given	not given	

Some serious deviations occur for CH_3 groups coupled to primary carbon atoms. In this case especially the tensor component parallel (or nearly parallel) to the C-C bond is overestimated. This also leads to a general overestimation of the mean value of the chemical shift on the end of chains. However, comparing theory and experiment one should keep in mind that the experimental conditions often do not meet the situation of a calculation of an isolated molecule, and the conformations of the molecules under study was unknown.

In the case of strained molecules such as cyclic molecules with less than six ring atoms the hybrids forming the bonds do not lie along the connection line of the nuclei any longer since the bond angles do not correspond with the hybridization of the atoms. This fact can be described by bent bonds.

To account for this phenomenon the hybrids used in the bond polarization calculations were described in agreement with an idea taken from the maximum overlap method (17). Atomic hybrids with a rigid relative orientation were optimized so that the overlap of the hybrids that form a bond is at maximum. The estimated tensors for the unpolarized and polarized contribution were oriented along the direction of these optimized hybrids instead of the direct connection line of the atoms. This refinement of the bond polarization approach reduced the otherwise large errors of the tensor components in some of the strained molecules as demonstrated in Table II.

Table II. Principal values of the chemical shift tensor calculated with uncorrected and corrected hybrids

molecule	group	uncorrected hybrids	corrected hybrids	experiment	*ab initio*	ref.
cubane	CH	22/54/54	50/36/36	62/35/35	75/25/25	(14)
cyclopentane	CH_2	40/35/24	42/32/23	49/21/12	45/22/4	(13)
cyclopentene	α-CH_2	42/26/32	43/34/24	52/30/12	50/29/3	(13)
	β-CH_2	39/23/33	41/32/22	39/22/7	49/18/5	(13)
	CH	214/141/16	235/124/17	235/118/39	not given	(16)
cyclobutane	CH_2	41/14/45	41/33/23	39/23/14	38/26/-6	(13)
cyclobutene	CH_2	38/18/45	44/35/25	43/33/23	60/37/11	(13)
	CH	245/119/15	237/130/14	244/138/30±5	not given	(16)
cyclopropane	CH_2	35/3/62	40/33/24	22/2/-36	42/-1/-10	(13)

In the cases of cyclopentane and cyclopentene the bond angles do not significantly differ from the ideal tetrahedron angle, therefore the correction is minimal. A clear improvement was achieved in the cases of cyclobutane and cubane. The calculation for the CH_2 groups using the uncorrected hybrids resulted in a different orientation of the δ_{11}, δ_{22} and δ_{33} components compared to the *ab initio* results. The largest shift component was calculated to be perpendicular to the HCH plane of the CH_2 group whereas the IGLO results report this as the smallest component. With optimized hybrids the bond polarization model gives similar results as the *ab initio* method.

However, this correction affected mainly the orientation of the individual chemical shift contributions of the bonds. The changes in the isotropic shift values were smaller and could be attributed to the geometry dependence of the polarization. The isotropic shift contribution from unpolarized bonds δ^{ai} is constant within the framework of this semi-empirical approach and remains unchanged by this correction. Therefore it must be assumed that the electron distribution around the carbon atom still corresponds to the standard hybridization. Consequently the bond polarization model fails as soon as the deviation from a regular hybridization causes considerable changes of the isotropic shift value. Apparently this is the case when the bond angles on the carbon are smaller than 90°. This corresponds with the results of Facelli *et al.* who reported the angular dependence of the CC bond component of the chemical shift tensor based on IGLO calculations (13).

The results for cyclopropane are given as an example for the poor correlation in these highly strained molecules. These extreme molecular structures have an electron distribution that cannot be adequately described using the approximations of the bond polarization model and must be subjected to *ab initio* calculations.

For the demonstration of the capabilities of this method we chose the Buckminster fullerene C_{60}. Since this molecule consists only of carbon atoms the net atomic charges of all atoms and hence the polarization are zero. The calculated chemical shift parameters depend only on the parameters of the unpolarized bonds and the π-bond order. The π-bond order determines the extent of the π parameters contribution.

The resulting principal values of the ^{13}C chemical shift tensors of the C_{60} carbons are δ_{11} = 228 ppm, δ_{22} = 178 ppm, and δ_{33} = -3 ppm. Tycko *et al.* reportet the experimental values δ_{11} = 213 ppm, δ_{22} = 182 ppm, and δ_{33} = 33 ppm obtained from low temperature measurements of a powder pattern spectrum (18). However, the spectra have a low signal to noise ratio and a wide slope so that a larger error for the experimental value can be assumed. The chemical shift anisotropy of 217 ppm corresponds quite well with the spectral range of about 200 ppm reported by Kerkoud *et al.* for low temperature single crystal measurements (19).

The δ_{11} and δ_{22} axes lie approximately in the plane of the pentagon, with δ_{22} bisecting the the bonds on its edge. This means δ_{22} points approximately along the edge between the two hexagons but deviates by about 36° from the bond direction due to the non planar arrangement of the bonds.

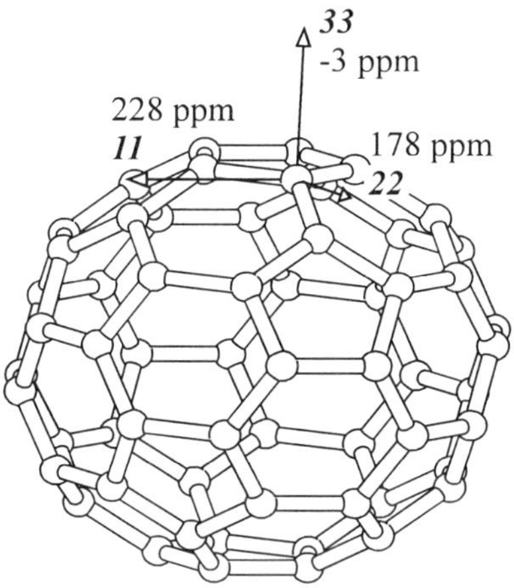

Figure 2. Model of a C_{60} molecule with principal axes of the calculated ^{13}C chemical shift tensor

Consequently, the orientation of the tensor is determined by the somewhat higher π character of the shorter pentagon edges in this molecule. Assuming a bond length of 1.40 Å (average of several experimental and theoretical values taken from (20)) the C-C bond valence of the pentagon edge is 1.54 according to the formula of O'Keefe and Brese (21) whereas the valence of the edge between two hexagons is 1.35 (bond length 1.45 Å). The orientation of the other tensor axes are sufficiently defined through the mirror plane in this molecule.

On a Pentium II PC (CPU-Speed 233 Mhz) the calculation of all ^{13}C shift tensors (without taking advantage of the high symmetry of the molecule) took only 0.4 sec whereas *ab initio* methods may take up to several days depending on the basis set.

Conclusion

The semi-empirical bond polarization model is a powerful tool for the calculation of ^{13}C chemical shift tensors. For most molecules the errors of this model are in the same order of magnitude as the errors of *ab initio* methods, under the condition that the surrounding of the carbon is not too much deformed by small bond angles. A great advantage of the model is that bond polarization calculations are very fast. The chemical shift tensors of small molecules can be estimated in fractions of a second. There is also virtually no limit for the size of the molecule. Systems with a few thousand atoms can be calculated with a standard PC within a few minutes. Possible applications are repetitive calculations during molecular dynamics simulations for the interpretation of dynamic effects on ^{13}C chemical shift distribution.

The bond polarization formula can also be used to refine molecular structures with a molecular force field that uses NMR data as an additional target function by minimizing the differences between calculated and experimental chemical shift data.

Acknowledgment

We thank J. Facelli for support and many helpful discussions. Ulrich Fleischer is thanked for supplying IGLO *ab initio* results.
This work was supported by the Deutsche Forschungsgemeinschaft.

REFERENCES

1. Kutzelnigg, W. *J. Molec.Structure* **1989,** *202,* 11.
2. Hansen, A. E.; Bouman, T. D. in NATO Advanced Research Workshop "Nuclear Magnetic Shieldings and Molecular Structure", NATO ASI C386, Kluwer Academic: Dordrecht/Norwell, Massachusetts, 1982.
3. Wollinski, K.; Hinton, J. F.; Pulay, P. *J. Am. Chem. Soc.* **1990,** *112,* 8251.
4. De Dios, A. C.; Laws, D. D.; Oldfield, E. *J. Am. Chem. Soc.* **1994,** *116,* 7784.
5. Sternberg, U.; Prieß, W. *J. Magn. Reson.* **1997,** *125,* 8.
6. Sternberg, U. *Mol. Phys.* 1988, 63, 249.
7. Dewar, M. J. S.; De Lano, C. *J. Am. Chem. Soc.* **1969,** *91,* 789.
8. Root, D. M.; Landix, C. R.; Cleveland, T. *J. Am. Chem. Soc.* **1993,** *115,* 4201.
9. Smith, A. Dissertation, Darmstadt, Germany, 1989.
10. Halgren, T. A. *J. Am. Chem. Soc.* **1992,** *114,* 7827.
11. Sternberg, U.; Koch, F.-T.; Möllhoff, M. *J. Comp. Chem.* **1994,** *15,* 524.
12. Solum, M. S.; Facelli, J. C.; Michl, J.; Grant, D. M. *J. Am. Chem. Soc.* **1986,** 108, 6464.

13. Facelli, J. C.; Orendt, A. M.; Beeler, A. J.; Solum, M. S.; Depke, G.; Malsch, K. D.; Downing, J. W.; Murthy, P. S.; Grant, D. M.; Michl, J. *J. Am. Chem. Soc.* **1985,** *107,* 6749.
14. Facelli, J. C.; Orendt, A. M.; Solum, M. S.; Depke, G.; Grant, D. M.; Michl, J. *J. Am. Chem. Soc.* **1986,** *108,* 4268.
15. Zilm, K. W.; Conlin, R. T.; Grant, D. M.; Michl, J. *J. Am. Chem. Soc.* **1980,** *102,* 6672.
16. Beeler, A. J.; Orendt, A. M.; Grant, D. M.; Cutts, P. W.; Michl, J.; Zilm, K. W.; Downing, J. W.; Facelli, J. C.; Schindler, M. S.; Kutzelnigg, W. *J. Am. Chem. Soc.* **1984,** *106,* 7672.
17. Eckert-Maksić, M.; Maksić, Z. *J. Mol. Struct.* **1974,** *22,* 445.
18. Tycko, R.; Dabbagh, G.; Fleming, R. M.; Haddon, R. C.; Makhija, A. V.; Zahurak, S. M. *Phys. Rev. Lett.* **1991,** *67,* 1886.
19. Kerkoud, R.; Auban-Senzier, P.; Godard, J.; Jerome, D.; Lambert, J.-M.; Bernier, P. C. R. *Acad. Sci.* Paris **1994,** *318,* 159.
20. Bohnen, K.-P.; Heid, R.; Ho, K.-M.; Chan, C. T. *Phys. Rev.* **1995,** *B51,* 5805.
21. O'Keefe, M.; Brese, N. E. *J. Am. Chem. Soc.* **1991,** *113,* 3226.

Chapter 7

Covering the Entire Periodic Table: Relativistic Density Functional Calculations of NMR Chemical Shifts in Diamagnetic Actinide Compounds

Georg Schreckenbach[1], Stephen K. Wolff[2], and Tom Ziegler[2]

[1]Theoretical Division (MS B268) and Seaborg Institute for Transactinium Science, Los Alamos National Laboratory, Los Alamos, NM 87545
[2]Department of Chemistry, University of Calgary, 2500 University Drive, NW, Calgary, Alberta T2N 1N4, Canada

Results from relativistic density functional calculations of nuclear magnetic resonance (NMR) shieldings and chemical shifts in diamagnetic uranium compounds are presented for the first time, thus extending the range of theoretically accessible systems to the heaviest part of the periodic table. Molecules covered include $UF_5(OCH_3)$, mixed uranium (VI) fluorides chlorides, $UF_nCl_{(6-n)}$, and a uranyl (VI) anion, $[UO_2(OH)_4]^{2-}$. Two relativistic methods are used, the "quasi-relativistic" method that employs a Pauli-Hamiltonian and the "zeroth order regular approximation" for relativistic effects (ZORA). The two relativistic methods are compared, and ZORA turns out to be more reliable for some of the test cases. Relativistic effects, including scalar relativistic and spin-orbit/Fermi contact effects, are discussed as they manifest themselves in the calculated NMR shieldings and chemical shifts.

1. Introduction

The chemistry of the actinide elements is still one of the most challenging areas for electronic structure calculations (*1*). There are several problems that have to be solved in one way or another. First, the actinide elements comprise the heaviest part of the periodic table. Hence, there are very many electrons that have to be dealt with, most of them occupying inert core shells. Second, scalar and spin-orbit relativistic effects have to be included for even a qualitative understanding of the f-block chemistry. Third, correlation effects resulting from the interaction between the electrons are at least equally important. Fourth, the *5f*, *6d*, *7s*, and *6p* orbitals are comparatively close in energy and spatial extent and can all participate in bonding. In addition, for the *5f* orbitals there are typically degeneracies or near degeneracies present in a molecule, which makes it difficult to achieve convergence in the self-consistent solution of the quantum-mechanical equations or to describe excited states.

Density functional theory (DFT) has been used extensively for the theoretical study of actinide complexes, making it one of the few first-principles methods that can treat large actinide complexes with a reasonable accuracy. This is because DFT treats electron correlation effects implicitly and hence efficiently. In addition, it is comparatively straightforward to include relativistic effects as well. The application of DFT to the chemistry of the actinides has been reviewed (*2*).

In the last few years, DFT has also become one of the prime methods for the study of nuclear magnetic resonance (NMR) chemical shifts in transition metal complexes and other large molecules. DFT calculations of NMR chemical shifts have been reviewed (*3,4*).

In this chapter, we discuss the application of our DFT-GIAO approach for the calculation of NMR shieldings and chemical shifts (*5-10*) to the heaviest part of the periodic table. (GIAO stands for "gauge including atomic orbitals". GIAOs have been introduced by London and Ditchfield (*11,12*). They are also called "London orbitals" (*11*) in the literature.) To the best of our knowledge, NMR chemical shifts in *f*-block compounds have never been studied by first principle theoretical methods before. Here, we will concentrate on diamagnetic (i.e., formally f^0) compounds of uranium.

This chapter is organized as follows. After the introduction, we will briefly present the relativistic methods used, the Pauli-Hamiltonian (*13*) and the more modern "zeroth order regular approximation" for relativistic effects (ZORA) (*14-16*). We refer the reader to the literature for a more comprehensive discussion of the theory (*3,5-10*) and of issues surrounding the application of DFT to magnetic properties (*3*). This section is followed by some computational details (section 3) and by a qualitative discussion of relativistic effects in calculated NMR shieldings and chemical shifts, section 4. We then compare the ZORA and Pauli approaches to relativity, using calculated ^{13}C chemical shifts in halomethanes as a representative example. Section 6 presents preliminary results on both ligand and metal NMR chemical shifts and shieldings in diamagnetic uranium compounds. We conclude the chapter with a discussion and outlook.

2. Some Aspects of the Theory: Approximate Relativistic Methods

Within the DFT framework, we apply two different approaches to deal with relativistic effects, the so-called quasi-relativistic (QR) method (*13*) and the more modern "Zeroth Order Regular Approximation for Relativistic Effects" (ZORA) (*14-16*). The QR method is also known as the Pauli approach.

Non-relativistic DFT is based on an exact energy expression (*17*) that can be written as

$$E^{NR}[\rho] = \sum_{i}^{n} \int d\vec{r}\, \Psi_i^* \left(\frac{p^2}{2} + V_N \right) \Psi_i + \frac{1}{2} \int d\vec{r}_1 d\vec{r}_2 \frac{\rho(\vec{r}_1)\rho(\vec{r}_2)}{|\vec{r}_1 - \vec{r}_2|} + E_{XC}[\rho] \quad (1)$$

where we have used atomic units. In equation 1, $\rho = \sum_i^n \Psi_i^* \Psi_i$ is the electronic density of the system, the $\{\Psi_i\}$ form a set of n orthonormal one-electron functions (Kohn-Sham orbitals (*18*)), V_N is the external, nuclear potential, and $\vec{p}$ is the electronic momentum operator. Hence, the first integral represents the kinetic and potential energy of a model system with the same density but without electron-electron interaction. The second term is the classical Coulomb interaction of the electron density with itself. E_{XC}, the exchange-correlation (XC) energy, and E^{NR} are functionals of the density. The exact functional form for E_{XC} is unknown; it is defined through equation 1 (*19*), and some suitable approximation has to be chosen in any practical application of

DFT. From equation 1, the Kohn-Sham one-electron equations are usually derived (*18*).

We will show in the following how the non-relativistic formalism of equation 1 can be augmented to approximately account for relativistic effects.

QR Method. The first relativistic method is the so-called quasi-relativistic (QR) method. It has been developed by Snijders, Ziegler and co-workers (*13*). In this approach, a Pauli Hamiltonian is included into the self-consistent solution of the Kohn-Sham equations of DFT. The Pauli operator is in a DFT framework given by

$$h^{Pauli} = h^{MV} + h^{Dar} + h^{SO} \tag{2}$$

The first two terms, the mass-velocity and the Darwin operators, are called scalar relativistic terms since they do not involve the electron spin. They are given by

$$h^{MV} = -\frac{1}{8c^2} p^4 \tag{3}$$

and

$$h^{Dar} = \frac{1}{8c^2} \nabla^2 V_{KS} , \tag{4}$$

where c is the speed of light, and V_{KS} is the total one-electron Kohn-Sham potential. Using only these two scalar relativistic operators results in the scalar relativistic approximation. Finally, the spin-orbit operator is given by

$$h^{SO} = \frac{1}{4c^2} \vec{\sigma}_S \cdot \left[\vec{\nabla} V_{KS} \times \vec{p} \right] \tag{5}$$

where $\vec{\sigma}_S$ is the three-component vector of the Pauli spin matrices (not to be confused with the shielding tensor, $\vec{\vec{\sigma}}$!). In equations 4 and 5, V_{KS} is often approximated by the potential due to the nuclei and the core electronic density, which is a rather good approximation (*13*).

The non-relativistic energy expression of equation 1 is modified as follows:

$$E^{QR}\left[\rho^{QR}\right] = E^{NR}\left[\rho^{QR}\right] + \sum_{i}^{n} \int d\vec{r}\, \Psi_i^* h^{Pauli} \Psi_i \tag{6}$$

where the expression for E^{NR} of equation 1 is used but depending on the QR density ρ^{QR}. Hence, the Pauli operator is included self-consistently into the solution of the Kohn-Sham equations (*13*).

The Pauli operator of equations 2 to 5 has serious stability problems so that it should not, at least in principle, be used beyond first order perturbation theory (*20*). These problems are circumvented in the QR approach where the frozen core approximation (*21*) is used to exclude the highly relativistic core electrons from the variational treatment in molecular calculations. Thus, the core electronic density along with the respective potential are extracted from fully relativistic atomic Dirac-Slater calculations, and the core orbitals are kept frozen in subsequent molecular calculations.

We refer to the literature for a more detailed discussion of this point (*3,7*), as well as for the actual implementation within the framework of the DFT-GIAO method for NMR shieldings (*6,7,9*).

ZORA Method. Very recently, we began applying a new relativistic approach, the ZORA method of van Lenthe and co-workers (*14-16*). The ZORA Hamiltonian is given by

$$h^{ZORA} = \vec{\sigma}_S \cdot \vec{p} \frac{K}{2} \vec{\sigma}_S \cdot \vec{p} + V \tag{7}$$

where K is defined as

$$K = \left(1 - V / 2c^2\right)^{-1}. \tag{8}$$

Within the Kohn-Sham formalism of DFT, V is substituted by the Kohn-Sham potential V_{KS}. Expanding equation 7 a little, we note that spin-orbit effects are implicitly included in the ZORA Hamiltonian:

$$h^{ZORA} = \vec{p} \cdot \frac{K}{2} \vec{p} + V + \frac{\vec{\sigma}_S}{2} \cdot \left(\vec{\nabla} K \times \vec{p}\right), \tag{9}$$

and by neglecting or including the last term, respectively, both scalar and spin-orbit calculations are possible. ZORA avoids the mentioned stability problems of the Pauli Hamiltonian. Consequently, both frozen core and all-electron calculations are possible.

The ZORA formalism can be extended by a simple scaling procedure for the one-electron energies, thus yielding improved agreement with the fully relativistic Dirac procedure, for the core orbitals in particular. We use this scaled ZORA approach in this study (*10*).

We refer the reader to the literature for details of the DFT-ZORA approach (*14-16*) and its application to NMR calculations (*10*).

3. Computational Details

In this chapter, we will use both the shielding and chemical shift terminologies. The experimentally used chemical shift δ follows from the theoretically calculated absolute shielding σ by:

$$\delta = \sigma_{ref} - \sigma \tag{10}$$

where σ_{ref} is the absolute shielding of the reference compound.

The NMR shielding tensor, $\vec{\vec{\sigma}}$, is calculated in the usual way as an energy derivative (22) using the relativistic QR and ZORA energy expressions, respectively (*5-7,9,10*). The gauge problem of magnetic properties is solved by employing GIAOs as basis functions (*5,11,12*). The resulting shielding tensor has three contributions, the diamagnetic, paramagnetic, and spin-orbit (spin-orbit/Fermi contact in the Pauli case) shieldings. The diamagnetic shielding depends only on the zero order, unperturbed electronic density whereas the paramagnetic shielding contains the magnetically perturbed molecular orbitals.

The calculation of NMR shielding tensors based on DFT and the GIAO approach has been implemented into the Amsterdam Density Functional code ADF (*21,23-27*). The non-relativistic as well as scalar relativistic (Pauli) implementations are the work of Schreckenbach and Ziegler (*5-7*) whereas the spin-orbit (Pauli) and ZORA NMR approaches were implemented by Wolff *et al.* (*9,10*). For the mathematical and technical details, the reader is referred to the literature.

We use Slater type basis sets that are of triple-ζ quality in the valence region. These basis sets are augmented by two (all elements in ZORA calculations; elements H - Kr in QR calculations; ADF standard basis V) or one (all other elements in QR calculations; ADF standard basis IV) sets of polarization functions. For Pauli (QR) calculations, we use the frozen core approach (*21*). The (frozen) core orbitals are described by a double-ζ Slater type basis whereas the basis for the valence orbitals is of single-ζ quality in the core region. ZORA calculations were performed with all-electron basis sets that are of double-ζ quality in the core region. (All-electron and/or double-ζ core basis sets cannot be used with the Pauli operator (*7*). This, of course, limits the accuracy of the method as we will see below.)

We use generalized gradient approximations (GGA) to the exchange-correlation (XC) functional of DFT for all NMR calculations that are reported here. All current dependent terms (*28*) in the XC functional are neglected. The $M(CO)_6$ and $[MO_4]^{2-}$, M = Cr, Mo, W, calculations (*7*) employed the BP86 functional (*29,30*). All other calculations were performed with the PW91 GGA (*31*).

We use relativistically optimized geometries throughout. (See, e.g., ref. (*32*) for an example of the optimization procedure that was used for the uranium compounds.) The transition metal complexes $M(CO)_6$ and $[MO_4]^{2-}$, M = Cr, Mo, W, are the only exception; experimental geometries have been used in these cases (*7*).

4. Relativistic Effects

In this section, we will use examples from our own work to qualitatively discuss the effects of special relativity on calculated NMR shieldings and chemical shifts.

Relativistic effects on calculated NMR shieldings and chemical shifts have sometimes been divided into "direct" and "indirect" effects. According to this point of view, indirect effects are those that result from relativistic changes of the molecular geometry (the well-known relativistic bond contraction (*33*) in particular) whereas direct effects refer to a fixed geometry.

In this chapter, we use exclusively relativistically optimized or experimental geometries. Hence, we concentrate on direct relativistic effects only. They can be separated into scalar and spin-orbit/Fermi contact effects. In addition, there are, in both cases, core and valence contributions.

Scalar Relativistic Effects. Scalar relativistic effects are due to the spin-free operators of equations 3 and 4 or 9. Table I illustrates these effects for transition metal carbonyls $M(CO)_6$, M = Cr, Mo, W (*7*). We have calculated the absolute shielding and the chemical shift, taken relative to the respective tetraoxide anion $[MO_4]^{2-}$. Comparing non-relativistic and scalar relativistic calculations, we observe a strong change due to relativity of up to some 1300 ppm in the absolute shielding of $[WO_4]^{2-}$. A large part of these effects cancels out in relative chemical shifts, and this can be attributed to core effects. Thus, it is well known that relativity contracts *s* and *p* type atomic orbitals (*33*). Such a contraction, of the *1s* orbital in particular, will result in a strong increase of the diamagnetic shielding (*7*). This increase occurs in any compound of, e.g., tungsten, and it is mostly invariant against changes in the chemical environment of the heavy nucleus. Thus, such core effects cancel out in relative chemical shifts, Table I.

Table I. Metal Shieldings and Chemical Shifts[a] in Transition Metal Carbonyls and in $[WO_4]^{2-}$.

Compound	Calculated Metal Shielding (ppm)		Metal NMR Chemical Shift[b] (ppm)		
			Calculated		
	Non-Relativistic	Scalar Relativistic	Non-Relativistic	Scalar Relativistic	Experiment
$Cr(CO)_6$	-507	-449	-1,866	-1,846	-1,795
$Mo(CO)_6$	1,452	1,720	-1,814	-1,804	-1,857
$W(CO)_6$	4,892	5,767	-4,050	-3,615	-3,505
$[WO_4]^{2-}$	841	2,152	0	0	0

[a]Ref. (*7*).
[b]The chemical shifts are taken relative to the respective metal oxides $[MO_4]^{2-}$.

The part that survives in the chemical shift is a paramagnetic effect, and it can be attributed to changes in the orbital energies. Thus, *s* and *p* type atomic orbitals are typically stabilized whereas *d* and *f* type orbitals are destabilized and expanded (*33*), with similar effects in the resulting molecular orbitals. These relativistic changes alter the energy separation and hence the magnitude of interaction between occupied and virtual orbitals. Consequently, they influence the (paramagnetic part of the) shielding *and* the chemical shift, Table I.

Spin-Orbit/Fermi Contact Effects. While scalar relativistic effects seem to be sufficient for some systems like the metal carbonyls of Table I (even though it has been speculated (*9*) that spin-orbit might improve the agreement with experiment even further), there are other cases where this is not the case. We have chosen as an example the proton NMR absolute shielding in hydrogen halides HX, X = F, Cl, Br, I (*7,9*), Figure 1. This series has also been studied by other authors (*34-38*), and it may well be the most prominent example for spin-orbit effects on NMR shieldings and chemical shifts.

We note from Figure 1 that scalar relativistic calculations (*7*) are entirely unable to reproduce the experimental trend for X = Br and I. Indeed, scalar relativistic and non-relativistic calculations gave almost identical results in this case (*7*). However, with the inclusion of spin-orbit/Fermi contact operators, the experimental trend is reproduced nicely (*9*).

The mechanism for spin-orbit effects on calculated NMR shieldings has recently been elucidated (*37*). Thus, spin-orbit splitting in the presence of the external magnetic field induces spin polarization in the electronic density near the heavy nucleus. This can also be shown rigorously (*9*). The induced spin-polarization is transferred through the bond to the NMR nucleus (the proton in our example) where it is picked up by means of a Fermi contact mechanism. An analogy with the indirect spin-spin coupling effects has been proposed (*37*): Spin polarization is, in the spin-spin coupling case, produced by a second nuclear magnetic moment through Fermi contact interactions but it is in either case picked up at the NMR nucleus by one and the same mechanism. One important consequence of this picture is that spin-orbit effects will only be relevant for systems that possess a strong bond contribution from *s* orbitals at the NMR nucleus -- otherwise, Fermi contact interactions are not relevant.

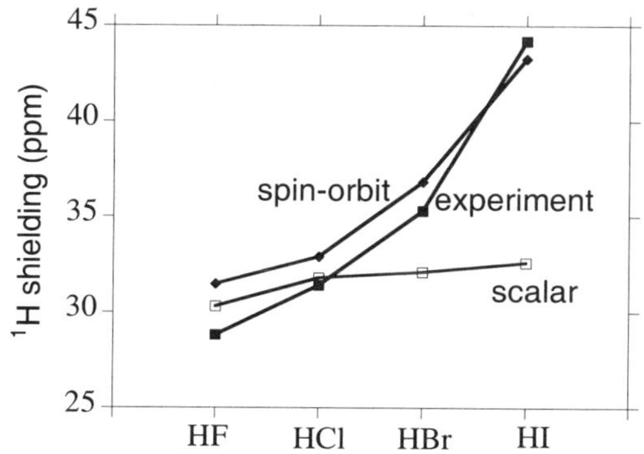

Figure 1. ¹H absolute shielding in HX, X = F, Cl, Br, I. The figure illustrates the importance of spin-orbit/Fermi contact effects in these systems (*9*); scalar relativistic calculation (*7*) are unable to reproduce the experimental trend.

5. Comparison of the QR (Pauli) and ZORA Methods

The older, Pauli-Hamiltonian based QR method has recently been compared to the more modern ZORA approach (*10*). The comparison has been done based on methane derivatives. We have summarized the results in Table II. Both the ZORA and QR methods agree well for molecules containing atoms no heavier than Cl. This shouldn't be surprising since relativistic effects are still small in such molecules. However, at least for this sample of molecules, the ZORA method is clearly superior for molecules containing heavy nuclei like Br or I. This is reflected in the mean absolute deviation between theory and experiment of 9.2 ppm (ZORA) and 15.6 ppm (QR), respectively (*10*). Note that some of the same systems have also been studied by other authors (*35-37*).

Table II. Comparison of ^{13}C Chemical Shifts in Methane Derivatives from ZORA and Pauli Spin-Orbit Calculations[a].

	Chemical Shift (ppm)[b]		
	Calculated		
Molecule	ZORA Spin-Orbit	Pauli Spin-Orbit	Experiment[c]
CH_4	-8.7	-10.3	-1.8
CH_3F	76.8	75.4	75.7
CH_3Cl	28.6	26.9	25.2
CH_3Br	10.7	6.0	9.7
CH_3I	-18.1	-34.2	-22.0
CF_4	132.5	131.3	119.9
CCl_4	122.1	119.6	96.7
CBr_4	-5.2	-58.5	-28.5
CI_4	-286.9	-241.8	-292.0

[a]Ref. (*10*).

[b]Relative to tetramethylsilane, TMS.

[c]Cited from ref. (*10*).

The deviation between the ZORA and the QR approaches may be due to the different basis sets used (double-ζ vs. single-ζ core type basis for ZORA and QR, respectively, cf. section 3 above), at least partly, rather than to the different treatment of relativity alone.

The comparison between the ZORA and QR methods is still rather limited (*10*), Table II. We are currently in the process of carrying out a more thorough comparison of the two approaches. The ZORA approach is clearly superior to the QR method on theoretical grounds since it avoids -- rather than circumvents -- the fundamental stability problems of the Pauli operator.

6. Results: Diamagnetic Uranium Compounds

We have used the Pauli formalism in its scalar and spin-orbit versions as well as the all-electron ZORA approach to calculate NMR shieldings and chemical shifts in various diamagnetic (formally f^0) uranium compounds. Preliminary results will be presented in the following.

^{19}F Chemical Shifts of the $UF_nCl_{(6-n)}$ Series. Uranium hexafluoride, UF_6, is perhaps the most studied and best characterized actinide molecule. The ^{19}F NMR chemical shift (taken relative to $CFCl_3$) of this molecule is presented in Table III for the scalar Pauli and all-electron spin-orbit ZORA approaches. In this table, we have also collected results for the whole series of the related uranium (VI) fluoride chlorides, $UF_nCl_{(6-n)}$, n =1 - 6. The ^{19}F NMR chemical shifts of all of these molecules have been measured in $CFCl_3$ solution (*39*). The solvent has also been used as internal standard. All mixed $UF_nCl_{(6-n)}$ species 'have been characterized virtually definitely' in these experiments (*39*).

Scalar Pauli computations and ZORA calculations with spin-orbit yield quite similar results, Table III, although the latter includes spin-orbit effects whereas the former does not. However, we note from a direct inspection of the ZORA calculations that the spin-orbit contributions are small (not shown in the table). The U-F bond is mostly p_σ- and p_π-type and should only have very little *s* character at the fluorine center. Hence, Fermi contact contributions are small, and spin-orbit is not relevant here, according to the qualitative discussion above.

Table III. Calculated and Experimental ^{19}F Chemical Shifts in $UF_nCl_{(6-n)}$, n = 1 - 6.

			^{19}F NMR Chemical Shift (ppm)[a]		
				Calculated	
Molecule	NMR System		Experiment[b]	Pauli Scalar	ZORA Spin-Orbit
UF_6	A_6		764.0	812.6	831.0
UF_5Cl	A_4X	A_4	762.0	792.6	813.3
		X	781.5	757.9	780.2
trans-UF_4Cl_2	A_4		755.5	782.5	805.0
cis-UF_4Cl_2	A_2X_2	A_2	760.0	779.2	803.3
		X_2	785.8	741.8	767.3
mer-UF_3Cl_3	A_2X	A_2	753.0	773.4	798.0
		X	782.6	731.1	759.0
fac-UF_3Cl_3	A_3		786.4[c]	727.9[c]	758.0[c]
trans-UF_2Cl_4	A_2		746.1	769.8	795.0
cis-UF_2Cl_4	A_2		781.0	724.1	755.3
$UFCl_5$	A	A	774.3	718.2	750.8

[a]Taken relative to $CFCl_3$. The calculated ^{19}F absolute shielding for (isolated) $CFCl_3$ is 130.7 ppm.

[b]Ref. (*39*).

[c]Averaged value; the three fluorine sites are not all equivalent in the optimized structure.

We see from Table III that the experimental chemical shifts are qualitatively reproduced by the calculations. Some of the remaining error might also be due to the calculated ^{19}F absolute shielding of the reference compound -- it is well known that ^{19}F NMR shieldings are notoriously difficult cases for DFT methods (*5,40,41*).

Inspecting the table, we observe that some experimental trends are reproduced by the calculations (for instance, the small increase in the chemical shift of *cis*-UF_4Cl_2 as compared to $UFCl_5$). However, other experimental trends are not reproduced by either theoretical method. The reasons for this somewhat disappointing result are not clear in the moment. However, solvation effects are expected to have a non-negligible influence on the electronic structure, and hence on the calculated chemical shifts, and probably also on the optimized geometries. Note that some of the molecules in the series have a permanent dipole moment whereas others don't, due to symmetry. This should lead to considerable differences in the electrostatic interactions with a polar solvent.

Chemical Shift Tensor. A particular strength of theoretical methods is that they can predict the entire chemical shift (shielding) tensor -- a property that is very difficult to measure experimentally. Further, the chemical shift tensor is thought to be a more stringent test of theoretical methods than the isotropic average alone. To the best of our knowledge, chemical shift tensors in actinide compounds have been measured only for the ^{19}F nucleus in UF_6 (*42*). The experimental anisotropy, i.e., the difference between parallel and orthogonal principal shielding tensor components, is 1210±30 ppm, which should be compared to the calculated values of 1217 ppm and 1270 ppm for the scalar Pauli and spin-orbit ZORA calculations, respectively. Hence, the experimental ^{19}F chemical shift tensor of UF_6 could be reproduced by either method, giving additional confidence in the theoretical approaches.

$U(OCH_3)F_5$. We have also investigated methoxy derivatives of UF_6. In Table IV, we present calculated and experimental (*43*) absolute shieldings for $U(OCH_3)F_5$ that has been chosen as a representative example. So far, scalar and spin-orbit Pauli (QR) calculations have been performed for this organometallic uranium (VI) system.

Table IV. Calculated and Experimental Absolute Shieldings in $U(OCH_3)F_5$.

	Absolute NMR Shielding (ppm)		
		Calculated (Pauli)	
Nucleus	Experiment[a]	Scalar	Spin-Orbit
^{1}H	20.3	25.2, 25.1, 27.7	19.4, 19.5, 20.1
^{17}O		-534.9	-567.7
^{19}F (A_4)	-402.8	-565.4, -563.5	-596.4, -597.2
		-608.4, -608.3	-641.9, -641.3
		-586.4 (average)	-619.2 (average)
^{19}F (X)	-465.9	-593.4	-624.9

[a]The experimental numbers from ref. (*43*) have been converted to absolute shielding using experimental absolute shielding scales.

From the table, we note that the experimental ^{19}F shielding (*43*) is again reproduced only qualitatively. Solvent effects are expected to be important here as well. We note in this connection that the U-O-Me bond is tilted with an optimized bond angle of 135.5°, leading to calculated ^{19}F shieldings that vary widely between the different fluorine sites, Table IV.

The proton NMR in $U(OCH_3)F_5$ is an example where spin-orbit effects cannot be neglected. Their inclusion leads to good agreement between theory and experiment (*43*), Table IV. It should not be too surprising that spin-orbit effects feature prominently in ^{1}H shieldings and chemical shifts since the proton will bind primarily through its *1s* orbital, and strong Fermi contact contributions can be expected.

Predicted ^{235}U Absolute Shieldings. Calculated uranium shieldings for selected molecules are collected in Table V.

Table V. Calculated ^{235}U Absolute Shieldings in Diamagnetic Uranium Compounds.

Molecule	Calculated ^{235}U NMR Absolute Shieldings (ppm)		
	Pauli Scalar	Pauli Spin-Orbit	ZORA Spin-Orbit
UF_6	-837	-1,170	678
UF_5Cl	-3,194	-3,452	-1,934
trans-UF_4Cl_2	-5,796	-6,165	-4,589
mer-UF_3Cl_3	-7,550	-7,889	-6,541
trans-UF_2Cl_4	-9,360	-9,809	-8,631
$UFCl_5$	-10,796	-11,169	-10,275
$UF_5(OCH_3)$	-225	-2,607	
$[UO_2(OH)_4]^{2-}$	-808	1,100	

Only once has an NMR signal of an actinide nucleus been observed, namely for UF_6 (*44*). Hence, ^{235}U relative chemical shifts (or any other actinide shifts) have not as yet been measured, due to an unfavorable signal-to-noise ratio combined with the large chemical shift range and the radioactivity of the actinides. It is hoped that the present calculations might help to guide future experiments, by narrowing down the magnetic field ranges that have to be scanned for a signal. The predicted chemical shift range is rather large indeed, and spans at least 12,000 ppm for the molecules studied (Pauli spin-orbit calculations.)

It follows from Table V that spin-orbit effects are relevant for the heavy metal shieldings and, since the spin-orbit contribution does not always have the same sign, for the relative chemical shifts. In this connection, it is interesting to note that the ZORA spin-orbit numbers are shifted as compared to their Pauli spin-orbit counterparts. This effect can be attributed, at least partly, to core contributions at the metal: while scalar contributions of the core orbitals are approximately accounted for by the frozen core approximation (*6,7*), spin-orbit contributions of the core orbitals are neglected. Hence, the more positive (diamagnetic) shieldings from the ZORA method are due to spin-orbit/Fermi contact contributions of the *s* orbitals in the uranium core.

7. Discussion and Conclusion

As should be evident from various chapters in this volume, the theoretical description of NMR shieldings and chemical shifts has seen a tremendous development during the past few years, and a number of methods are now at a point to be truly useful to experimental researchers as well as to answer fundamental questions. DFT based methods are a major aspect of these developments. The particular strength of DFT is perhaps its computational expedience, paired with accuracy and reliability. Hence, DFT applications are particularly useful for large and/or heavy systems including metal complexes or biological molecules.

Relativistic effects cannot be neglected if heavier systems are studied; we have discussed the major relativistic effects on calculated NMR shieldings and chemical shifts in this chapter. Besides relativistic effects, electron correlation has to be included for even a qualitatively correct treatment of transition metal or actinide complexes. So far, DFT based methods are about the only approaches that can handle both relativity and correlation, and DFT is, for the time being, the method of choice for these heavy element compounds. In this chapter, we have presented results from two relativistic DFT methods, the Pauli- (QR-) and ZORA approaches.

The two methods have been compared but this comparison is still rather limited. The ZORA method has a better theoretical foundation than the Pauli approach (section 2), allowing, among other things, the use of larger basis sets (section 3). ZORA was found to be the more accurate method for the halomethanes studied (section 5). However, no clear preference for either method could be established yet for the actinide compounds.

We have, for the first time, presented calculated shieldings and chemical shifts in compounds of *f*-block elements. Thus, the entire periodic table is now becoming accessible to the theoretical study of NMR parameters. Relativistic effects are essential for a correct description of actinide compounds. We noted, in particular, that the relativistic spin-orbit/Fermi contact contribution cannot be neglected for the ^{235}U shielding nor for the shieldings of the neighboring ligands.

Experimental data is available for ligand chemical shifts. We achieved qualitative and in some, but not all cases also quantitative agreement between theory and experiment. Nevertheless, this gives confidence in the predictive power of the DFT-GIAO methods, as one of the major goals of the present studies is to use calculated actinide metal chemical shieldings to guide future experiments. It is hoped that experimental ^{235}U NMR can be established in this way.

The remaining differences between theory and experiment may be due to a number of reasons. First, basis set issues have to be explored further. Thus, it is well known from the lower part of the periodic table that polarization functions are necessary for chemical shift calculations. Currently, *g*-type polarization functions on uranium cannot yet be tested for technical reasons. Second, calculated NMR shieldings and chemical shifts are known to be sensitive to the reference geometry. Here, we have used optimized geometries. In cases where experimental structures are available we found that bond lengths tend to be somewhat overestimated by the current theoretical techniques (*2*). Third, solvation effects cannot yet be modeled realistically by our DFT-GIAO method. They are thought to be of major importance, both, indirectly by way of their influence on the optimized structures and directly through changes in the electronic structure in going from vacuum to solution.

Limitations and possible future directions of DFT-NMR methods have been discussed elsewhere (*3*). In addition to the issues discussed in this paper (*3*), it will be necessary to develop the theory and computational methods for paramagnetic molecules. Such paramagnetic molecules are prevalent in actinide chemistry, in particular in the organometallic chemistry of these elements. Typically, the unpaired electron(s) occupy various, formally non-bonding *f* levels. In many cases, ligand NMR signals have been obtained for such species (e.g., refs. (*45-47*)). Current theory is, however, unable to model the chemical shift in such open shell systems. Thus, for theory to be truly useful in this area of chemistry (as well as for transition metal complexes), it becomes necessary to extend existing NMR methods towards open shell systems. This requires that we treat the interaction of the magnetic perturbation with the spin and orbital angular momentum on an equal footing.

Acknowledgments

G. S. would like to thank P. J. Hay, R. L. Martin, D. W. Keogh, and D. L. Clark for inspiring discussions and encouragement. S. K. W. and T. Z. are grateful to E. J. Baerends and E. van Lenthe for discussions on ZORA-NMR calculations. G. S. acknowledges funding from the Laboratory Directed Research and Development program of the Los Alamos National Laboratory, operated by the University of California under contract to the US Department of Energy (W-7405-ENG-36), and from the Seaborg Institute for Transactinium Science. S. K. W. and T. Z. acknowledge financial support from the National Science and Engineering Research Council of Canada (NSERC), as well as from the Petroleum Research Fund, administered by the American Chemical Society (Grant ACS-PRF No. 31205-AC3).

Literature Cited

(1) Pepper, M.; Bursten, B. E. *Chem. Rev.* **1991**, *91*, 719.
(2) Schreckenbach, G.; Hay, P. J.; Martin, R. L. *J. Comput. Chem.* **1999**, in print.
(3) Schreckenbach, G.; Ziegler, T. *Theor. Chem. Acc.* **1998**, *99*, 71.
(4) Bühl, M.; Kaupp, M.; Malkin, V. G.; Malkin, O. L. *J. Comput. Chem.* **1999**, in print.
(5) Schreckenbach, G.; Ziegler, T. *J. Phys. Chem.* **1995**, *99*, 606.
(6) Schreckenbach, G.; Ziegler, T. *Int. J. Quantum Chem.* **1996**, *60*, 753.
(7) Schreckenbach, G.; Ziegler, T. *Int. J. Quantum Chem.* **1997**, *61*, 899.
(8) Schreckenbach, G.; Dickson, R. M.; Ruiz-Morales, Y.; Ziegler, T. In *Chemical Applications of Density Functional Theory, ACS Symposium Series 629*; Laird, B. B.; Ross, R. B. and Ziegler, T., Ed.; American Chemical Society: Washington DC, 1996; pp 328-341.
(9) Wolff, S. K.; Ziegler, T. *J. Chem. Phys.* **1998**, *109*, 895.
(10) Wolff, S. K.; Ziegler, T.; van Lenthe, E.; Baerends, E. J. *J. Chem. Phys.* **1999**, submitted.
(11) London, F. *J. Phys. Radium* **1937**, *8*, 397.
(12) Ditchfield, R. *Mol. Phys.* **1974**, *27*, 789.
(13) Ziegler, T.; Tschinke, V.; Baerends, E. J.; Snijders, J. G.; Ravenek, W. *J. Phys. Chem.* **1989**, *93*, 3050.
(14) van Lenthe, E.; Baerends, E. J.; Snijders, J. G. *J. Chem. Phys.* **1993**, *99*, 4597.
(15) van Lenthe, E.; Baerends, E. J.; Snijders, J. G. *J. Chem. Phys.* **1994**, *101*, 9783.
(16) van Lenthe, E.; van Leeuwen, R.; Baerends, E. J. *Int. J. Quantum Chem.* **1996**, *57*, 281.
(17) Hohenberg, P.; Kohn, W. *Phys. Rev.* **1964**, *136*, B864.
(18) Kohn, W.; Sham, L. J. *Phys. Rev.* **1965**, *140*, A1133.
(19) Becke, A. D. In *Modern Electronic Structure Theory, Part II*; Yarkony, D. R., Ed.; World Scientific: Singapore, 1995; pp 1022-1046.
(20) Kutzelnigg, W. *Z. Phys. D* **1990**, *15*, 27.
(21) Baerends, E. J.; Ellis, D. E.; Ros, P. *Chem. Phys.* **1973**, *2*, 41.
(22) Helgaker, T.; Jaszunski, M.; Ruud, K. *Chem. Rev.* **1999**, accepted.
(23) Baerends, E. J.; Ros, P. *Chem. Phys.* **1973**, *2*, 52.
(24) Baerends, E. J. Ph.D. Thesis, Free University, Amsterdam, The Netherlands, 1973.
(25) te Velde, G.; Baerends, E. J. *J. Comp. Phys.* **1992**, *99*, 84.
(26) Fonseca Guerra, C.; Visser, O.; Snijders, J. G.; te Velde, G.; Baerends, E. J. In *Methods and Techniques in Computational Chemistry METECC-95*; Clementi, E. and Corongiu, C., Ed.; Cagliari, 1995; pp 305-395.
(27) *ADF, 2.3*, Theoretical Chemistry, Vrije Universiteit, Amsterdam, The Netherlands, **1997**.
(28) Rajagopal, A. K. *J. Phys. C* **1978**, *11*, L943.
(29) Becke, A. D. *Phys. Rev. A* **1988**, *38*, 3098.
(30) Perdew, J. *Phys. Rev. B* **1986**, *33*, 8822.
(31) Perdew, J.; Wang, Y. *Phys. Rev. B* **1992**, *45*, 13244.
(32) Schreckenbach, G.; Hay, P. J.; Martin, R. L. *Inorg. Chem.* **1998**, *37*, 4442.
(33) Pyykkö, P. *Chem. Rev.* **1988**, *88*, 563.
(34) Pyykkö, P.; Görling, A.; Rösch, N. *Mol. Phys.* **1987**, *61*, 195.
(35) Nakatsuji, H.; Takashima, H.; Hada, M. *Chem. Phys. Lett.* **1995**, *233*, 95.
(36) Malkin, V. G.; Malkina, O. L.; Salahub, D. R. *Chem. Phys. Lett.* **1996**, *261*, 335.

(37) Kaupp, M.; Malkina, O. L.; Malkin, V. G.; Pyykkö, P. *Chem. Eur. J.* **1998**, *4*, 118.
(38) Vaara, J.; Ruud, K.; Vahtras, O.; Ågren, H.; Jokisaari, J. *J. Chem. Phys.* **1998**, *109*, 1212.
(39) Downs, A. J.; Gardner, C. J. *J. Chem. Soc., Dalton Trans.* **1984**, 2127.
(40) Malkin, V. G.; Malkina, O. L.; Casida, M. E.; Salahub, D. R. *J. Am. Chem. Soc.* **1994**, *116*, 5898.
(41) Lee, A. M.; Handy, N. C.; Colwell, S. M. *J. Chem. Phys.* **1995**, *103*, 10095.
(42) Zeer, E. P.; Falaleev, O. V.; Zobov, V. E. *Chem. Phys. Lett.* **1983**, *100*, 24.
(43) Cuellar, E., A.; Marks, T. J. *Inorg. Chem.* **1981**, *20*, 2129.
(44) Le Bail, H.; Chachaty, C.; Rigny, P.; Bougon, R. *J. Physique Lett.* **1983**, *44*, 1017.
(45) Luke, W. D.; Streitwieser, A. In *Lanthanide and Actinide Chemistry and Spectroscopy, ACS Symposium Series 131*; Edelstein, N. M., Ed.; American Chemical Society: Washington DC, 1980; pp 93-140.
(46) Streitwieser, A. *Inorg. Chim. Acta* **1984**, *94*, 171.
(47) Fischer, R. D. In *Fundamental and Technological Aspects of Organo-f-Element Chemistry, NATO ASI C155*; Marks, T. J. and Fragalà, I. L., Ed.; D. Reidel: Dordrecht, The Netherlands, 1985; pp 277-326.

Chapter 8

The Effect of Electron Correlation on the ^{19}F Chemical Shifts in Fluorobenzenes

Peter B. Karadakov, Graham A. Webb, and James A. England

Department of Chemistry, University of Surrey, Guildford, Surrey GU2 5XH, United Kingdom

Ab initio Hartree-Fock (HF) level calculations using gauge-including atomic orbitals (GIAOs) are known to provide a remarkably good description of the experimental ^{19}F chemical shieldings in fluorobenzenes. However, the HF approach does not account for any correlation effects, the inclusion of which can lead to significant changes both in the optimized molecular geometry and in the wavefunction, and consequently, in the calculated chemical shielding values. We present and discuss the results of second- and fourth-order Møller-Plesset (MP2 and MP4), '6 in 6' complete-active space self-consistent field (CASSCF) and other correlated geometry optimizations for the members of the fluorobenzene series, followed by HF-GIAO, '6 in 6' CASSCF-GIAO and MP2-GIAO shielding calculations.

There is a massive and constantly expanding body of evidence indicating that inclusion of electron correlation effects is essential for obtaining quantitatively and, in many cases, even qualitatively correct results from quantum-chemical calculations. When considering electron correlation, it has become fairly common to distinguish between dynamic and nondynamic correlation effects. The need to identify these two types of correlation effects stems from a simplifying assumption, very common to chemistry, according to which the electrons in a small number of 'active' orbitals only participate in most molecular processes, or determine the local properties of a fragment or an electron subsystem (*e.g.* the π electrons) in a much larger molecule. This assumption, known as the core-valence separation (*1*), suggests that it should be sufficient to provide correlation for these 'active' (or valence) orbitals only, and to keep all other 'inactive' (or core) orbitals doubly-occupied, just as in a closed-shell Hartree-Fock (HF) wavefunction. The most common type of 'active' space wavefunction is the '*N* in *M*' complete-active-space self-consistent field (CASSCF, or just CAS for brevity) construction, which involves all spin-adapted configurations that can be generated by distributing *N* electrons among *M* orbitals. As a rule, the nondynamic correlation effects are considered to be those accounted for by the CASSCF wavefunction; dynamic correlation designates the further difference between the CASSCF wavefunction and the exact one.

Normally, inclusion of nondynamic correlation effects is expected to be sufficient for the proper qualitative description of bond-breaking and bond-making processes, or of systems which are not well described by closed- or open-shell HF wavefunctions (an example is provided by singlet biradicals such as square cyclobutadiene and various dehydrobenzenes). However, as it has been shown recently by Borden and Davidson (*2*) (see also refs. *3* and *4*), there are important cases in which wavefunctions that include nondynamic correlation for the active electrons only (such as CASSCF) give results which are quantitatively and sometimes even qualitatively incorrect. Their examples include calculations on excited states, radical cations, potential energy surfaces for pericyclic reactions and dehydrobenzenes, in which the inclusion of dynamic correlation is shown to be of crucial importance for achieving qualitatively correct results. The standard way of treating dynamic correlation effects usually involves some variant of many-body perturbation theory (MBPT), and the most frequently used approaches employ Møller-Plesset constructions of various orders (MP*n*), based on a HF reference wavefunction. A closed-shell HF reference can be a bad choice for systems for which the nondynamic correlation is important (see above). This has prompted the development of more complicated MBPT approaches based on a CASSCF reference, *e.g.* CASPT2N (*3*) and CASMP2 (*4*).

Until the beginning of this decade, the inclusion of correlation effects was not considered to be of considerable importance for the accurate calculation of nuclear magnetic resonance (NMR) chemical shielding constants. The main reason for this is in the fact that HF-level calculations using GIAOs (gauge-including atomic orbitals) or the IGLO (individual gauges for localized orbitals) approach in many cases produce results that are in excellent agreement with experiment. One convincing example is provided by the extensive calculations of ^{19}F shielding constants in fluorobenzenes performed by de Dios and Oldfield (*5*). It is, however, natural to expect that for systems where nondynamic and dynamic correlation effects are important for the proper description of the molecular geometry or of excited state energies, they will be equally important for the accurate calculation of NMR shieldings. For example, the multi-configuration IGLO (MC-IGLO) approach developed by van Wüllen and Kutzelnigg (*6*) is capable of treating the ^{13}C shieldings in cyclobutadiene in a consistent way both for rectangular and square geometries, whereas the single-configuration IGLO calculation diverges at the square geometry, at which the molecule is considered to be antiaromatic.

It is currently possible to perform gauge-origin-independent NMR shielding calculations using several types of correlated wavefunctions. The already mentioned MC-IGLO approach of van Wüllen and Kutzelnigg (*6*), as well as the MC-GIAO method of Ruud *et al.* (*7*) allow inclusion of nondynamic correlation effects. Dynamic correlation can be treated either through a MP2-GIAO (*8–10*), MP3-GIAO (*11*), MP4-GIAO (*12*) construction, or through a coupled-cluster (CC) ansatz involving GIAOs, *e.g.* the CC with singles and doubles (CCSD)-GIAO approach developed by Gauss and Stanton (*13*). Other possibilities to include correlation effects, without making the distinction between nondynamic and dynamic correlation, are presented by the combination of second-order polarization propagator approximation (SOPPA) method and LORG (localized orbital/local origin) which results in SOLO (second order LORG) (*14*), and by density functional (DFT) theory in its sum over states (SOS)-DFT-IGLO (*15, 16*) and GIAO variants (*17*).

Most of the existing applications of correlated approaches for calculating NMR shielding tensors target small molecules and do not make a distinction between nondynamic and dynamic correlation effects. In fact, the effect of electronic correlation on the calculated NMR shieldings can be regarded as two-fold: Firstly, the use of a correlated wavefunction directly influences the calculated values of all properties, including NMR shieldings, and secondly, if one makes use of an optimized, rather than an experimental geometry, the introduction of correlation usually modifies the optimized molecular geometry to an extent that can have an appreciable effect on the calculated shieldings.

In this paper we present a detailed investigation of the effects of electron correlation on the geometry and ^{19}F shielding constants in mono-, di-, tri-, tetra-, penta- and hexa-substituted fluorobenzenes. This is a group of medium-size molecules which have been studied extensively at the HF-GIAO level (*5*). All fluorobenzenes are planar and, in analogy with benzene, one can expect to obtain reasonably good results by introducing the σ-π separation and treating only the π electrons as active within an appropriate CASSCF ansatz (*i.e.* accounting for the nondynamic correlation only). On the other hand, all fluorobenzenes are described sufficiently well by the closed-shell HF wavefunction to allow use of this wavefunction as a reference in approaches including dynamic correlation effects.

Computational Procedure

The calculations included in this article were performed using three *ab initio* packages: GAUSSIAN94 (*18*), DALTON (*19*) and ACES II (*20*). All geometry optimizations at the HF, CASSCF, MP2, MP4(SDQ) (MP4 with single, double and quadruple substitutions), QCISD (quadratic configuration interaction including single and double substitutions) and DFT levels of theory were carried out within GAUSSIAN94, under the 'Tight' convergence criteria (thresholds of 1.5×10^{-5}, 10^{-5}, 6×10^{-5} and 4×10^{-5} for the maximum force, RMS (root mean square) force, maximum displacement and RMS displacement, respectively). This package was also used to produce the HF-GIAO and DFT-GIAO NMR shielding values. The CASSCF-GIAO shieldings were calculated using DALTON, while their MP2-GIAO counterparts come from ACES II.

The basis sets used in the calculations were taken from the GAUSSIAN94 basis set library. All geometry optimizations employed the 6-31G** basis. The most accurate shielding calculations presented in this article involve, in a locally-dense context, 6-311++G(2d,2p) sets of basis functions on the fluorines, while the carbons and hydrogens are described within the 6-31G** basis.

Results and Discussion

Optimized Geometries. As the elements of the NMR shielding tensor for a nucleus represent local properties, the most important geometrical parameter influencing the value of the ^{19}F shielding constant can be expected to be the length of the bond which attaches the fluorine to the benzene ring. The optimized values of the carbon-fluorine bond lengths in all fluorobenzenes calculated at the HF, CASSCF(6,6) (6 electrons in 6 active orbitals) and MP2 levels of theory are collected in Table I.

Table I. Optimized carbon-fluorine bond distances in mono-, di-, tri-, tetra-, penta- and hexa-substituted fluorobenzenes calculated at the HF, CASSCF(6,6) and MP2 levels of theory. The basis set used for all optimizations is 6-31G**. All distances in Å. F_1, F_2, etc denote the fluorines connected to carbon atoms number 1, 2, etc. See text for additional CASSCF(8,7), MP4(SDQ), QCISD and DFT results for C_6H_5F.

CF Bond/Method	HF	CAS(6,6)	MP2
C_6H_5F	1.331	1.335	1.358
o-$C_6H_4F_2$	1.325	1.327	1.351
m-$C_6H_4F_2$	1.327	1.331	1.355
p-$C_6H_4F_2$	1.331	1.334	1.357
1,2,3-$C_6H_3F_3$ (CF_1)	1.322	1.325	1.349
1,2,3-$C_6H_3F_3$ (CF_2)	1.320	1.321	1.345
1,2,4-$C_6H_3F_3$ (CF_1)	1.325	1.326	1.350
1,2,4-$C_6H_3F_3$ (CF_2)	1.321	1.324	1.348
1,2,4-$C_6H_3F_3$ (CF_4)	1.327	1.330	1.354
1,3,5-$C_6H_3F_3$	1.322	1.328	1.353
1,2,3,4-$C_6H_2F_4$ (CF_1)	1.322	1.324	1.349
1,2,3,4-$C_6H_2F_4$ (CF_2)	1.317	1.318	1.343
1,2,3,5-$C_6H_2F_4$ (CF_1)	1.318	1.322	1.347
1,2,3,5-$C_6H_2F_4$ (CF_2)	1.321	1.320	1.344
1,2,3,5-$C_6H_2F_4$ (CF_5)	1.323	1.328	1.352
1,2,4,5-$C_6H_2F_4$	1.322	1.324	1.348
C_6HF_5 (CF_1)	1.318	1.321	1.347
C_6HF_5 (CF_2)	1.317	1.318	1.343
C_6HF_5 (CF_3)	1.313	1.316	1.342
C_6F_6	1.314	1.316	1.342

In the case of the simplest fluorobenzene, C_6H_5F, we performed additional geometry optimizations using a CASSCF(8,7) (8 electrons in 7 orbitals) wavefunction which adds the F π lone pair to the active space, MP4(SDQ) perturbation theory, the QCISD method, and a DFT approach involving the B3LYP exchange-correlation potential, all within the 6-31G** basis. CASSCF(8,7) produces a CF bond length of 1.334 Å, virtually indistinguishable from the CASSCF(6,6) value in Table I, while the MP4(SDQ), QCISD and B3LYP results are 1.357 Å, 1.357 Å and 1.351 Å, respectively. The very small differences between the MP2, MP4(SDQ) and QCISD values indicate that the MP2 level of theory yields geometries which are, to all intents and purposes, as good as those produced by the most advanced methods for which GAUSSIAN94 offers analytical gradients [MP4(SDQ) and QCISD]. Therefore, for the remaining fluorobenzenes we decided not to go beyond MP2 in the geometry optimizations. Each of the geometry optimizations we performed involved a single molecule only which means that the results should be compared to gas-phase measurements. The CF bond length in C_6H_5F, 1.358 Å, calculated at the MP2/6-31G** level of theory, compares very favourably with the experimental microwave measurement of 1.354±0.006 Å (*21*). In the case of fluorobenzene, the HF and π-active space CASSCF methods significantly underestimate the carbon-fluorine bond length, while the MP2, MP4(SDQ), QCISD and B3LYP (DFT) results are in much better agreement between themselves and with experiment. This suggests that dynamic correlation influences the ground-state geometry of C_6H_5F much more than its nondynamic counterpart. As the CASSCF(6,6) and CASSCF(8,7) wavefunction include most of the π-space correlation effects, it is also reasonable to assume that the correlation effects of importance for the ground-state geometry are predominantly within the σ-framework. As the system of σ bonds does not change much throughout the fluorobenzene series, it can be expected that the variation in the CF bond lengths will not be very significant. This is fully supported by the data in Table I: The longest MP2/6-31G** CF bond is 1.358 Å, in fluorobenzene, while the shortest one is 1.342 Å, in C_6F_6.

NMR Shielding Calculations. The influence of the combination of basis set, wavefunction type and theoretical geometry on ^{19}F shielding in monofluorobenzene were investigated in much greater detail than for the other fluorobenzenes. The results of the calculations are presented in Figures 1–3.

Figure 1 shows the ^{19}F absolute chemical shielding values calculated by means of the HF-GIAO approach using a 6-31G** basis on all atoms and a locally-dense basis (6-31G** on C and H, and 6-311++G(2d,2p) on F) and geometries optimized at the HF, CASSCF(6,6), MP2, MP4(SDQ) and QCISD levels of theory (all within the 6-31G** basis). Use of the larger basis leads to an almost uniform decrease of all shieldings by 8–9 ppm. As the MP2, MP4(SDQ) and QCISD optimized geometries are very similar, the corresponding ^{19}F HF-GIAO shielding values differ by very little. While the switching from the HF-optimized to the CASSCF(6,6)-optimized geometry leads to a decrease of the ^{19}F shielding by less than 2 ppm, the transition from the HF-optimized geometry to one of its MP2, MP4(SDQ) or QCISD-level counterparts has a more pronounced effect, a decrease of about 5 ppm.

Figure 2 (in which the HF//HF results present in Figure 1 are repeated for comparison purposes) illustrates the effect of including different types of electron correlation on the ^{19}F absolute chemical shielding. The numbers are clearly separated in three groups: A set of CASSCF-GIAO shieldings, a set of DFT-GIAO shieldings and an MP2-GIAO value. One possible reason for the unexpectedly low DFT-GIAO values is in the absence of a magnetic-field dependent term in the B3LYP exchange-correlation potential (for an interesting study of the effect of the form of the exchange-correlation potential on the DFT-GIAO shieldings, see ref. *22*).

The most interesting data are collected in Figure 3 which shows the deviations of the theoretical ^{19}F isotropic shielding values relative to $CFCl_3$ from the corresponding experimental value of 113.12 ppm (*23*). The notable absence of CASSCF-GIAO results from this figure follows from the fact that it is not possible to perform a CASSCF(6,6) calculation on $CFCl_3$ within an active space analogous to the π-active space for C_6H_5F. While the HF-GIAO results obtained at MP2, MP4(SDQ)

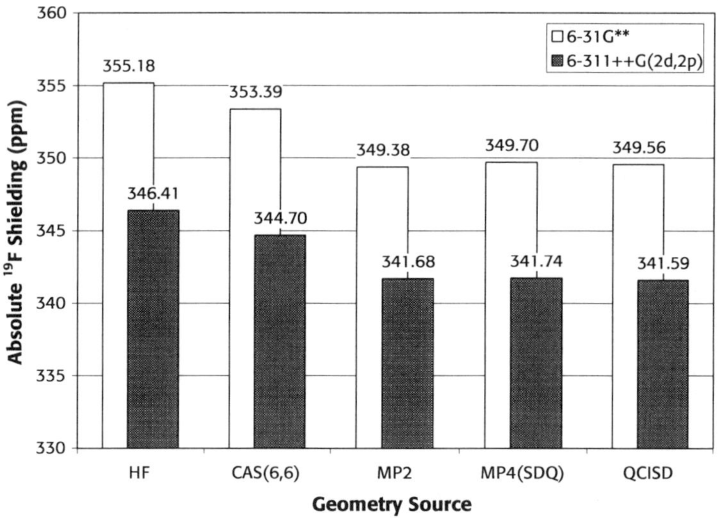

Figure 1. HF-GIAO ^{19}F absolute chemical shieldings for C_6H_5F (in ppm) using (*i*) a 6-31G** basis on all atoms and (*ii*) a locally-dense basis (6-31G** on C and H, and 6-311++G(2d,2p) on F) and geometries optimized at the HF, CASSCF(6,6), MP2, MP4(SDQ) and QCISD levels of theory (all within the 6-31G** basis).

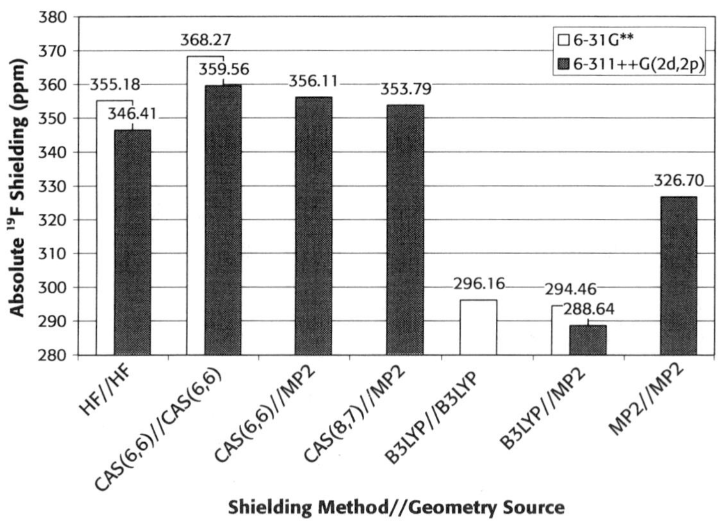

Figure 2. ^{19}F absolute chemical shieldings for C_6H_5F (in ppm) corresponding to different combination between a correlated GIAO wavefunction and a theoretical geometry. Basis set details as in Figure 1.

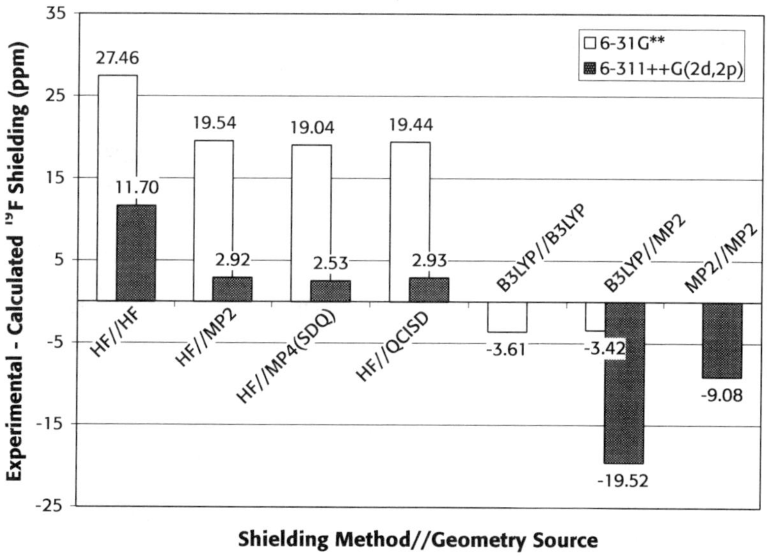

Figure 3. Deviations of the theoretical ^{19}F isotropic shielding values for C_6H_5F (relative to $CFCl_3$) from the corresponding experimental chemical shift of 113.12 ppm (*23*).

and QCISD geometries are remarkably accurate, the inclusion of electron correlation in the MP2-GIAO calculation leads to an overestimation of the experimental value. This feature of the MP2-GIAO calculations has also been observed by other authors (see *e.g.* ref. *24*). The DFT-GIAO results exhibit an unexpected anomaly: The extension of the basis impairs, rather that improves, their quality. Of course, it cannot be expected that recipés known to work well in HF-GIAO calculations will work equally well when transferred, without further modification, to a DFT-GIAO context.

Table II contains a selection of our theoretical results for the ^{19}F absolute and relative chemical shieldings for all fluorobenzenes, together with experimental data. The results for *o*-$C_6H_4F_2$, *m*-$C_6H_4F_2$ and *p*-$C_6H_4F_2$ have much in common. In all three cases the variation of the theoretical geometry changes the relative shielding calculated at the HF-GIAO level by not more than *ca.* 2 ppm. This is a consequence of the fact that the relative shieldings are given with respect to C_6H_5F, and the effects of electron correlation on the geometries of monofluorobenzene and its disubstituted counterparts are not unsimilar. For *m*- and *p*-difluorobenzene the experimental value is bracketed by the HF//MP2 and MP2//MP2 theoretical results, while in the case of *o*-fluorobenzene it is underestimated by both the HF//MP2 (slightly) and MP2//MP2 (even more so) values. This suggests that a more accurate theoretical estimate of these relative shieldings would require a higher-level MP*n*-GIAO approach.

The numbers for the trifluorobenzenes show that the HF//MP2 and, especially, the MP2//MP2 values are in much better agreement with the experimental data than the remaining theoretical estimates. 1,3,5-$C_6H_3F_3$ is an exception: All HF-GIAO-level relative shieldings are very close together and *ca.* 5 ppm below the experimental value.

For all tetrafluorobenzenes the most accurate results for the relative shieldings come again from either the HF//MP2 or MP2//MP2 calculations. It should be noted that for 1,2,3,5-$C_6H_2F_4$, for which the relative shieldings are given with respect to a different standard, $CFCl_3$, the HF//MP2 values are much more accurate than their MP2//MP2 counterparts. This suggests that the better accuracy of the MP2//MP2 relative shieldings with respect to C_6H_5F comes from the cancellation of similar MP2-GIAO errors in the calculations for the substance and for the standard.

Due to program limitations within DALTON and ACES II, in the cases of pentafluorobenzene and hexafluorobenzene we could calculate HF-GIAO-level shieldings only when using the locally dense basis set. The relative shieldings for C_6HF_5 are the only ones for which we observe better agreement with experimental data at the HF//HF, rather than at the HF//MP2 level. However, for C_6F_6 the HF//MP2 approach once again produces the most accurate relative shielding value.

It could appear interesting to apply Chestnut's approximate infinite-order perturbation-theory prescription for isotropic NMR chemical shieldings (*25*) in order to extrapolate our HF//MP2 and MP2//MP2 results to higher orders of perturbation theory. However, due to the fact that reference MP3-GIAO and MP4-GIAO results are not available, even for C_6H_5F, we did not feel that the use of ratios, derived from MP*n*-GIAO calculations on smaller moleules, would be sufficiently well justified.

Conclusions

Electron correlation has a nonnegligible effect on the calculated ^{19}F NMR shielding

Table II. ^{19}F absolute and relative chemical shieldings (in brackets, with respect to C_6H_5F) for mono-, di-, tri-, tetra-, penta- and hexa-substituted fluorobenzenes calculated using different combinations of a correlated GIAO wavefunction and a theoretical geometry. The experimental chemical shifts were taken from refs. *23* and *26*. All values in ppm. A locally dense basis set (6-31G** on C and H, and 6-311++G(2d,2p) on F) was used throughout. In this basis, the CASSCF(6,6)-GIAO and MP2-GIAO calculations for C_6HF_5 and C_6F_6 require more than 256 basis functions and cannot be performed by the current versions of DALTON and ACES II.

F Atom/Method	HF//HF	HF//CAS(6,6)	HF//MP2	CAS(6,6)//MP2	MP2//MP2	Exp
C_6H_5F	346.41	344.70	341.68	356.11	326.70	
o-$C_6H_4F_2$	373.29(26.88)	371.27(26.57)	366.53(24.85)	362.08(5.97)[a]	349.57(22.87)	25.9
m-$C_6H_4F_2$	340.92(-5.49)	339.33(-5.37)	336.30(-5.38)	353.42(-6.14)[b]	324.08(-2.62)	-3.15
p-$C_6H_4F_2$	354.63(8.22)	352.66(7.96)	348.75(7.07)	359.61(3.50)	332.27(5.57)	6.6
1,2,3-$C_6H_3F_3$ (F_1)	367.28(20.87)	365.31(20.61)	360.67(18.93)	373.89(17.78)	347.52(20.82)	19.3
1,2,3-$C_6H_3F_3$ (F_2)	398.94(52.53)	396.47(51.77)	389.54(47.86)	387.41(31.30)	370.75(44.05)	46.1
1,2,4-$C_6H_3F_3$ (F_1)	380.85(34.44)	378.61(33.91)	372.84(31.16)	376.31(20.20)	353.87(27.17)	27.5
1,2,4-$C_6H_3F_3$ (F_2)	366.54(20.13)	364.72(15.17)	359.87(18.19)	369.91(13.80)	345.24(18.54)	17.6
1,2,4-$C_6H_3F_3$ (F_4)	349.00(2.59)	347.19(2.49)	343.15(1.47)	356.43(0.32)	328.80(2.10)	2.8
1,3,5-$C_6H_3F_3$	335.93(-10.48)	334.49(-10.21)	331.55(-10.13)	352.31(-3.8)[c]	322.31(-4.39)	-5.5
1,2,3,4-$C_6H_2F_4$ (F_1)	374.99(28.58)	372.84(28.14)	367.14(25.46)	373.57(17.46)	351.77(25.07)	24.40
1,2,3,4-$C_6H_2F_4$ (F_2)	391.78(45.37)	389.48(44.78)	382.61(40.93)	386.03(29.92)	367.68(40.98)	39.88
1,2,3,5-$C_6H_2F_4$ (F_1)	361.61(15.20)	359.84(15.14)	355.24(13.56)	368.52(12.41)	344.80(18.10)	11.3
1,2,3,5-$C_6H_2F_4$ (F_2)	406.48(60.07)	403.82(59.12)	395.77(54.09)	392.03(35.92)	375.29(48.59)	43.2
1,2,3,5-$C_6H_2F_4$ (F_5)	344.74(-1.67)	343.11(-1.59)	339.05(-2.63)	354.72(-1.39)	327.83(1.13)	-5.5
1,2,4,5-$C_6H_2F_4$	375.42(29.01)	373.36(28.66)	367.27(25.59)	366.76(10.65)	350.45(23.75)	26.7
C_6HF_5 (F_1)	365.90(19.49)	363.98(19.28)	357.79(16.11)			23.39
C_6HF_5 (F_2)	391.58(45.17)	389.16(44.46)	380.79(39.11)			46.00
C_6HF_5 (F_3)	378.02(31.61)	375.96(31.26)	369.03(27.35)			37.56
C_6F_6	397.32(50.91)	394.89(50.19)	386.83(45.09)			46.44

[a] The unrealistic CAS(6,6)//MP2 values for o-$C_6H_4F_2$ are a consequence of the fact that the CASSCF procedure in DALTON converges to a local minimum that is above the lowest-energy CASSCF solution.

[b] This is a CAS(6,6)//CAS(6,6) result.

[c] The CASSCF(6,6) wavefunction for 1,3,5-$C_6H_3F_3$ exhibits a $D_{3h} \rightarrow C_{2v}$ symmetry breaking, due to which there are two different ^{19}F shieldings. The entry for the second one is 351.27(-4.84).

values for all mono-, di-, tri-, tetra-, penta- and hexa-substituted fluorobenzenes. All fluorobenzenes are reasonably well described by a closed-shell HF wavefunction and, as a consequence, the dynamic correlation effects are much more important than non-dynamic correlation. In fact, the π-active space CASSCF calculations offer little, if any improvement over the closed-shell HF results, both in the geometry optimizations and in the subsequent GIAO calculations. They can be plagued by problems associated with difficulties in the choice of the active space, convergence to undesired local minima and spatial symmetry breakings. However, in a different context, namely for molecular systems, for which the closed-shell HF wavefunction is inadequate, a CASSCF-GIAO, MCSCF-GIAO or MC-IGLO procedure may offer the only way forward.

The inclusion of dynamic correlation effects through Møller-Plesset constructions of various orders or QCISD has a pronounced effect on the optimized ground-state geometries of all fluorobenzenes. As indicated by the more extensive calculations on C_6H_5F, second-order MP theory (within a sufficiently flexible basis) provides quantitatively correct descriptions of the fluorobenzene geometries which are unlikely to change to an appreciable extent upon further improvement of the wavefunction. As a rule, the better quality MP2-level theoretical geometries, on their own, bring about a significant improvement of the accuracy of the HF-GIAO calculations performed on top of these geometries. However, the direct inclusion of dynamic correlation in the shielding calculations through the MP2-GIAO approach does not lead to a universal further improvement of the calculated shieldings. The fact that the experimental shieldings are often bracketed by the HF//MP2 and MP2//MP2 theoretical values indicates that the perturbation treatment is targeting the experimental value, but may require higher orders of perturbation theory to achieve a complete agreement with it.

DFT incorporating a B3LYP exchange-correlation potential produces a reasonably good ground-state geometry for C_6H_5F, but the decline of the accuracy of the DFT-GIAO shieldings with the increase of the size of the basis set is disappointing.

The results presented in this paper strongly suggest that, at the present moment, the most efficient and sufficiently accurate approach for calculating ^{19}F NMR shielding values in fluorobenzenes and other related molecules is a HF-GIAO calculation on top of a ground-state geometry optimized at the MP2 level of theory.

Literature Cited

1. McWeeny, R. *Proc. Roy. Soc. Lond., Ser. A* **1954**, *223*, 306.
2. Borden, W. T.; Davidson, E. R. *Acc. Chem. Res.* **1996**, *29*, 67.
3. Hrovat, D. A.; Morokuma, K.; Borden, W. T. *J. Am. Chem. Soc.* **1994**, *116*, 1072.
4. Kozlowski, P. M.; Dupuis, M.; Davidson, E. R. *J. Am. Chem. Soc.* **1995**, *117*, 774.
5. de Dios, A. C.; Oldfield, E. *J. Am. Chem. Soc.* **1994**, *116*, 7453.
6. van Wüllen, C.; Kutzelnigg, W. *Chem. Phys. Lett.* **1993**, *205*, 563.
7. Ruud, K.; Helgaker, T.; Kobayashi, R.; Jørgensen, P.; Bak, K. L.; Jensen H. J. Aa. *J. Am. Chem. Soc.* **1994**, *100*, 8178.

8. Vauthier, E. C.; Comenau, M.; Odiot, S.; Eliszar, S. *Can. J. Chem.* **1988**, *66*, 1781.
9. Gauss, J. *Chem. Phys. Lett.* **1992**, *191*, 614.
10. Gauss, J. *J. Chem. Phys.* **1993**, *99*, 3629.
11. Fukui, H.; Baba, T.; Matsuda, H.; Miura, K. *J. Chem. Phys.* **1994**, *100*, 6608.
12. Gauss, J. *J. Chem. Phys.* **1994**, *229*, 198.
13. Gauss, J.; Stanton, J. *J. Chem. Phys.* **1995**, *103*, 3561.
14. Bouman, T. D.; Hansen, A. E. *Chem. Phys. Lett.* **1990**, *175*, 292.
15. Malkin, V. G.; Malkina, O. L.; Salahub, D. R. *Chem. Phys. Lett.* **1993**, *204*, 87.
16. Malkin, V. G.; Malkina, O. L.; Casida, M. E.; Salahub, D. R. *J. Am. Chem. Soc.* **1994**, *116*, 5898.
17. Cheeseman, J. R.; Trucks, G. W.; Keith, T. A.; Frisch, M. *J. Chem. Phys.* **1996**, *104*, 5497.
18. Frisch, M. J.; Trucks, G. W.; Schlegel, H. B.; Gill, P. M. W.; Johnson, B. G.; Robb, M. A.; Cheeseman, J. R.; Keith, T.; Petersson, G. A.; Montgomery, J. A.; Raghavachari, K.; Al-Laham, M. A.; Zakrzewski, V. G.; Ortiz, J. V.; Foresman, J. B.; Cioslowski, J.; Stefanov, B. B.; Nanayakkara, A.; Challacombe, M.; Peng, C. Y.; Ayala, P. Y.; Chen, W.; Wong, M. W.; Andres, J. L.; Replogle, E. S.; Gomperts, R.; Martin, R. L.; Fox, D. J.; Binkley, J. S.; Defrees, D. J.; Baker, J.; Stewart, J. P.; Head-Gordon, M.; Gonzalez, C.; Pople, J. A. *Gaussian 94, Revision D.4* Gaussian, Inc.: Pittsburgh, PA, 1995.
19. Helgaker, T.; Jensen, H. J. Aa.; Jørgensen, P. ; Olsen, J.; Ruud, K.; Ågren, H.; Anderersen, T.; K. L. Bak; V. Bakken; Christiansen, O.; Dahle, P.; Dalskov, E. K.; Enevoldsen, T.; Fernandez, B.; Heiberg, H.; Hettema, H.; Jonsson, D.; Kirpekar, S.; Kobayashi, R.; Koch, H.; Mikkelsen, K. V.; Norman, P.; Packer, M. J.; Saue, T.; Taylor, P. R.; Vahtras, O. *Dalton, An Electronic Structure Program, Release 1.0 (1997)* .
20. Stanton, J.; Gauss, J.; Watts, J.; Lauderdale, W. *Input Manual for ACES II Release 2.0* Quantum Theory Project, Departments of Chemistry and Physics, University of Florida, Gainesville, 1998.
21. Nygaard, L.; Bojesen, I.; Pedersen, T.; Rastrup-Andersen, J. *J. Mol. Structure* **1986**, *2*, 209.
22. Lee, A. M.; Handy, N. C.; Colwell, S. M. *J. Chem. Phys.* **1995**, *103*, 10095.
23. Dungan, C. H.; van Wazer, J. *Compilation of Reported F^{19} NMR Chemical Shifts, 1951 to Mid-1967* Wiley, N.Y., 1970.
24. Fukui, H. *Progr. Nucl. Magn. Res. Spectr.* **1997**, *31*, 317.
25. Chesnut, D. B. *Chem. Phys. Let.* **1995**, *246*, 235.
26. Emsley, J. W.; Phillips, L. in *Progress in Nuclear Magnetic Resonance Spectroscopy, Vol. 7*, edited by Emsley, J. W.; Feeney, J.; Sutcliffe, L. H., page 2, Pergamon Press, Oxford, 1971.

Chapter 9

^{17}O NMR Chemical Shifts in Peptides

S. Kuroki[1], K. Yamauchi[2], Hiromichi Kurosu[3], S. Ando[1], Isao Ando[1], A. Shoji[4], and T. Ozaki[4]

[1]Department of Chemistry and Materials Science, Tokyo Institute of Technology, Ookayama, Meguro-ku, Tokyo 152–8852, Japan
[2]Solid NMR Application, Bruker Japan Company Ltd., Ninomiya, Tsukuba-shi, Ibaraki 305–0015, Japan
[3]Department of Human Life and Environment, Nara Women's University, Kita-uoya, Nishimachi, Nara 630–8263, Japan
[4]Department of Biological Sciences, Gunma University, Tenjin-cho, Kiryu-shi, Gunma 376–8515, Japan

The O-17 NMR spectra of L-alanine containing polypeptides in the solid state were measured chemical shift, quadrupolar coupling constant and asymmetric parameter information obtained. The relationship between the hydrogen-bonding structure and these NMR parameters was investigated. Theoretical calculations of the electric field gradients and chemical shielding values of hydrogen-bonded carbonyl oxygens in peptides were also carried out, and it was found that there exists a relationship between hydrogen-bonding structure and the NMR parameters such as quadrupolar coupling constants and chemical shieldings.

The oxygen atom is one of the more important atoms involved with hydrogen-bonding in polymers such as peptides and polypeptides. Nevertheless, few solid-state ^{17}O NMR studies of such polymers have been carried out. This is due to the very weak sensitivity of solid-state ^{17}O NMR measurements, due to the two following reasons. First, the ^{17}O nucleus has a very low natural abundance, 0.037 %. Second, the ^{17}O nuclear spin quantum number(I) is 5/2, which implies a quadrupolar nucleus, and so the ^{17}O signal is broadened by nuclear quadrupolar effects in the solid. On the other hand, solution-state ^{17}O NMR spectroscopy has been successfully employed to elucidate a number of structural problems in organic chemistry, because the ^{17}O signal becomes relatively sharp due to the removal of the quadrupolar interaction by fast isotropic reorientation in solution. For example, as the oxygen atom is directly associated with the formation of a hydrogen bond, hydrogen bonding for the carbonyl group in various compounds often results in large low frequency shifts of the carbonyl ^{17}O NMR signal. From these results, solution-state ^{17}O NMR has been established as a means for structural characterization.

In such situations, it can be expected that solid-state ^{17}O NMR will provide a deep insight into understanding the hydrogen-bonding structures of solid polymers, such as peptides and polypeptides(1). High-resolution ^{17}O NMR spectra of solid polyglycine(PGI(β-sheet) and PGII(3_1helix)), glycylglycine(GlyGly), glycylglycine nitrate (GlyGly•HNO_3), have been measured previously, in order to obtain three kinds of NMR parameters: the chemical shift(δ), quadrupolar coupling constant(e^2qQ/h) and asymmetry parameter(η), in order to understand the relationship between these NMR parameters and the hydrogen-bonding structure(2). From these observed ^{17}O carbonyl NMR spectra, it is found that e^2qQ/h values decrease linearly with a decrease in the hydrogen bond length. This indicates that it is possible to determine the hydrogen bond length through the observation of e^2qQ/h values. The chemical shift values in peptides and in polypeptides move to low frequency with a decrease in the hydrogen bond length. However, there is a difference in the chemical shift value between peptides and polypeptides, which may come from differences in molecular packing. From these experimental findings, it was demonstrated that ^{17}O NMR spectroscopy is a useful mean for elucidating the hydrogen-bonding structure in solid peptides and polypeptides.

In addution, it was also demonstrated that the quantum-chemical approach of calculating the quadrupolar coupling constants and the chemical shieldings will provide systematic information on the conformation and electronic state. The theoretical calculations of the electric field gradients of hydrogen-bonded carbonyl oxygen were successfully carried out by the FPT-MNDO-PM3 method. From these calculated results, it was found that the calculated e^2qQ/h value decreases with a decrease in the hydrogen bond length, and the chemical shielding values move to low frequency with a decrease in the hydrogen bond length. These calculations explain the experimental findings obtained in the NMR measurements. Also, the directions of the principal axes of the electric field gradient tensor and chemical shielding of the carbonyl oxygen were obtained with resonable accuracy(3).

In addition to the glycine-containing polypeptides and peptides, we show here new results on the ^{17}O NMR chemical shift and quadrupolar coupling constant of poly(L-alanine)s in both α-helix and β-sheet forms, in the solid state. There results help clarify the relationship between hydrogen-bond length and e^2qQ/h and chemical shift parameters.

Experimental Section

Sample Preparations. 10% ^{17}O-labeled the L-alanine methyl ester in $Na^{17}OH$/methanol solution, where $Na^{17}OH$ was prepared by reaction of 20% ^{17}O-labeled water with Na metal. Further, L-Ala N-carboxyanhydride(NCA) was prepared using 10% ^{17}O-L-alanine. 10% ^{17}O-labeled poly(L-alanine)[(L-Ala)$_n$] was prepared by heterogeneous polymerization of 10% ^{17}O-labeled NCA in acetonitrile by using n-butylamine as the initiator. The conformational characterizations of these polymers were made on the basis of ^{13}C CP/MAS(cross polarization/magic angle spinning) by using reference data of ^{13}C chemical shift values for various polypeptides, which are associated with the main-chain conformation. [L-Ala]$_n$ with A/I=100 takes α-helix form, and [L-Ala]$_n$ with A/I=5 takes β-sheet forms, respectively.

Solid state ^{17}O NMR Measurements. Static ^{17}O CP NMR spectra were recorded with a JEOL GSX-270(6.3T) spectrometer and a JEOL GSX-500(11.7T) spectrometer operating at 36.6 and 67.8 MHz, respectively, with a CP/MAS accessory at room temperature. In the CP static methods, $Mg(^{17}OH)_2$ was used for ^{1}H-^{17}O CP matching($\gamma_H B_H = 3\gamma_o B_o$). The ^{1}H $\pi/2$ pulse length was 5μs and the ^{17}O $\pi/2$ pulse length was 5μs for a solid sample(which corresponded to 15μs for a solution sample). The ^{1}H decoupling field strength was 50 kHz and repetition time was 5s. According to the previous work(2) on ^{17}O CP NMR experiments for polyglycines and its peptides, 9ms was used as the appropriate contact time. The ^{17}O chemical shifts were calibrated through external liquid water(δ= 0 ppm). Further, ^{17}O MAS NMR spectra were recorded with a Bruker AVANCE400(9.4T) spectrometer, a Bruker AVANCE 500(11.7T) spectrometer, and a Bruker AVANCE800(18.8T) spectrometer with a high-speed MAS accessory operating at 54.2, 67.8, and 108.6MHz. The sample spinning rate is 15kHz at 54.2 and 67.8MHz, and 25kHz at 108.6MHz, respectively. These spectra were obtained by single $\pi/2$ pulse with ^{1}H decoupling experiments.

Solid State ^{17}O NMR Spectral Analysis. The ^{17}O nucleus has a 5/2 spin and so has a quadrupolar moment. This means that static and MAS ^{17}O NMR spectrum contains information about quadrupolar interactions and the chemical shift. In order to obtain the quadrupolar coupling constant and chemical shift separately, we have carried out spectral analysis with theoretical calculation taking into account quadrupolar interactions and chemical shift interactions.

Results and Discussion

^{17}O NMR spectra of poly(L-alanine)s: Figures 1 and 2 shows the static ^{17}O CP NMR spectra of solid (L-Ala)$_n$[A/I=100] with an α-helix form, and solid (L-Ala)$_n$[A/I=5] with a β-sheet form at 67.8 MHz(a) and 36.6 MHz(b), respectively(4). The spectrum at 36.6 MHz consists of two major split signals, but that at 67.8 MHz has one major signal overlapping with two signals. Such large variation comes from the quadrupolar interaction because the appearance of the spectrum depends on the NMR frequency. If the NMR frequency is extremely high, the influence of quadrupolar interactions may be neglected in the spectrum. From these static spectra we can obtain the NMR parameters such as chemical shift tensors(δ_{11}, δ_{22}, and δ_{33}), quadrupolar coupling constant(e^2qQ/h), asymmetry parameter of electric field gradient(η),and the Euler angle(α,β,γ) between the principal axes of the quadrupolar tensor and the chemical shift tensor by theoretical simulation, but it is difficult to determine these eight parameters exactly at the same time. For the case of MAS spectra, we can obtain only three NMR parameters such as isotopic chemical shift(δ_{iso}), e^2qQ/h , and η. It is not so difficult to determine these three NMR parameters from MAS spectra compared with the case of static spectra. But until now, the typical sample spinning rate is below 10kHz, it was not enough to separate the centerband signal and sideband signals of ^{17}O NMR spectrum so that identical quadrupolar MAS signal could not be observed. Very recently the high speed MAS technique has been developed and we can observed the centerband signal of ^{17}O NMR signal separated from the sideband signals easily. Figures 3 and 4 shows the

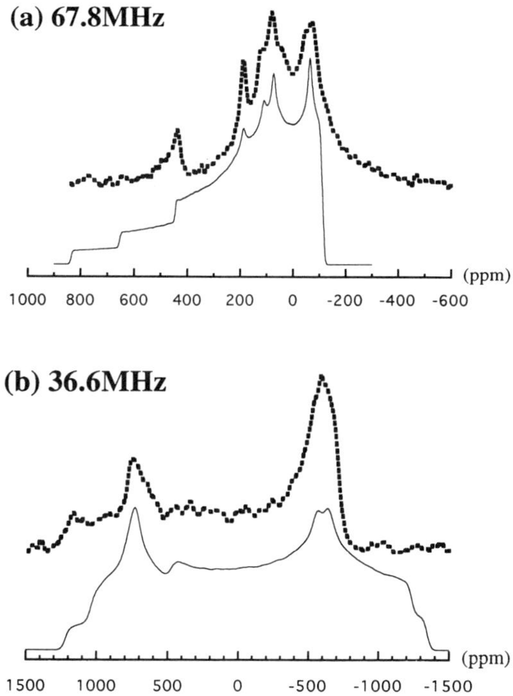

Figure 1. 67.8MHz (a) and 36.6MHz (b) ^{17}O CP static NMR spectra of $(L\text{-}Ala)_n$ with the α-helix form (A/I=100) in the solid state.

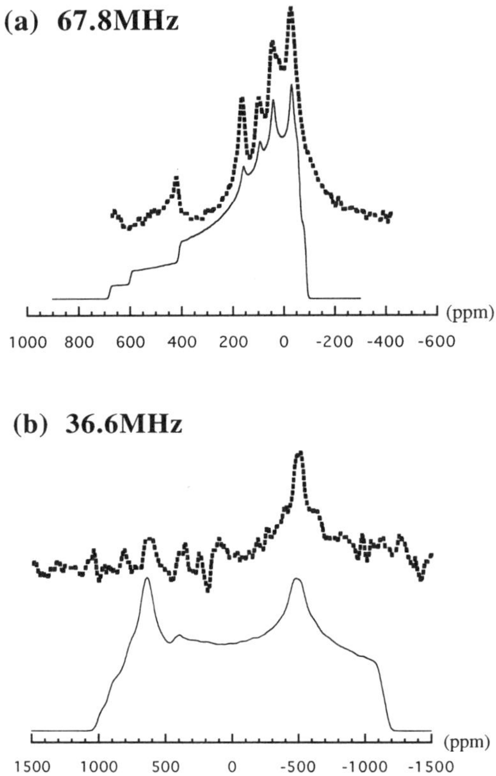

Figure 2. 67.8MHz (a) and 36.6MHz (b) ^{17}O CP static NMR spectra of $(L\text{-}Ala)_n$ with the β-sheet form (A/I=5) in solid state.

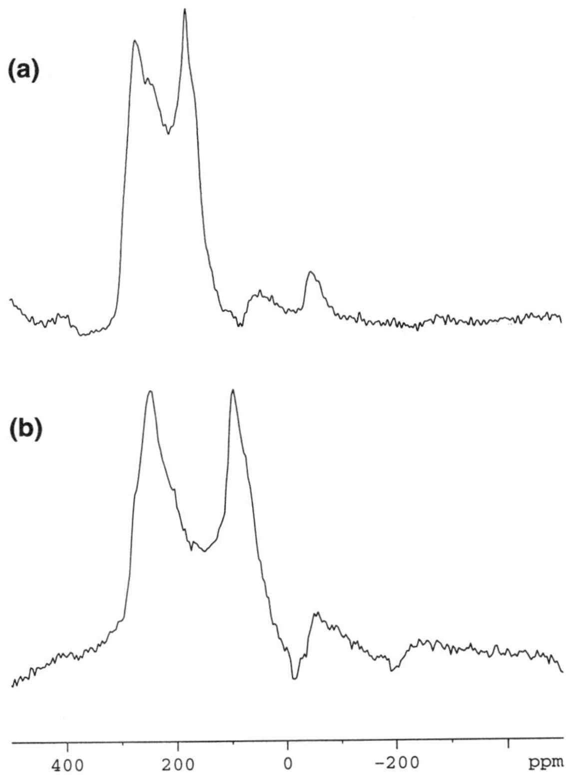

Figure 3. ^{17}O MAS NMR spectra of solid (L-Ala)$_n$[A/I=100] with an α-helix form at 67.8 MHz(a) and 54.2 MHz(b) spun at 15kHz.

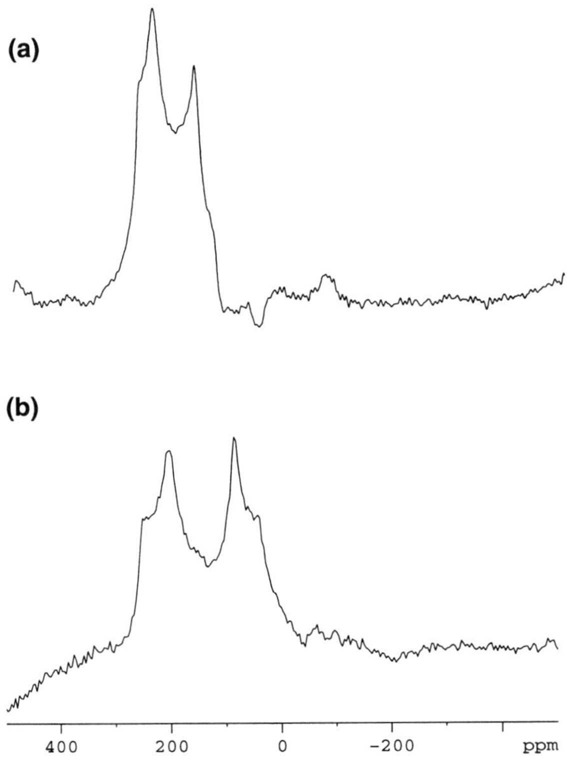

Figure 4. ^{17}O MAS NMR spectra of solid (L-Ala)$_n$[A/I=5] with a β-sheet form at 67.8 MHz(a) and 54.2 MHz(b) spun at 15kHz.

^{17}O MAS NMR spectra of solid (L-Ala)$_n$[A/I=100] with an α-helix form, and solid (L-Ala)$_n$[A/I=5] with a β-sheet form at 67.8 MHz(a) and 54.2 MHz(b) spun at 15kHz, respectively. At 54.2MHz, the centerband signal was not separated from the sideband signal both poly(L-alanine)s, though at 67.8 MHz the centerband signal was separated from the sideband signals. The centerband linewidth at 54.2MHz was broadened compared with that at 67.8MHz. The centerband linewidth is proportional to ω_Q^2/ω_L. ω_L is the Larmor frequency and ω_Q is proportional to e^2qQ/h. Therefor, it can be said that high-frequency measurement is needed for a ^{17}O MAS experiment simultaneously with high-speed sample spinning. In addition, the ^{17}O MAS NMR spectra spun at 25kHz of solid (L-Ala)$_n$[A/I=100] with an α-helix form, and solid (L-Ala)$_n$[A/I=5] with a β-sheet form at 108.6 MHz are shown in Figure 5. In these spectra, the centerband signal was completely separated from the sideband signals. The obtained NMR parameters for the α-helix form and the β-sheet form of poly(L-alanine)s are shown in Table I, together with the NMR parameters obtained from polyglycines I(β-sheet form) and II(3_1-helix form) as reported previously(2). Table I also shows the hydrogen-bond length N...O of poly(L-alanine)s and polyglycines obtained by X-ray analysis. It was found that e^2qQ/h value decreases with a decrease in the hydrogen bond length, and the chemical shift values move low frequency with a decrease in the hydrogen bond length. This results coincided with glycine residue containing peptides and polypeptides results(2).

Electric field gradient and oxygen nuclear shielding calculations of the carbonyl oxygen of the model hydrogen-bonding system; The theoretical approach of calculating the quadropular coupling constants and the chemical shieldings on the basis of quantum chemistry provides systematic information on the hydrogen-bonding structure and its electronic state. The theoretical calculations of the electric field gradients and of hydrogen-bonded carbonyl oxygen of glycine residue were successfully carried out by the FPT-MNDO-PM3 method(3). Figures 6 and 7 shows these calculated results, and it was found that the calculated e^2qQ/h value decreases with a decrease in the hydrogen bond length, and the chemical shielding values move low frequency with a decrease in the hydrogen bond length. These calculations explain the experimental finding obtained in the NMR measurement. Further, in this study we calculated the oxygen chemical shielding values by the ab-initio GIAO-CHF method using model compound. The ^{13}C chemical shielding calculated by the ab-initio GIAO-CHF method reasonably explained the experimental findings(6, 7). Figure 8 shows the calculated oxygen chemical shielding against the hydrogen bond length N...O. The chemical shielding values moved low frequency with a decrease in the hydrogen bond length. This result coincided with the results obtained by the FPT-MNDO-PM3 method and explain the experimental finding obtained in the NMR measurement.

Summary

We can successfully measure ^{17}O MAS NMR spectra of poly(L-Ala)s using high-field and high-speed magic-angle spinning technique. From the ^{17}O MAS spectra three kinds of NMR parameters such as chemical shift, quadrupolar coupling constant and asymmetric parameter were obtained exactly and understood the relationship

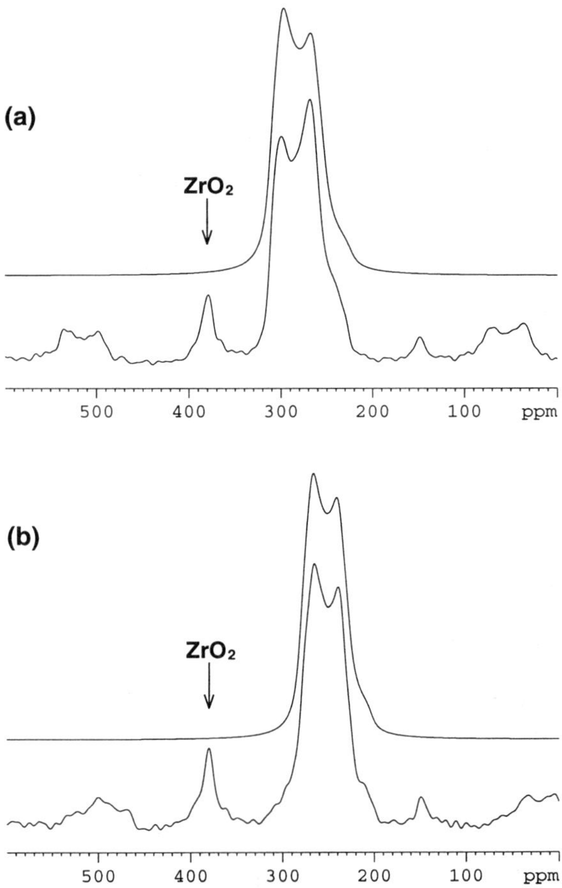

Figure 5. ^{17}O MAS NMR spectra spun at 25kHz of solid (L-Ala)$_n$[A/I=100] with an α-helix form(a), and solid (L-Ala)$_n$[A/I=5] with a β-sheet form(b) at 108.6 MHz with theoretically simulated spectra.

Table I. The determined NMR parameters of $(L\text{-}Ala)_n$ in this work together with those of $(Gly)_n$.

Sample		e^2qQ/h(MHz)	η	δ_{iso}(ppm)	$R_{N...O}$(Å)
$(L\text{-}Ala)_n$	α-helix	8.59	0.28	319	2.87
	β-sheet	8.04	0.28	286	2.83
$(Gly)_n$	PG II	8.30	0.29	288	2.73
	PG I	8.55	0.26	299	2.95

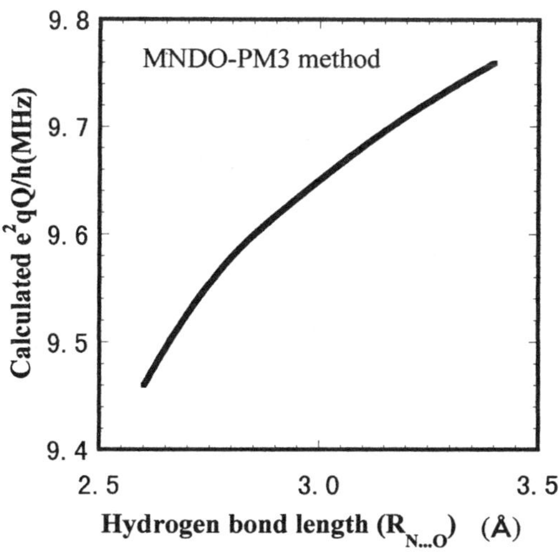

Figure 6. Plots of the calculated e^2qQ/h using the FPT-MNDO-PM3 method against the hydrogen bond length.

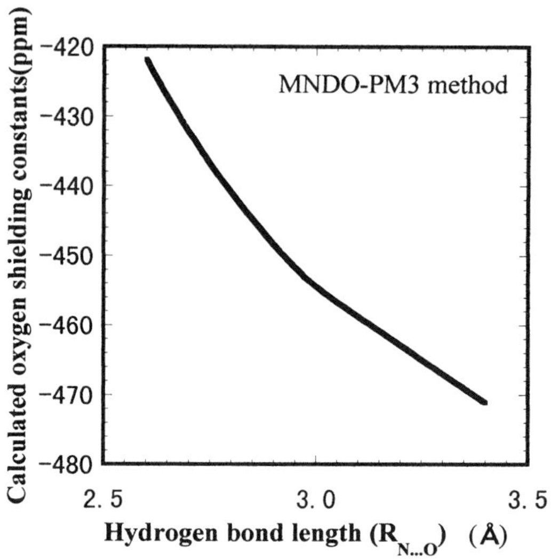

Figure 7. Plots of the calculated oxygen chemical shielding using the FPT-MNDO-PM3 method against the hydrogen bond length.

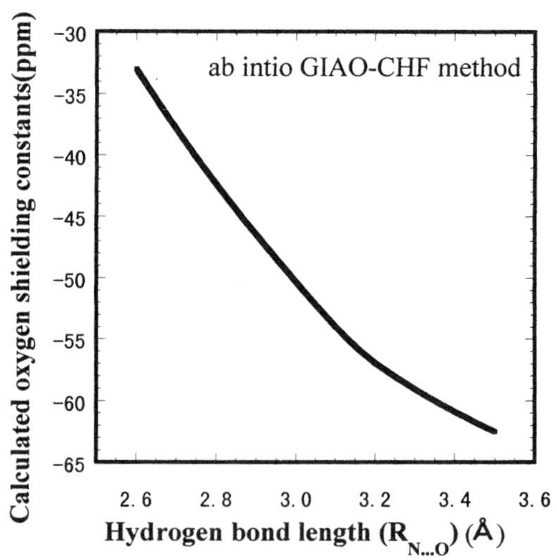

Figure 8. Plots of the calculated oxygen chemical shielding using the GIAO-CHF method against the hydrogen bond length.

between the hydrogen-bonding structure and these NMR parameters. Further, the theoretical calculations of the electric field gradients and the chemical shielding values of hydrogen-bonded carbonyl oxygens in peptides were carried out by the FPT-MNDO-PM3 and ab-inito GIAO-CHF methods, and it was found that there is a relationship between the hydrogen-bonding structures and the NMR parameters such as quadrupolar coupling constants and the chemical shielding values.

References

1. Tsuchiya.K.; Takahashi.A; Takeda,N.; Asakawa,N.; Kuroki,S.; Ando,I.; Shoji,A.; Ozaki,T. *J.Mol.Struct.* **1995**, 350, 233-240.
2. Kuroki.S; Takahashi,A.; Ando,I,; Shoji,A.; Ozaki,T. *J.Mol.Struct.* **1994,** 323, 197-208.
3. Kuroki,S.; Ando,S.; Ando,I. *Chemical Physics*, **1995**, 195, 107-116.
4. Takahashi,A.; Kuroki,S.; Ando,I.; Ozaki,T.; Shoji,A. *J.Mol.Struct.* **1998,** 442, 195-199.
5. Yamauchi,K.; Kuroki,S.; Ando,I.; Ozaki,T.; Shoji, A. *Chem.Phys.Lett.* to be submitted.
6. Asakawa, N.; Kurosu,H.; Ando, I; Shoji, A.; Ozaki, T. *J.Mol.Struct.* **1994,** 317, 119-129.
7. Asakawa,N.;Kameda,T.;Kuroki,S.;Kurosu,H.;Ando,S.;Ando,I.;Shoji,A. *Annual Reports on NMR Spectroscopy*, **1998,** 35, 56-137.

Chapter 10

A Conformational Study of the L-Alanine Residue in Polypeptides by Ab Initio ^{13}C NMR Shielding Calculation

Hiromichi Kurosu[1], Kouji Fukuyama[2], Shigeki Kuroki[2], and Isao Ando[2]

[1]Department of Textile and Apparel Science, Nara Women's University, Kita-uoya, Nishimachi, Nara 630–8263, Japan
[2]Department of Polymer Chemistry, Tokyo Institute of Technology, 2–12–1 Ookayama, Meguro-ku, Tokyo 152, Japan

The ^{13}C chemical shift contour map for the C_β carbon of the L-alanine residue in peptides and polypeptides was made as a function of the dihedral angles(Φ, Ψ) by using the experimental data. Also, the corresponding calculated map was made by using the ab initio coupled Hartree-Fock method with the gauge included atomic orbitals(GIAO-CHF). From these results, it was found that the calculated map explains the chemical shift behavior of the α-helix and β-sheet forms in poly(L-alanine) and some proteins. This suggests that the calculated map is applicable to the structural analysis of proteins with complicated structure.

Solid-state high-resolution NMR spectroscopy, combined with quantum chemistry, is able to provide detailed information on the electronic and stereochemical structures of molecules[1]. Quantum-chemical calculations produce three principal values of the nuclear shielding tensor. These principal values have more detailed information about structure of molecules as compared with isotropic chemical shifts. Comparison of the observed chemical shift and chemical shift tensor with the calculated shieldings, produced by quantum-chemical calculations, permits deep insight into the structures of the molecules under investigation.

So far, we have investigated higher-order structure of polypeptides by solid-state high-resolution NMR not only using experimental but also theoretical methods[2-4]. The chem cal shifts can be characterized by variations in the electronic states of the local conformation as defined by the dihedral angles(Φ,Ψ). Ando et al. have calculated contour map for the C_β carbons of an alanine dipeptide by using the FPT INDO method within the semi-empirical MO framework. The calculated map reasonably predicts the experimental version. This shows that the chemical shift behavior of the L-alanine residue C_β-carbonyl carbons in the copolymer of L-alanine can be understood by the conformation-dependent chemical shift[5]. It is well known that not only dihedral angle of the main chain but also hydrogen bond angle affect on the chemical shift and chemical shift tensor behavior of the peptides and polypeptides. In this paper, however, we focus on the dihedral angle of the main chain since it plays major roll in determining the chemical shift and chemical shift tensor behavior of peptides and polypeptides.

In this paper, we aim to elucidate the correlation between the chemical shift tensors and the conformation of peptides, by carrying out NMR shielding calculations using the ab initio GIAO-CHF MO with 6-31G basis set, in order to understand the ^{13}C chemical shift and chemical shift tensor behavior of the peptides and polypeptides.

^{13}C NMR measurements

Solid state ^{13}C NMR measurements were performed on a JEOL GSX-270 spectrometer operating at 67.8 MHz equipped with a CP/MAS accessory. The field strength of the ^{1}H decoupling was 1.2 mT, contact time 2 ms, repetition time 5s, and spectral width 27.0 kHz. 8K data points were used. Samples was placed in a cylindrical rotor and spun at about 4 kHz. The ^{13}C chemical shifts were calibrated indirectly through the adamantane peak(29.5 ppm relative to tetramethylsilane, TMS, $(CH_3)_4Si$).

^{13}C NMR shielding calculation

The ^{13}C NMR shieldings are calculated by ab initio GIAO-CHF method with the 6-31G basis set. N-acetyl-N'-methyl-L-alanine amide (Ac-L-Ala-NH-Me) as shown in Figure 1 was employed as a model molecule to investigate the effect of dihedral angles of the main chain. The Gaussian 94W package performed by a Pentium II 266MHz processor was used for the calculation. The geometrical parameters used here were Momany's standard values[6].

Results and discussion

Table 1 shows the observed ^{13}C chemical shifts and the main-chain conformation for series of peptides that contain the L-alanine residue. Figure 2 shows the contour map of calculated ^{13}C NMR shieldings with the observed ^{13}C NMR chemical shifts for C_β carbon of the L-alanine residue. It should be noted that the negative sign of shielding σ indicates deshielding and shielding variations can be compared with the experimental chemical shift δ, where a positive sign denotes deshielding. This figure shows the dependence on the dihedral angles(Φ,ψ) of the calculated isotropic ^{13}C chemical shift for the L-alanine residue C_β-carbon. The observed ^{13}C isotropic chemical shifts for the α_R-helix and β-sheet in poly(Ala) are 15.7 and 21.0 ppm, respectively. The calculated isotropic ^{13}C NMR shieldings for the C_β-carbon are 188.0 ppm for the main-chain dihedral angles which correspond to the α_R-helix conformation ((Φ,Ψ)= (-47.0,-57.2)), 186.5 ppm for the β-sheet conformation ((Φ,Ψ)= (-142.0,145.0)). The calculated ^{13}C shielding contour map reproduce well the experimental results. This means that the change of the main-chain dihedral angles dominates the ^{13}C isotropic chemical shift behavior of the L-alanine residue C_β-carbon.

In order to obtain further insight into the observed C_β-carbon chemical shift

Table 1 Observed ^{13}C NMR chemical shifts of the L-alanine residue C_β-carbons for peptides and polypeptides containing the L-alanine residues from ^{13}C CP/MAS NMR measurements, and their main chain dihedral angles and hydrogen-bond length for the L-alanine residues determined by X-ray diffraction studies.

Sample	^{13}C chemical shift/ppm	Dihedral angle/deg		Hydrogen-bond length/Å
		Φ	Ψ	
Z-Ala-Pro[7]	15.6	-76.1	153.0	3.01
Boc-Ala-Pro[8]	17.2	-95.3	153.6	3.12
Boc-Ala-Alib[9]	17.3	-66.3	-24.1	3.13
Boc-Ala-Gly[10]	18.7	-76.0	164.3	2.97
Boc-Ala[11]	18.3	-76.0	-2.6	3.16
	18.3	-76.0	177.7	
Ac-Ala-Aib-OMe[12]	22.5	-129.8	121.4	
	22.5	134.0	122.2	
Ac-Ala-NH_2[13]	20.0	-124.7	127.9	2.96
Ac-Ala-NHMe[14]	18.8	-84.3	159.0	2.92
Ac-Ala.HCl[15]	18.5	-153.5	162.7	
Gly-Ala[16]	16.4	-78.4	150.0	
Gly-Ala.HCl[17]	16.3	-72.4	179.8	
Asp-Ala[18]	17.7	-134.6	83.0	2.89
Ala-<u>Ala</u>-Ala[19]	20.0	-145.7	145.5	
Gly-Ala-Ile[20]	12.1	-70.4	75.4	
Rnase H[21]	21.3	-144.6	134.7	
	19.7	-160.5	150.4	
	19.3	-55.2	-46.1	
	15.3	-71.3	-35.4	
	17.2	-65.8	-49.2	
	17.2	-81.0	-11.5	
	16.0	-62.4	-22.9	
	15.8	-53.5	-49.9	
	16.5	-72.2	-25.6	
	16.4	-66.5	123.7	

Table 1 *Continued*

Sample	^{13}C chemical shift/ppm	Dihedral angle/deg		Hydrogen-bond length/Å
		Φ	Ψ	
	16.6	-53.7	-57.3	
	15.7	-67.4	-38.4	
	15.3	-67.6	-41.7	
	16.0	-60.5	-22.9	
BPTI[21]	18.5	-76.2	172.0	
	17.5	-62.3	-28.0	
	19.0	-82.9	-23.1	
	18.5	-62.5	151.1	
	16.0	-61.7	-35.9	
poly(Ala) α_R-helix[22]	15.7	-47.0	-57.2	
poly(Ala) β-sheet[22]	21.0	-142.0	145.0	

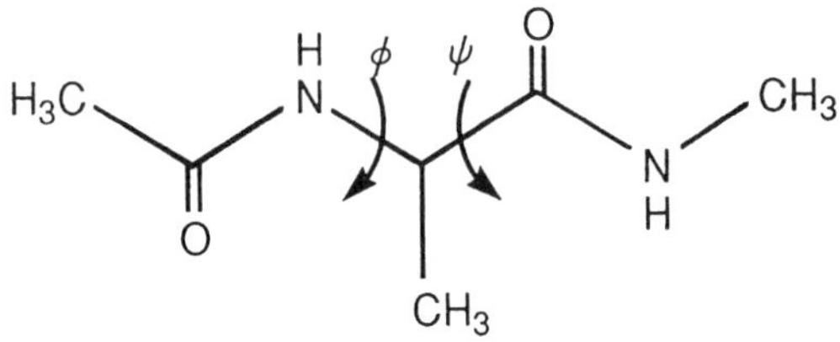

Figure 1. Structure of N-acetyl-N′-methyl-L-alanine amide used in NMR shielding calculations.

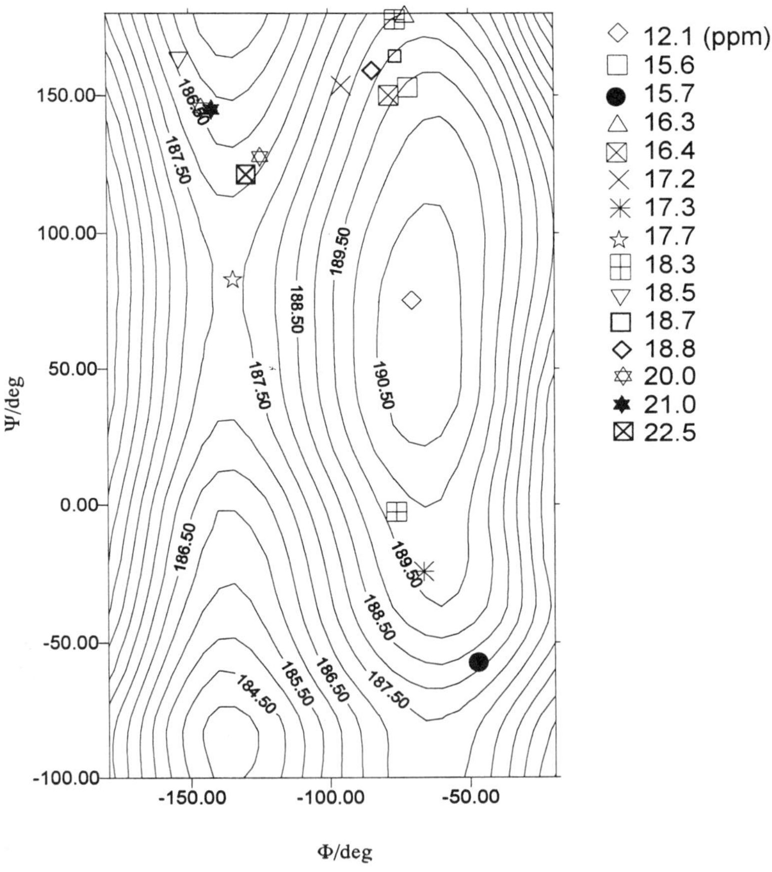

Figure 2. The calculated contour map of ^{13}C NMR shieldings with the observed ^{13}C NMR chemical shifts for C_β carbon of the L-alanine residue.

behavior, we carried out the principal values of the ^{13}C NMR shielding tensor calculations. Table 2 shows the observed principal values of ^{13}C chemical shifts for a series of peptides containing the L-alanine residue. The contour map of the calculated principal values of the ^{13}C shielding(σ_{11}, σ_{22} and σ_{33}, defined from the least to the most shielding) together with the observed principal values are shown in Figure 3. This contour map of σ_{11} shows that σ_{11} is very sensitive to change of the dihedral angles and reproduces the experimental data successfully. The observed σ_{22} values of Z-Ala-Pro, Boc-Ala-Gly, Boc-Ala-Pro and Ac-Ala-NH2 are 17.4, 17.7, 18.1 and 22.2 ppm, respectively. As seen from Figure 3(b), the calculated principal values of σ_{22} for Z-Ala-Pro, Boc-Ala-Gly, Boc-Ala-Pro and Ac-Ala-NH2 are 189.3, 188.2, 187.9 and 185.2 ppm, respectively. This shows that the calculated contour map of σ_{22} explains well the behavior of the observed σ_{22} principal values. As seen from Figure 3(c), the principal value of σ_{33} is not so sensitive to change of the dihedral angles(Φ,Ψ). From above results, it was found that σ_{11} and σ_{22} is quite sensitive to change of the dihedral angles(Φ,Ψ).

From these results, it was found that the calculated map explains well the chemical shift behavior of the α-helix and β-sheet forms in poly(L-Ala) and some proteins. This suggests that the calculated map is applicable to the structural analysis of proteins with complicated structure.

References

1. I.Ando, T.Yamanobe, H.Kurosu and G.A.Webb, *Ann.Rept.NMR Spectrosc.*, **22**, 205(1989).
2. H.Saito and I.Ando, *Ann.Rept.NMR Spectrosc.*, **21**, 210(1989).
3. N.Asakawa, T.Kameda, S.Kuroki, H.Kurosu, S.Ando, I.Ando and A.Shoji, *Ann.Rept.NMR Spectrosc.*, **35**, 55, (1998).
4. I.Ando and T.Asakura(ed.), *Solid State NMR of Polymers*, Elsevier Science, Amsterdam,1998.
5. I.Ando, H.Saito, R.Tabeta, A.Shoji and T.Ozaki, *Macromolecules*, **17**, 457(1984).
6. F.Momany, R.F.MacGuire, J.F.Yan and H.A.Sceraga, *J.Phys.Chem.*, **75**,2286(1971).
7. K.Panneerselvam, K.K.Chacko, K.R.Veena, *Acta Cryst.*,**C46**, 81 (1990).
8. M.E.Kamwaya, O.Oster, H.Brandaczek, *Acta Cryst.*, **B38**,172(1982).
9. R.Bosch, K.P.Voges, G.Iung, W.Winter, *Acta Cryst.*, **C39**, 481(1983).
10. E.Subramanian, J.J.Sahayamary, *Int.J.Peptide Protein Res.*, **41**,319 (1993).
11. L.V.Meervelt, K.D.Wael, T.Z.Huyskens, *J.Mol.Struct.*, **356**, 183(1995).
12. G.Valle, M.Crisma, F.Formaggio, C.Toniolo, G.Jung, Liebigs *Ann.Chem.*, 1055(1987).
13. R.Puliti, *Acta Cryst.*, **C45**, 1554(1989).
14. Y.Harada, Y.Iitaka, *Acta Cryst.*, **B30**, 1452(1974).
15. Y.Tokuma, T.Ashida, M.Kakudo, *Acta Cryst.*, **B25**, 1367(1969).
16. A.H.-J.Wang, I.C.Paul, *Cryst.Struct.Comm.*, **8**, 269(1979).
17. P.S.Naganathan, K.Venkatesan, *Acta Cryst.*, **B28**, 552(1972).
18. C.H.Gorbitz, *Acta Chemica Scandinavia*, **B41**, 679(1987).
19. J.K.Fawcett, N.Camerman, *Acta Cryst.*, **B31**, 658(1975).
20. K.Go, S.Chaturvedi, R.Parthasarathy, *Biopolymers*, **32**, 107(1992).
21. A.Wagner, D.Bruehwiler, *Biochemistry*, **25**, 5939(1986).
22. S.Tuzi, PhD thesis, Tokyo Institute of Technology(1989).

Table 2 Observed principal values of ^{13}C NMR chemical shifts of the L-alanine residue C_β-carbons for peptides containing the L-alanine residues from ^{13}C CP/MAS slow-spinning NMR measurements.

Sample	σ_{11}	σ_{22}	σ_{33} /ppm
Z-Ala-Pro	27.1	17.4	2.3
Boc-Ala-Pro	32.3	18.1	1.2
Boc-Ala-Aib	29.5	19.4	3.0
Boc-Ala-Gly	34.3	17.7	3.8
Ac-Ala-NH2	35.7	22.2	2.1
Poly(Ala) *)	25.8	18.6	2.4

*)α_R-helix

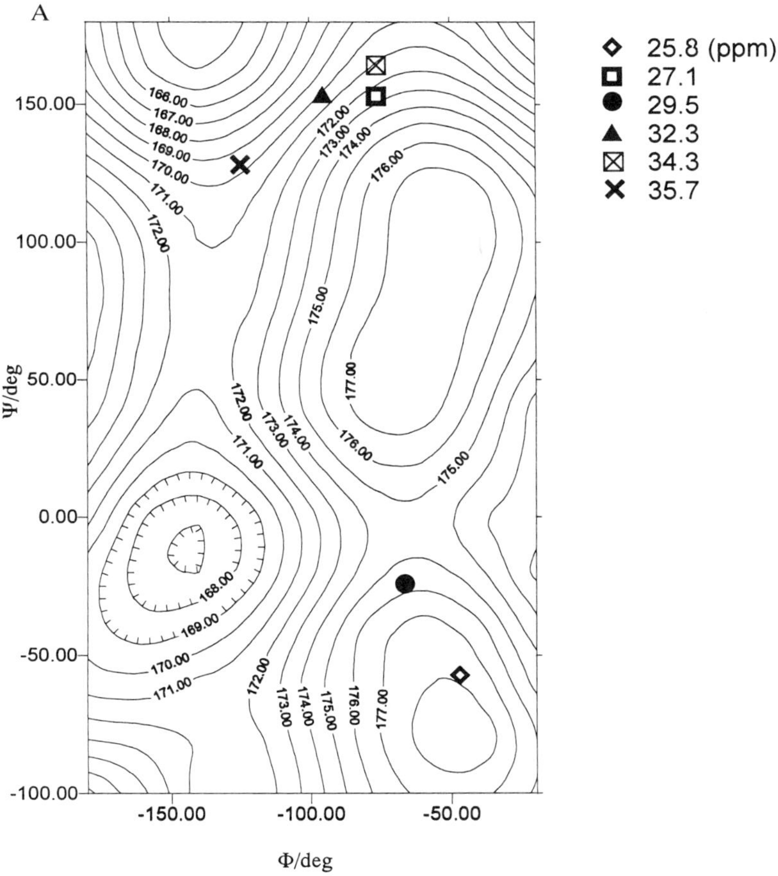

Figure 3. The calculated contour map of ^{13}C NMR principal values with the observed ^{13}C NMR principal values for C_β carbon of the L-alanine residue. (a) σ_{11}, (b) σ_{22} and (c) σ_{33}.

Continued on next page.

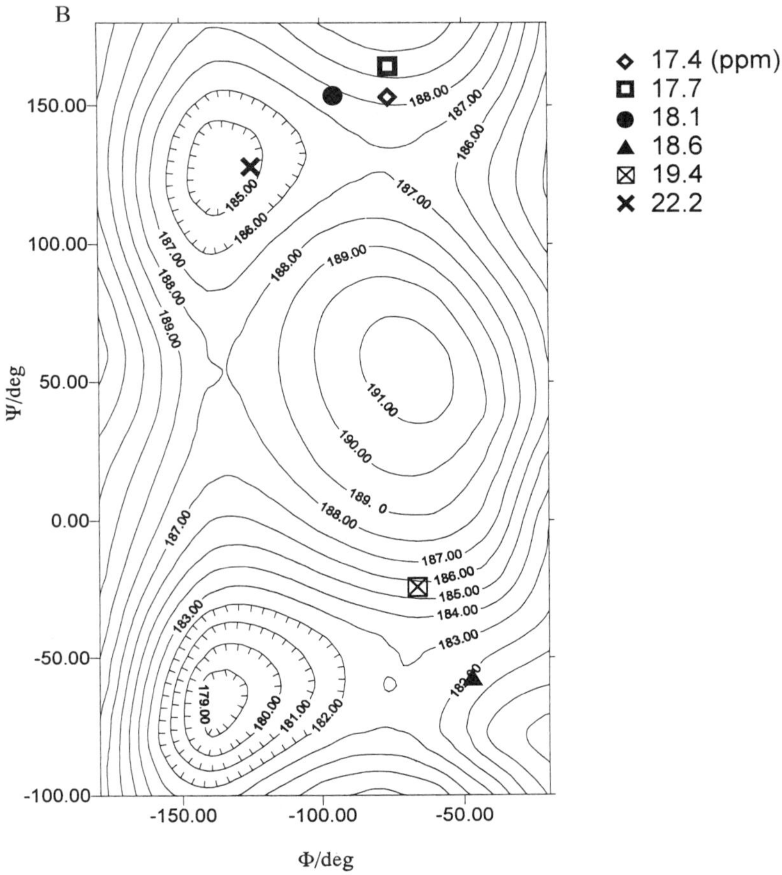

Figure 3. *Continued.*

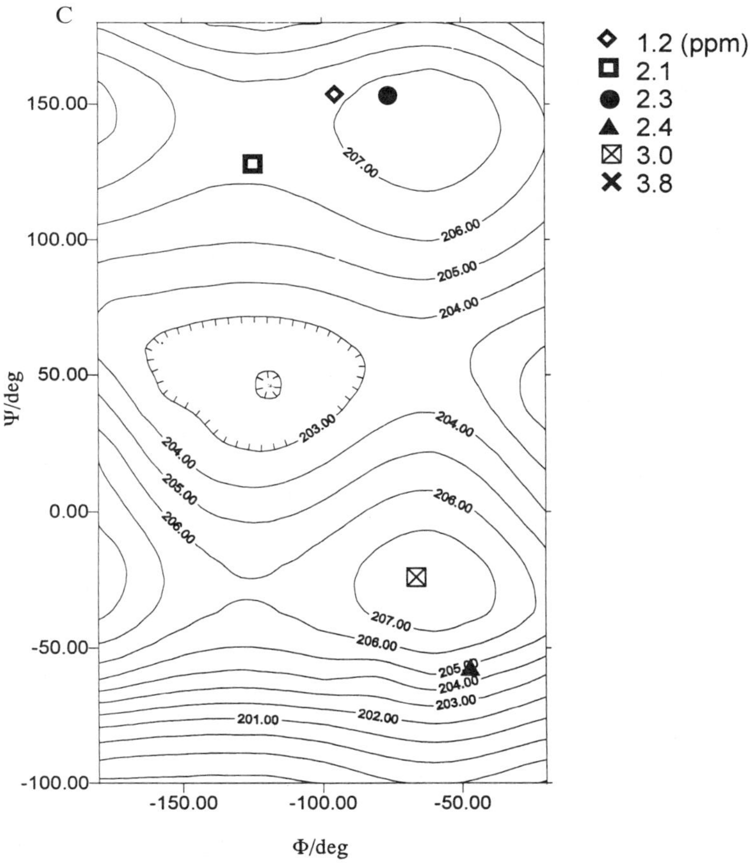

Figure 3. *Continued.*

Chapter 11

^{13}C Chemical Shift–Conformation Relationship in the Chromophores of Rhodopsin and Bacteriorhodopsin

Minoru Sakurai, Mitsuhito Wada, Hirohiko Houjou, Naoki Asakawa, and Yoshio Inoue

Department of Biomolecular Engineering, Tokyo Institute of Technology, 4259 Nagatsutacha, Midori-ku, Yokohama 226–8501, Japan

The chromophores of rhodopsin and bacteriorhodopsin are 11-*cis*- and *all-trans*-retinal Schiff bases, respectively. Upon binding to the proteins, their unsaturated carbons show anomalous ^{13}C chemical shifts compared with those of corresponding model compounds. This indicates the occurrence of interactions between the chromophore and its surrounding protein matrix. *Ab inito* shielding calculation reveals that the major part of such anomalous shifts originates in the conformational change of the chromophore.

Ab inito chemical shielding calculation, combined with solid-state NMR experiment, is a powerful tool to elucidate the structures of active sites in proteins and enzymes. A landmark study in such an approach is a series of our works (1-3) in which the ^{13}C chemical shifts of the chromophores of retinal proteins, obtained mainly by Griffin's group (4-8), were successfully interpreted using *ab initio* shielding calculation. Photoreceptive protein such as rhodopsin (Rh) or bacteriorhodopsin (bR) possesses a retinal isomer bound to a lysine residue via the protonated Schiff base linkage (9). Rh, present in the rod cell, is responsible for vertebrate dim-light vision. Rh possesses a 11-*cis* retinal (Figure 1), which is isomerized into the *all-trans* form by the absorption of photons, finally leading to signal transduction. On the other hand, bR exists in the purple membrane of *Halobacterium salinarium* and acts as a light-driven proton pump through a photocycle including the conversion of *all-trans*-retinal (Figure 1) into 13-*cis* isomer. In both proteins, the conformational state of retinal closely relates to the appearance of the biological function, especially to the regulation of their absorption maxima (10). For example, the twisting angle about the C6-C7 bond is an important factor to determine the color of these proteins. The ^{13}C NMR chemical shifts of the chromophores involve invaluable information about their conformational states and electronic structures in the protein matrix. Thus, an in-depth analysis of the chemical shifts enables us to better understand the functions of these proteins.

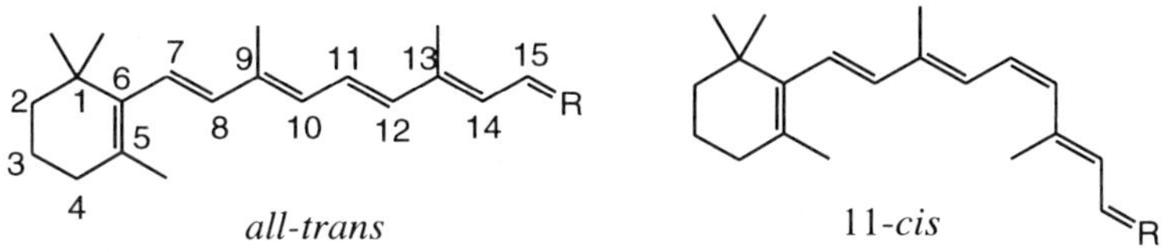

R: CH=O (retinal), R: $C{=}N^{+}H^{-}$ (protonated Schiff base)

Figure 1. Molecular structures of retinal and its Schiff bases.

At present, we know the ^{13}C chemical shift values for all the unsaturated carbons of the chromophores of Rh and bR before illumination. Figure 2 shows the chemical shift differences between the chromophores and their model compounds (free protonated retinal Schiff bases, hereafter abbreviated as PRSB). As for Rh, the chemical shifts of the carbons from C8 to C13 show significant downfield displacements relative to those for the model. As for bR, the chemical shifts of C5 and C8 are displaced significantly to downfield and upfield, respectively. These anomalous displacements suggest the occurrence of interactions between the chromophore and the surrounding protein matrix, including the resultant conformational change of the chromophore.

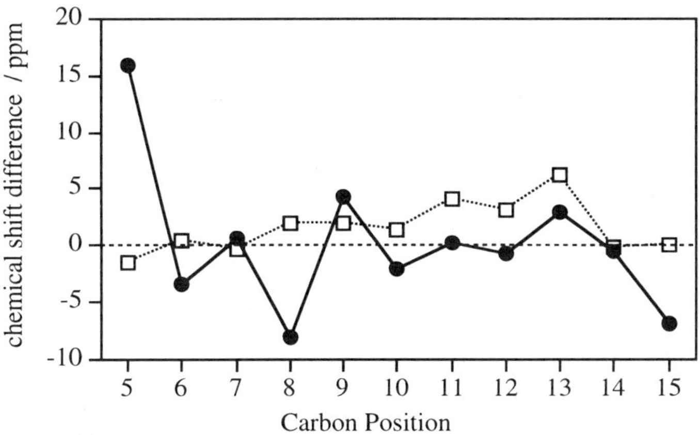

Figure 2. Plots of differences in chemical shift observed between the bR chromophore and *all-trans* PRSB chloride salt (circle), and between the Rh chromophore and 11-*cis* PRSB chloride salt (square).

Conformation dependence of ^{13}C chemical shifts have been theoretically studied for saturated cyclic or acyclic hydrocarbons (11-13). It is well known that the ^{13}C chemical shift of aliphatic carbon significantly shift to upfield when a methyl group is introduced at the γ-position with respect to the carbon of interest, called γ-steric effect. However, such an empirical rule deduced from aliphatic compounds may be insufficient to interpret the chemical shift data for conjugated compounds, because the shieldings of conjugated carbons would be influenced with changing π-electronic state, induced by rotation of a single bond. Therefore, a systematic study is newly required for conjugated systems such as retinal.

In this paper, we briefly describe some empirical rules found for conformation-dependent chemical shift displacements in conjugated systems. On the basis of these results, we interpret the anomalous chemical shift displacements of the chromophores of Rh and bR, leading to the determination of their conformations.

Calculation

The shielding calculations were carried out using two different progarams, RPAC9.0 (14) and GAUSSIAN94 (15). The so-called localized orbital/local origin (LORG) theory, developed by Hansen and Bouman (16), is implemented into the former program. Although three types of methods for shielding calculation are coded in the GAUSSIAN94 program, we adopted the gauge-invariant atomic orbital (GIAO) theory (17), because it is extensively used in actual calculations of NMR shielding. All shielding calculations were performed at Hartree-Fock level.

According to a recent study by us (18), the 6-31G* or 6-31G** basis sets are needed to accurately reproduce the observed chemical shifts for all the conjugated carbons in retinal and its Schiff bases. However, even in the 4-31G level of calculation, there have been found good correlation between the calculated and observed data except for carbonyl and imine carbons. Here, the 4-31G and 6-31G* basis sets were used for the calculations of retinal and its model compounds (shorter linear polyene such as diene), respectively. The calculated and experimental shieldings were converted to a methane (Figures 3, 6 and 7) or tetramethylsilane (Figure 9) reference and the positive sign indicates deshielding.

The geometries of all the molecules studied were optimized using the same basis sets as in the shielding calculation. Then, the initial geometries of *all-trans* and 11-*cis* retinals were taken from the corresponding crystal structures (19, 20).

Accuracy of the Shielding Calculation

First, we compared the 4-31G shieldings with the experimental ones for the unsaturated carbons of *all-trans* retinal (1) and its Schiff bases (2). It was confirmed that the calculation completely reproduces the relative order of all the carbon peaks: the standard deviation was 3.3 ppm for *all-trans* retinal. In general, the principal values of a shielding tensor involves more detailed information on the conformational and electronic states of a molecule than the isotropic shielding. Next, in order to more severely check the accuracy of the calculation, the principal values for each shielding tensor were compared with the corresponding experimental values (1). As shown in Figure 3, some characteristic features are found in the experimental data. As a whole, each tensor component changes with a zigzag pattern along the conjugated chain, called "odd-even effect". Among the three principal components, the σ_{11} behaves most regularly except for the position of C15. The σ_{22} and σ_{33}

components exhibit some irregularities in the middle positions of the conjugated chain. Clearly, the calculation completely reproduces these features.

The odd-even effect on the σ_{11} component almost disappears when the methyl groups on C9 and C13 are deleted, but those on the σ_{22} and σ_{33} components remain (1). This fact implies that the σ_{11} and the other components are dominated by different factors from each other, as will be discussed in the next section.

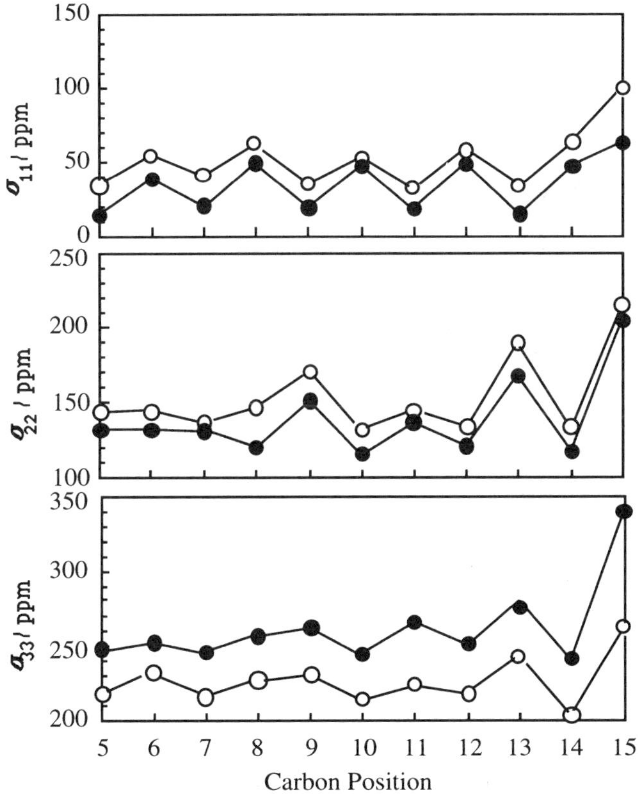

Figure 3. Comparison of the 4-31G (solid circle) and experimental (open circle) principal values for the unsaturated carbons of *all-trans*-retinal.

Shielding Tensor Analysis

The essential features of the shielding tensor of a conjugated carbon could be deduced from the calculation for minimal models, namely, diene and triene compounds. Here, using a few examples (Figure 4), we explain how each of the three principal values is influenced by the twist of a single bond. For all the unsaturated carbons studied, the σ_{11} is aligned perpendicular to the plane made by the trans zigzag chain, whereas the σ_{22} and σ_{33} components (in-plane elements) are roughly parallel and perpendicular to the relevant double bond, respectively. During twisting of the C2-C3 bond, the direction of each principal axis was almost kept constant within 5°.

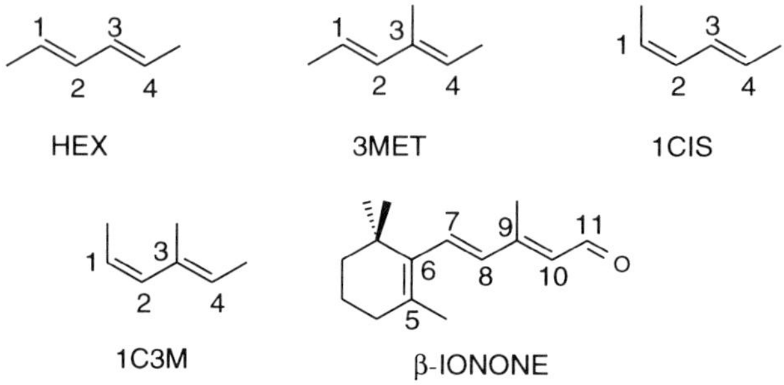

Figure 4. Molecular Structures of Model Compounds.

Out-of-plane component (σ_{11}). In order to explain the conformation dependence of σ_{11}, we take up the γ-steric effect , as an example, induced by the rotation of the C2-C3 bond and particularly give our attention to the C1 shielding of 3MET and 1C3M. Then, it is convenient to take HEX as a reference compound, because it has no substituent at the γ-position with respect to C1. The net contribution of the steric interaction, denoted as Δ3MET (or Δ1C3M), is obtained by subtracting the C1 shielding of HEX from that of 3MET (or 1C3M).

As can be seen from Figure 5, the isotropic shielding of C1 of 3MET shifts to upfield when this carbon stays in eclipsed conformation with respect to the γ-methyl group, namely, $\varphi_{2\text{-}3}$= 180°. A more interesting finding is that this shift predominantly originates from a large upfield shift of σ_{11}. This finding agrees with the experimental result that for sp^2 carbon, the γ-effect is reflected on the out-of plane element of the principal values (21). And it is also consistent with the result of a recent theoretical study, which indicated that for sp^3 carbon the γ-effect originates from the change of the principal value whose direction is perpendicular to the plane in which the steric interaction forces operate (22). It can be thus said that the behavior of the σ_{11} component governs the γ-effect.

On the other hand, the isotropic shielding of 1C3M certainly shifts to upfield when the angle $\varphi_{2\text{-}3}$ is around 180°, but the σ_{11} component does not. This indicates that the change in the isotropic shielding may not be ascribed to the γ-effect mentioned above. In fact, the behavior of σ_{11} is rather similar to that for the C1 carbon of 1CIS (see the plot of Δ1CIS), where the γ-methyl carbon is not present. The behavior of Δ1C3M and Δ1CIS both should be understood as a result of the *trans-cis* isomerization of the C1=C2 double bond. On the basis of these results, we can provide a more limited definition for the γ-effect in the conjugated systems: the true γ-effect on a CH carbon occurs only when the C-H vector directs towards the same side of the γ-methyl group. This statement implies that a quaternary carbon (non-protonated carbon) is insensitive to the γ-effect, as shown by the data for the C1 of 1C3M and for the C3 of 1C3M (data not shown).

Similar to the γ-effect, other steric effects such as β- and δ-effects are also dominated by the σ_{11} component (3). In particular, the β-effect is characterized as a large downfield shift: for example, the C2 shielding of 1C3M exhibits a downfield shift of about 12 ppm as a result of a steric interaction with the C3 methyl group.

In summary, the σ_{11} value of the shielding tensor is a measure of the chemical shift change originating in steric interactions.

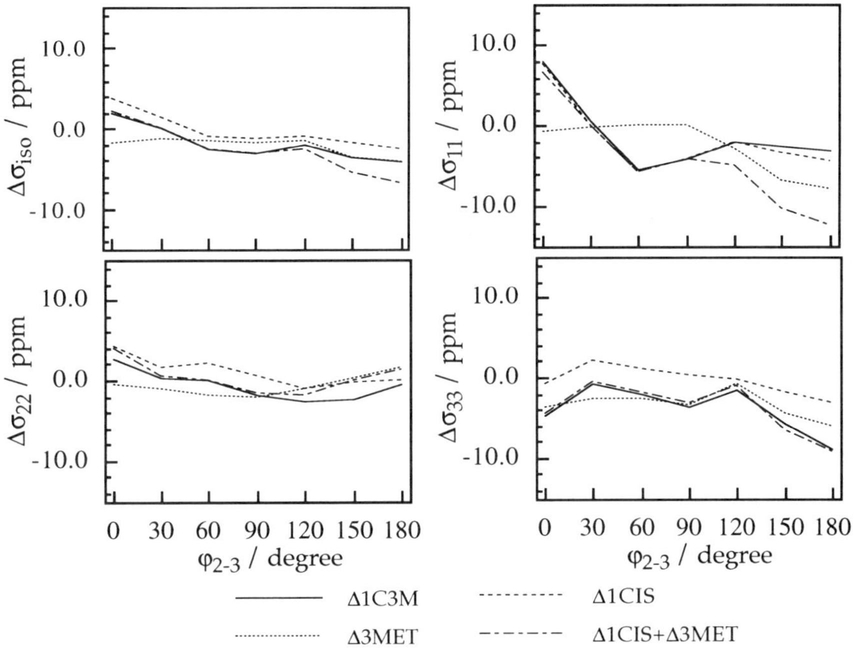

Figure 5. Conformation dependence of the C1 shieldings in Δ3MET, Δ1CIS, and Δ1C3M.

In-plane components (σ_{22} and σ_{33}). The feature of conformation dependence of the σ_{22} and σ_{33} components is better understood by taking β-IONONE as an example. Here, the conformation-dependent shifts of C5 and C8 are examined. As can be seen from Figure 6, the isotropic shieldings of both carbons exhibit clear periodic dependence on the twist angle about the C6-C7 bond,φ_{6-7}. For example, the σ-φ_{6-7} curve of C5 appears to be approximated by a sine wave whose period is 180°. The curve for C8 shifts in phase by a half period relative to that for C5. These curves are expressed by using the Fourier series of expansion as follows:

$$\sigma_{C5} = 138.4 - 1.466 \cos \varphi_{6-7} + 4.238 \cos 2\varphi_{6-7} \quad (1)$$

$$\sigma_{C8} = 142.4 - 2.729 \cos \varphi_{6-7} - 2.368 \cos 2\varphi_{6-7} \quad (2)$$

In Figure 7, the principal values for the C5-shielding tensor are plotted against the angle φ_{6-7}. It is evident that the σ_{22} component dominantly contributes to causing the periodic angular dependence of the C5 shielding. The σ_{33} component also shows an angular dependence with less-clear periodicity. In contrast to the in-plane components, σ_{11} is scarcely affected by a conformational change about the C6-C7 bond because C5 is a quaternary carbon. The angular dependence of the σ_{22} and σ_{33} can be represented by means of the Fourier series of expansion as follows:

$$\sigma_{22} = 143.3 + 9.680 \cos 2\varphi_{6\text{-}7} \tag{3}$$

$$\sigma_{33} = 258.2 - 6.643 \cos \varphi_{6\text{-}7} + 2.831 \cos 2\varphi_{6\text{-}7} \tag{4}$$

where the higher terms whose coefficients are less than 0.5 ppm are neglected. From these equations, the σ_{22} and σ_{33} components are mainly responsible for the cos 2φ and cosφ terms in equation 1, respectively. Consequently, the change in the isotropic shielding of C5 can be described by means of that in the σ_{22} and σ_{33} shieldings. The similar result was obtained for the C8 shielding (1).

The behavior of σ_{22} and σ_{33} could be explained in terms of the so-called resonance effect. When the molecular structure is planar, positive charge should be induced on the odd-numbered carbons (C5, C7, C9 and C11), while the even-numbered carbons should be negatively-charged.

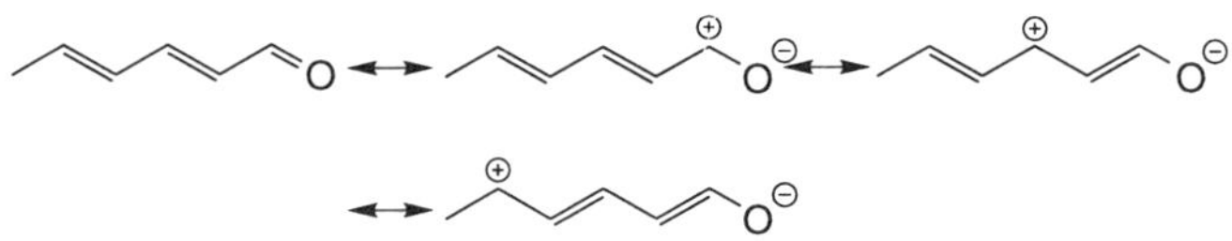

With a rotation of the C6-C7 bond, the contribution of the resonance structure is reduced. Consequently, the positive charge density on the odd-numbered carbons decreases, causing an upfield shift to the shielding. This successfully explains the fact that the C5 shielding exhibits a concave-type curve in the region from 0° to 180° and that from 180° to 360°. Inversely, the shieldings of the even-numbered carbons such as C8 exhibit convex curves. Therefore, it is concluded that σ_{22} and σ_{33} change as a result of the change in the electronic structure of the conjugated system.

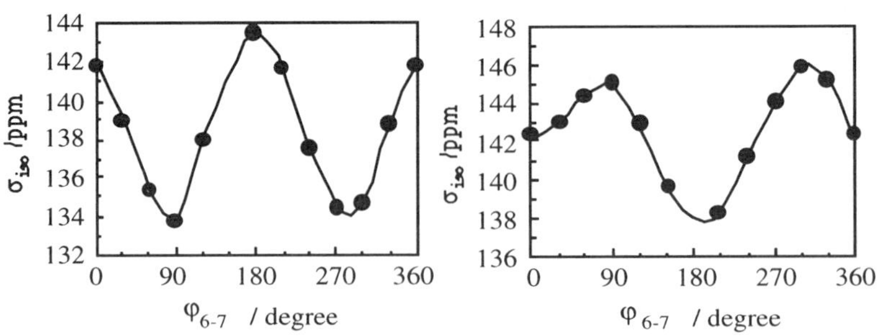

Figure 6. Conformation dependence of the isotropic shieldings of C5 (left) and C8 (right) of β-IONONE.

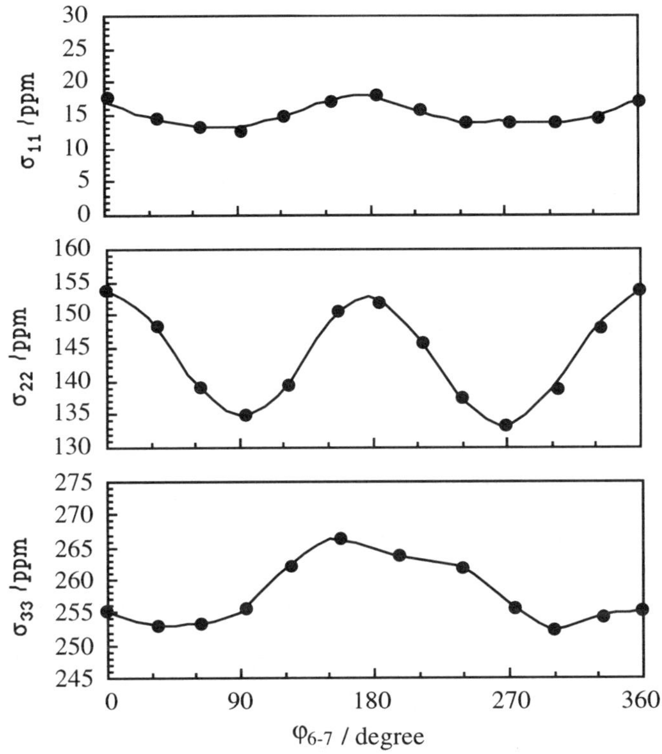

Figure 7. Changes of the principal values of the C5 shielding.

In the above example, we have not given attention to the shieldings of the carbon atoms directly participating in the twisting bond C6-C7. To understand the conformation dependence of the C6 and C7 shieldings, a special treatment is required. For simplicity, we consider a four π-electron system such as HEX. It is now assumed that the x, y and z axes coincide with the direction of the principal axes for σ_{33}, σ_{22} and σ_{11}, respectively. Then, the rotation of the C2-C3 bond modifies the p_z orbitals of these carbons, resulting in a distortion of the π-orbitals, which is represented by the mixing of the p_z orbital with p_{x-y} orbital defined as

$$p_{x-y} = \frac{1}{2}\left(p_x - \sqrt{3}\, p_y\right) \tag{5}$$

whose direction is perpendicular to both the z-axis and the single bond. Using the mixing parameter λ, the distortion of the p_z orbital can be described as follows:

$$p_z \rightarrow \frac{1}{\sqrt{1+\lambda^2}}\left(p_z \pm \lambda p_{x-y}\right) \tag{6}$$

where the choice of the double sign depends on the symmetry of the π-orbital in question. According to Ramsey's theory for the NMR shielding (23), we can derive the approximate expression of the paramagnetic shielding term for both C2 and C3 carbons as follows (3):

$$\sigma_{xx}^{\text{para}} = 3\,\sigma_{yy}^{\text{para}} = -\frac{1}{c^2}\Delta E^{-1}\left\langle r^{-3}\right\rangle \frac{3\lambda^2}{4\left(1+\lambda^2\right)^2} \qquad (7)$$

$$\sigma_{zz}^{\text{para}} = 0 \qquad (8)$$

where ΔE is the average π–π^* excitation energy of the molecule. According to these equations, the orbital modification induced by rotation of the C2-C3 bond affects mainly the terms of σ_{xx} and σ_{yy}, namely, σ_{33} and σ_{22}, in the diene system. Since the value of λ increases with a rotation of the bond, an appreciable amount of downfield shift would be induced for these tensor elements.

In summary, the in-plane components (σ_{33} and σ_{22}) of the shielding tensor depend on the electronic structure changes of the conjugated systems, induced by the inductive effect and the π-modification effect mentioned above.

Additivity of Miscellaneous Effects. We have discussed several factors affecting conformation-dependent changes of the shielding tensor, including the β-, γ- and δ-steric effects, the *cis-trans* isomerization effect, the inductive effect, and the π-orbital modification effect. Basically, these mechanisms additively operate for each of the principal values (3). A typical example is shown in Figure 5, where the sum of Δ1CIS and Δ3MET is compared with Δ1C3M. For each principal value, the simple sum is in good agreement with the latter value. In general, the only situation in which the additivity is disturbed is in the conformational range where the γ-effect is acting predominantly (3).

The Conformation of the Chromophores of Rh and bR

As described in the introduction, the absorption maxima of Rh and bR depend on the conformation of their chromophores, especially on the twisting angle about the C6-C7 bond in bR. As for Rh, the chemical shifts of C11, C12 and C13 undergo relatively large displacements (Figure 2), implying that the twisting angle of the C12-C13 single bond may be perturbed from the protein environment. We can determine the conformations about these single bonds on the basis of results of the shielding calculation.

The conformation about the C6-C7 bond. The periodic angular dependence, as observed in the C5 and C8 shieldings of β-IONONE, were also reproduced for the corresponding carbon shieldings of model compounds with the same length of the conjugated chain as PRSB (2). Thus, this may be a common behavior among the molecules possessing the β-ionone moiety, including the Rh and bR chromophores. An important finding from Figure 6 is that there is the apparent phase shift between the periodic functions for C5 and C8 (see equations 1 and 2). As a consequence, the difference between the C5 and C8 shieldings also shows a periodic dependence against the angle $\varphi_{6\text{-}7}$ as shown in Figure 8. This fortunate result enables us to estimate the twist angle about the C6-C7 bond through only an observation of the chemical shift difference between C5 and C8.

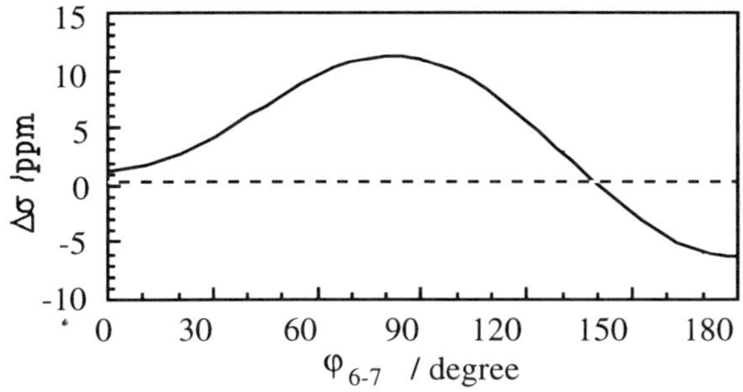

Figure 8. Conformation dependence of $\Delta\sigma_{C8\text{-}C5}$ obtained by subtracting equation (1) from equation (2).

Table 1 summarizes the experimental $\varphi_{6\text{-}7}$ angles and the calculated and observed shielding differences $\Delta\sigma_{C8\text{-}C5}$, obtained by subtracting the value of the C5 shielding from that of the C8 shielding. Hereafter, for the sake of convenience, the conformation about the C6-C7 bond is classified into the following four types: planar 6-*s-cis* ($\varphi_{6\text{-}7}$ = 0° -30°), skewed 6-*s-cis* (30° - 90°), skewed 6-*s-trans* (90° -150°°), and planar 6-*s-trans* (150° -180°). Before discussing the chromophores of Rh and bR, we compare the calculated and observed data for several model compounds listed in Table 1. Except for 6-*s-trans*-retinoic acid, all the other models have skewed *s-cis* conformations about the C6-C7 bond in the crystalline state, and their $\Delta\sigma_{C8\text{-}C5}$ values exhibit relatively large plus ones. In contrast, 6-*s-trans* retinoic acid exhibit a minus value for $\Delta\sigma_{C8\text{-}C5}$. These characteristic tendencies are reproduced well by the calculation. It is thus concluded that Figure 8 is applicable to a wide variety of retinal derivatives.

Table 1. Chemical Shift Difference between the C5 and C8

		$\Delta\sigma_{C8\text{-}C5}$ /ppm	
	$\varphi_{6\text{-}7}$ [a]	obsd	calcd
6-*s-cis*-retinoic acid	41.2[b]	10.1[e]	6.1
retinol acetate	57.9[c]	10.2[e]	9.0
6-*s-cis*-13-*cis*-retinal	65.4[c]	12.2[e]	9.9
all-trans-retinal	62.0[d]	6.6[e]	9.6
6-*s-trans*-retinoic acid	165.8[b]	-5.0[e]	-6.0
rhodopsin		8.9[f]	
bacteriorhodopsin		-12.1[g]	

[a] Experimental dihedral angle in degree. [b] Ref 24 . [c] Ref 25 . [d] Ref 19. [e] Ref 21. [f] Ref. 5. [g] Ref. 7.

Next, we predict the C6-C7 conformations of the bR and Rh chromophores. As shown in Table 1, in bR the observed values of $\Delta\sigma_{C8\text{-}C5}$ is -12.1 ppm. Figure 8 shows that the value of $\Delta\sigma_{C8\text{-}C5}$ has a minus sign only in the planar *trans* form. It can be thus concluded that the chromophore of bR takes the planar 6-*s-trans* conformation around the C6-C7 bond. This is consistent with the results from recent NMR studies (26, 27). On the other hand, the observed value of $\Delta\sigma_{C8\text{-}C5}$ is 8.9 ppm for Rh. On the basis of Figure 8, there are two solutions for the $\varphi_{6\text{-}7}$ angle, namely 58° and 135°. This indicates that the chromophore of Rh takes a highly skewed conformation about the C6-C7 bond.

The conformation about the C12-C13 Bond in Rh. Here, we describe the calculated results for 11-*cis*-retinal, which is a better model of the Rh chromophore. Among the largely perturbed carbons shown in Figure 2, particular attention is given to the C12 carbon, because its principal values have been experimentally obtained (5, 28). Experimental data are summarized in Table 2, where for Rh Mollevanger's data (28) are listed and the data for *all-trans* PRSB are added as a reference. A notable feature is that the σ_{11} and σ_{22} components largely shift relative to those of the reference (see $\Delta\sigma$ values). Special care must be, however, taken in analyzing the tensor data for this carbon, because it is pertinent to the rotating bond (C12-C13) in question. As described in the section of tensor analysis, the σ_{22} and σ_{33} components of such a type of carbon undergo simultaneously several types of perturbations, including the π-orbital modification effect, inductive effect, etc. In addition, electrostatic interactions between the chromophore and the protein environment should reflect mainly on these components, because they modify the electronic structure of the conjugated system. On the other hand, the σ_{11} component is dominated by steric interactions depending on the conformation. It is thus appropriate to analyze the σ_{11} for the present purpose.

Table 2. Experimental chemical shifts of C12 (in ppm)

	σ_{iso}	σ_{11}	σ_{22}	σ_{33}
all-trans PRSB[a]	134.3	58	33	212
Rh[b]	133.5	41	49	209
$\Delta\sigma$[c]	-0.8	-17	16	3

[a] Ref. 5. [b]Ref. 28. [c]Obtained by subtracting the data for *all-trans* PRSB from those for Rh.

The quantity $\Delta\sigma$ includes mainly two contributions: namely, the effects of *trans-cis* isomerization of the C11=C12 double bond and of a rotation of the C12-C13 single bond, where the C12-C13 conformation of the reference was assumed to be planar *s-trans*, namely $\varphi_{12\text{-}13}$=180°. In order to examine the net contribution of the former effect, we compared the calculated shielding of C12 of *all-trans*-retinal and that for 11-*cis*-retinal (Table 3). When the angle of $\varphi_{12\text{-}13}$ is kept at 180°, the σ_{11} component shifts to upfield by 2.7 ppm upon isomerization from *all-trans* to 11-*cis* form. 3MET and 1C3M (Figure 4) are the minimal models available for examining the *trans-cis* isomerization effect. As shown in Table 3,the σ_{11} of C2, corresponding to C12 of retinal, shifts to upfield by 6.5 ppm upon going from 3MET to 1C3M. Taken together with these results, the net contribution of the *trans-cis* isomerization on the σ_{11} value is not more than several ppm, which is insufficient to explain the observed value of $\Delta\sigma_{11}$.

As the C12-C13 bond of 11-*cis* retinal is rotated from the *s-trans* form, the σ_{11} shielding gradually shifts to upfield. As shown in Figure 9, the σ_{11} of C12 exhibits a concave-type profile with a minimum at $\varphi_{12\text{-}13}$ = 90-120°. When $\varphi_{12\text{-}13}$ equals 120°, the σ_{11} undergoes an upfield shift of 18.1 ppm, as measured from that of the *all-trans* isomer with $\varphi_{12\text{-}13}$ = 180°. A comparable amount of upfield shift (19.9 ppm) was also obtained for the model compound 1C3M. These values are in good agreement with the observed $\Delta\sigma_{11}$. Therefore, it can be concluded that the chromophore of Rh takes the skewed *s-trans* form around the C12-C13 bond.

When the C12-C13 takes the *s-trans* conformation, the C12 shielding should receive the β-steric effect from the C13 methyl. Similarly, the C2 shielding of 1C3M receives the effect from the C3 methyl. As mentioned in the section of tensor analysis, such a β-effect causes a large downfield shift (11-12 ppm) to the σ_{11} shielding of a carbon of interest. Thus, the rotation of the C12-C13 or C2-C3 bond should decrease the β-effect on the C12 or C2 shieldings, resulting in a large upfield shift. This explains the reason why these shieldings exhibit concave-type profiles.

Table 3. Conformation dependence of the calculated C12 shielding

	φ^{a}	$\sigma_{iso}{}^{b}$	$\sigma_{11}{}^{b}$	$\sigma_{22}{}^{b}$	$\sigma_{33}{}^{b}$
11-*cis* retinal	180	-4.2	-2.7	-2.7	-7.1
	120	-4.0	-18.1	+0.3	+5.9
1C3M	180	-2.2	-6.5	-3.0	-3.1
	120	-3.2	-19.9	+3.1	+7.0

[a] The dihedral angle (in degree) about the C12-C13 bond for 11-*cis* retinal and the C2-C3 bond for 1C3M. [b] The C12 shielding in ppm.

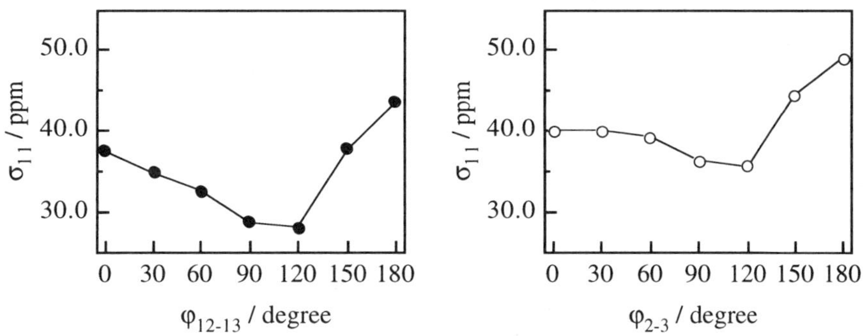

Figure 9. Conformation dependence of the C12 shielding of retinal (left) and the C2 shielding of 1C3M (right).

Conclusion

Throughout this chapter, we have focused on the conformation of the chromophore responsible for the color regulation of Rh and bR. The *ab inito* shielding calculation revealed that the major part of the anomalous chemical shifts observed for Rh and bR is caused by the conformational change of the chromophore on binding to the protein. Our calculation demonstrated that ^{13}C shielding tensor involves invaluable information about the conformational and electronic structures of a molecule. In addition, it should be noted that the shielding calculation serves us a new analytical procedure for experimental data. The calculation of β-IONONE is a good example, where it is revealed that the difference between the C5 and C8 shieldings involves information on the conformational state of the chromophore. An advantage of this procedure is that an external reference is not required in order to interpret the native data for the proteins. Larger scale calculation with explicitly taking into account the surrounding amino acids would enable us to obtain a more detailed picture of the chromophore-protein interaction in Rh and bR in the future.

Acknowledgment. The authors gratefully thank the Computer Center, Institute for Molecular Science, Okazaki, Japan, for the use of an NEC SX-3 supercomputer. They also thank the Computer Center, Tokyo Institute of Technology, Tokyo, Japan, for the use of a Cray C916 system.

Literature Cited

1) Wada, M.; Sakurai, M.; Inoue, Y.; Tamura, Y.; Watanabe, Y. *J. Am. Chem. Soc.* **1994**, *116*, 1537.
2) Sakurai, M.; Wada, M.; Inoue, Y.; Tamura, Y.; Watanabe, Y. *J. Am. Chem. Soc.* **1996**, *100*, 1957.
3) Houjou, H.; Sakurai, M.; Asakawa, N.; Inoue, Y.; Tamura, Y. *J. Am. Chem. Soc.* **1996**, *118*, 8904.
4) Smith, O.; Palings, I.; Copie, V.; Raleigh, D. P.; Courtin, J.; Pardoen, J. A.; Lugtenburg, J; Mathies, R. A.; Griffin, R. G. *Biochemistry* **1987**, *26*, 1606.
5) Smith, S. O.; Palings, I.; Miley, M. E.; Courtin, J.; de Groot, H.; Lugtenburg, J.; Mathies, R. A.; Griffin, R. G. *Biochemistry* **1990**, *29*, 8158.
6) Harbison, G. S.; Smith, S. O.; Pardoen, J. A.; Mulder, P. P. J.; Lugtenburg, J; Hertzfeld, J; Mathies, R. A.; Griffin, R. G. *Biochemistry* **1984**, *23*, 2662.
7) Harbison, G. S.; Smith, S. O.; Pardoen, J. A.; Courtin, J. M. L.; Lugtenburg, J; Hertzfeld, J; Mathies, R. A.; Griffin, R. G. *Biochemistry* **1985**, *24*, 6955.
8) Smith, S. O.; de Groot, H. J. M.; Gebharrd, R; Courtin, J. M. L.; Lugtenburg, J; Hertzfeld, J; Griffin, R. G. *Biochemistry* **1989**, *28*, 8897.
9) See, for a review: Birge, R. R. *Annu. Rev. Phys. Chem.* **1990**, *41*, 683.
10) van der Steen, R; Biesheuvel, P. L.; Mathis, R. A.; Lugtenburg, J. *J. Am. Chem. Soc.* **1986**, *108*, 6410.
11) Barfield, M.; Yamamura, S. H. *J. Am. Chem. Soc.* **1990**, *112*, 4747.
12) Barfield, M. *J. Am. Chem. Soc.* **1993**, *115*, 6916.
13) Barfield, M. *J. Am. Chem. Soc.* **1995**, *117*, 2862.
14) Bouman, T. D.; Hansen, Aa. E. RPAC Ver. 9.0; 1991.
15) Frisch, M.J.; Trucks, G.W.; Schlegel, H.B.; Gill, P.M.W.; Johnson, B.G.; Robb, M.A.; Cheeseman, J.R.; Keith, T.; Petersson, G.A.; Montgomery, J.A.;

Raghavachari, K.; Al-Laham, M.A.; Zakrzewski, V.G.; Ortiz, J.V.; Foresman, J.B.; Cioslowski, J.; Stefanov, B.B.; Nanayakkara, A.; Challacombe, M.; Peng, C.Y.; Ayala, P.Y.; Chen, W.; Wong, M.W.; Andres, J.L.; Replogle, E.S.; Gomperts, R.; Martin, R.L.; Fox, D.J.; Binkley, J.S.; Defrees, D.J.; Baker, J.; Stewart, J.P.; Head-Gordon, M.; Gonzalez, C.; Pople, J.A. GAUSSIAN94, Revision B.2; Gaussian, inc.: Pittsburgh PA, 1995.

16) Hansen, Aa. E; Bouman, T. D. *J. Chem. Phys.*, **1985**, *82*, 5035.

17) Ditchfield, R.; Ellis, P. D. *Topics in Carbon-13 NMR Spectroscopy*; Levy, G. C., Ed; John Wiley & Sons: New York, 1974; pp 18-34.

18) Wada, M.; Sakurai, M.; Inoue, Y.; Tamura, Y.; Watanabe, Y. *Magn. Reson. Chem.* **1995**, *33*, 453.

19) Hamanaka, T.; Mitsui, T.; Ashida, T.; Kakudo, M. *Acta Crystallogr.* **1972**, *B28*, 214.

20) Gilardi, R. D.; Karle, I. L.; Karle, J. *Acta Crystallogr.* **1972**, *B28*, 2605.

21) Harbison, G. S.; Mulder, P. J. J.; Perdon, H.; Lugtenberg, J.; Hertzdeld, J; Griffin, R. G. *J. Am. Chem. Soc.* **1985**, *107*, 4809.

22) Soderquist, A.; Facelli, J. C.; Horton, W. J.; Grant, D. M. *J. Am. Chem. Soc.* **1995**, *117*, 8441.

23) Ramsey, N. *Phys. Rev.* **1952**, *86*, 243.

24) Stam, C. H. *Acta Crystallogr.* **1972**, *B28*, 2936.

25) Simmons, C. J.; Liu, R. S. H.; Denny, M; Seff, K. *Acta Crystallogr.* **1981**, *B37*, 2197.

26) Ulrich, A. S.; Heyn, M. P.; Watt, A. *Biochemistry* **1992**, *31*, 10390.

27) Creuzet, F; McDrmott, A; Gebhard, R; van der Hoef, K.; Spijker-Assink, M. B.; Herzfeld, J.; Lugtenburg, J; Levitt, M. H.; Griffin, R. G. *Science* **1991**, *251*, 783.

28) Mollevanger, L. C. P. J.; Kentgens, A. P. M.; Padoen, J. A.; Courtin, J. M. L.; Veeman, W. S.; Lugtenburg, J; de Grip, W. *Eur. J. Biochem.*, **1987**, *163*, 9.

Chapter 12

Modeling of the ^{15}N and ^{13}C Chemical Shift Tensors in Purine

Julio C. Facelli[1], Jian Zhi Hu[2], Mark S. Solum[2,3], Ronald J. Pugmire[3], and David M. Grant[2]

[1]Center for High Performance Computing, [2]Department of Chemistry, and [3]Department of Chemical and Fuels Engineering, University of Utah, Salt Lake City, UT 84112

This paper presents quantum mechanical studies of the ^{15}N and ^{13}C chemical shifts in both the N7-H and N9-H tautomeric forms of purine. Quantum mechanical calculations of the chemical shifts were used to assign the NMR resonances and the spatial orientation of the principal axes of the chemical shift tensors. Calculations in purine and in a pyridine-methanol complex model provide insights on the importance of the intermolecular interactions on the chemical shifts of the nucleic acid bases.

The study of nucleic acid bases by NMR has been reported in a number of monographs (*1*), but very little data is available on the ^{13}C and ^{15}N NMR chemical shift tensors in these compounds. The low sensitivity of NMR spectroscopy and the long relaxation times exhibited by many of these compounds have posed the main impediments for these studies. The use of sample doping with free radical relaxation reagents, to reduce the relaxation times facilitating 2D multiple pulse experiment (*2, 3*), enables one to measure and analyze the principal values of the chemical shift tensors in natural abundance samples. In previous papers from this laboratory we have presented ^{15}N NMR chemical shift principal values for adenine, guanine, cytosine, thymine and uracil (*4, 5*).

The experimental data on the principal components of the ^{13}C and ^{15}N chemical shift tensors can be complemented by quantum mechanical calculations to gain information on the spatial assignment of the principal components. The calculations also provide valuable information for exploring the relationships between the NMR chemical shifts and the molecular geometry and environment (*6, 7*). Recent work from this laboratory has demonstrated that the inclusion of intermolecular interactions is necessary to reproduce accurately the ^{15}N experimental chemical shift principal values (*8*).

Purine can exist in two tautomeric forms, N7-H and N9-H depending on the state (*9, 10*). In the crystalline state, it exists in the N7-H form (11), but its N9-H form has been observed in the gas phase, matrix isolation, and in solution. Quantum mechanical calculations always predict that the N9-H form is the more stable tautomer for the isolated molecule (*12*).

This paper presents a study of the ^{15}N and ^{13}C chemical shifts in both tautomeric forms of purine. All the principal values of the ^{15}N and ^{13}C chemical shifts tensors were measured in a sample in which only the N7-H tautomeric form was present. In another sample, containing both tautomeric forms, the principal values of the quaternary carbons, the isotropic shifts of the remaining carbons and isotropic shifts of the nitrogens could also be obtained for the N9-H tautomeric form. The experimetal procedures are described in detail elsewhere.

Quantum chemical calculations employing different molecular models, some of which include the intermolecular interactions in the calculations, are used to guide the interpretation of the experimental results. The intermolecular interactions have been approximated by explicitly adding part of the nearest neighboring molecules into the calculations. Comparison of the results obtained when including the intermolecular interactions with those in the isolated molecule gives semi-quantitative information on the effects of intermolecular interactions on the principal values of the nitrogen shift tensors. The intermolecular interactions are dominated by the hydrogen bonding (HB) and calculations in a pyridine-methanol model complex provide important insights on how the chemical shift of aromatic nitrogens depends on the strength of the HB. These computational results are used in this paper to interpret the purine data.

Computational Procedures

All the chemical shift calculations and geometry optimizations were performed with the GAUSSIAN94 (*13*) suite of programs using the DFT approach and the D95** bases set, which includes polarization functions in all the atoms (*14*). The calculations were performed with an IBM SP and a SGI Origin 2000 using the parallel capabilities of these systems. The chemical shift calculations employed the method proposed by Cheeseman *et al.* (*15*) with the BLYP exchange-correlation functional (*16*) as implemented in the GAUSSIAN94 program (*13*). This method makes use of an efficient implementation (17) of the gauge-including atomic orbitals (GIAO) methods (*18,19*). The calculated ^{15}N chemical shielding values were converted to the shift scale by subtracting the shielding value of nitromethane, -135.8 ppm, obtained from the literature (*20*). The ^{13}C shifts were converted to the TMS scale by subtracting the estimated absolute chemical shieldings of TMS, 185.4 ppm, which was obtained from the calculated shielding in methane minus the 7 ppm reported as the difference between liquid TMS and gas phase methane (*21*).

Calculations were performed for a number of molecular structures corresponding to both tautomeric forms. Calculations for the isolated molecule used the molecular optimized geometries (OPT) obtained with the same computational method and basis set employed in the shielding calculations. For the isolated molecules the calculations predict that the N9-H tautomer is the most stable by 3.4 Kcal/mol. This value is in good agreement with a recent value of 3.8 Kcal/mol, which was obtained using the MP2 method and corrected by zero-point vibrations (*12*). This result is also in agreement with experimental results indicating that in the gas phase and in matrix isolation the N9-H form is the most stable (*12*). Additional calculations were performed utilizing the X-ray structure (*11*) of purine. These included calculations for the N7-H tautomeric form using the X-ray data without further changes (X-ray) and the X-ray data with the X-H bond distances optimized (Hopt).

To explore the importance of the intermolecular interactions model systems were constructed for both the N7-H and the N9-H tautomeric forms. These models, labeled HB:N9-H (**1**) and HB:N7-H (**2**) in the tables, are based on the X-ray structure of the N7-H tautomer partially surrounded by the nearest neighbor molecules. In the models the position of the nearest neighbor fragments was determined using the x-ray structure and the optimized X-H distances. The models and the numbering of the atoms in purine are shown in Figure 1.

Spectral Assignments

Table I presents the comparison between the calculated and experimental ^{15}N chemical shifts in purine. The assignment of the NMR resonances is straight-forward using the liquid state, in 7-methyl substituted purine (*22*), and calculated values. As expected, in the predominat N7-H form, N7 experiences a large up field shift while N9 shows a similar down field shift as a consequence of the migration of the proton from N9 in the liquid to N7 in the solid. The isotropic values for the N9-H form in the solid are quite similar to the values observed in the liquid, the largest difference between corresponding chemical shift values is 12 ppm.

The calculations reproduce quite well the isotropic shifts of both forms, but show larger discrepancies in the principal values of the chemical shifts of the N7-H tautomeric form. The lack of experimental values, for the principal components of the nitrogen chemical shifts in the N9-H tautomeric form, precludes a comparison of the calculated data with the experiment.

The assignment of the ^{13}C chemical shifts for the N7-H tautomeric form can be done by comparison with the calculated results and using the information obtained from the dipolar dephasing experiments. These experiments indicate that the peaks at 158 ppm and 123 ppm, in the N7-H tautomer, corresponds to the quaternary carbons. The large difference in their chemical shifts gives confidence for using the calculations to assign them to C4 and C5 respectively. The resonance at 153 ppm is well separated and can be assigned with confidence to C2. The other two resonances, at 141 ppm and 146 ppm, are quite close but the large difference in their δ_{33} components, 28 ppm and 56 ppm, respectively, gives confidence for using the calculated values to assign them to C6 and C8, respectively. The isotropic chemical shifts for the carbons in the N9-H tautomer are similar to those in the N7-H form, with the largest difference observed in C5 at 9 ppm. The rationale for this difference and the assignments, based on the tautomeric forms, was described earlier by Pugmire *et al* (23). The differences in the principal components, available only for the quaternary carbons, C4 and C5, are significantly larger than the differences observed in the isotropic shifts.

The similarity between the ^{13}C chemical shift tensors of the N7-H and N9-H forms can be rationalized by inspecting the optimized structures of both conformers. Figure 2 shows the optimized structures for the N7-H and N9-H tautomeric forms. It is apparent that most of the interatomic distances of equivalent atoms agree within a few thousands of an Angstrom. The bond angles, not shown in the figure, also are quite similar.

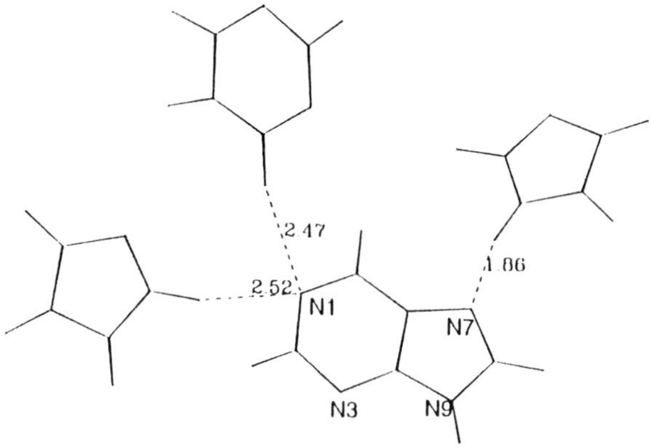

1

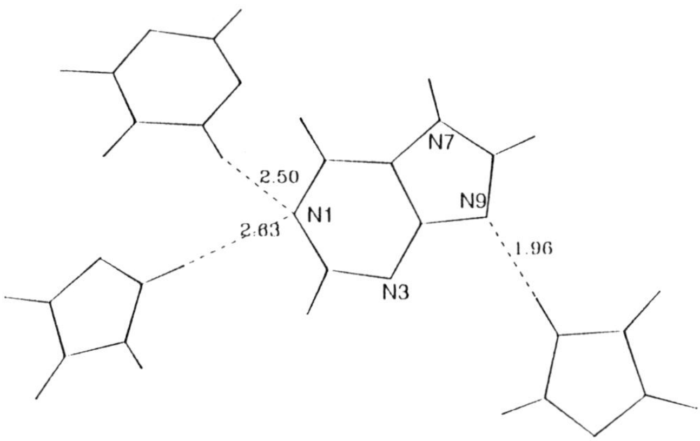

2

Figure 1: Numbering of the atoms in purine. The figure also shows the neighbouring fragments included in some of the calculations to simulate the intermolecular effects on the calculated chemical shifts of purine. The close approaches are indicated in Å. The calculations using these models are labeled HB:N9-H (**1**) and HB:H7-H (**2**) in the tables.

Table I: Experimental and Calculated ^{15}N Chemical Shifts principal values in purine. [a]

Model Form		Calculated OPT N7-H	Calculated OPT N9-H	Calculated HB N7-H	Calculated HB N9-H	Calculated X-ray N7-H	Calculated Hopt N7-H	Experimental N7-H[b]	Experimental N9-H[c]
N1	δ_{iso}	-81	-81	-99	-94	-99	-99	-102	-102(-104)
	δ_{11}	157	145	136	146	148	142	110	n/a
	δ_{22}	10	14	-19	-12	-14	-17	-5	n/a
	δ_{33}	-411	-404	-416	-416	-431	-421	-412	n/a
N3	δ_{iso}	-88	-118	-94	-105	-91	-92	-115	-122(-131)
	δ_{11}	137	74	122	101	130	126	87	n/a
	δ_{22}	-19	-39	-14	-26	-11	-12	-36	n/a
	δ_{33}	-381	-389	-391	-390	-392	-390	-395	n/a
N7	δ_{iso}	-241	-119	-255	-136	-271	-259	-226	-137(-140)
	δ_{11}	-159	87	-186	49	-195	-192	-135	n/a
	δ_{22}	-266	-76	-263	-105	-281	-270	-222	n/a
	δ_{33}	-297	-370	-314	-351	-338	-314	-322	n/a
N9	δ_{iso}	-113	-227	-134	-239	-119	-123	-150	-221(-233)
	δ_{11}	79	-146	19	-158	65	54	0	n/a
	δ_{22}	-60	-247	-81	-258	-61	-67	-98	n/a
	δ_{33}	-357	-288	-341	-302	-362	-356	-351	n/a

[a] Values in ppm referenced to nitromethane. Liquid chemical shifts (given between parenthesis) are for the methyl derivatives of the N7-H conformer from the literature (*22*).
[b] From a cupric chloride doped sample containing only the N7-H tautomeric form.
[c] From a BDPA/DPPH (melt/quench) doped sample containing both N7-H and N9-H tautomeric forms.

Table II: Experimental and calculated ^{13}C Chemical shift principal values in purine.[a]

Model		Calculated						Experimental	
		Opt.		HB		X-ray	Hopt		
Form		N7-H	N9-H	N7-H	N9-H	N7-H	N7-H	N7-H[b]	N9-H[c]
C2	δ_{iso}	158	157	156	153	154	155	~153	~152(154)
	δ_{11}	247	243	242	235	242	241	246	n/a
	δ_{22}	154	155	156	154	153	153	160	n/a
	δ_{33}	73	72	70	69	67	72	53	n/a
C4	δ_{iso}	162	153	157	153	159	157	158	152(155)
	δ_{11}	221	230	217	235	218	216	230	238
	δ_{22}	198	170	192	157	196	196	192	175
	δ_{33}	68	59	60	67	62	59	52	44
C5	δ_{iso}	122	132	118	123	118	118	123	132(128)
	δ_{11}	202	200	196	185	199	197	209	208
	δ_{22}	142	163	137	166	136	137	147	159
	δ_{33}	23	34	21	19	19	20	15	30
C6	δ_{iso}	137	149	136	142	129	133	141	~146(145)
	δ_{11}	218	235	212	221	210	210	226	n/a
	δ_{22}	150	164	155	162	145	146	169	n/a
	δ_{33}	44	49	41	43	31	43	28	n/a
C8	δ_{iso}	140	137	134	133	127	131	146	~146(148)
	δ_{11}	219	216	205	206	207	206	221	n/a
	δ_{22}	131	127	133	132	125	124	160	n/a
	δ_{33}	71	69	64	62	49	63	56	n/a

[a] Values in ppm from TMS. Liquid chemical shifts (given bewteen parenthesis) for the N7-H tautomeric form are from the literature (*23*).

[b] From a cupric chloride doped sample containing only the N7-H tautomeric form.

[c] From a BDPA/DPPH (melt/quench) doped sample containing both N7-H and N9-H tautomeric forms.

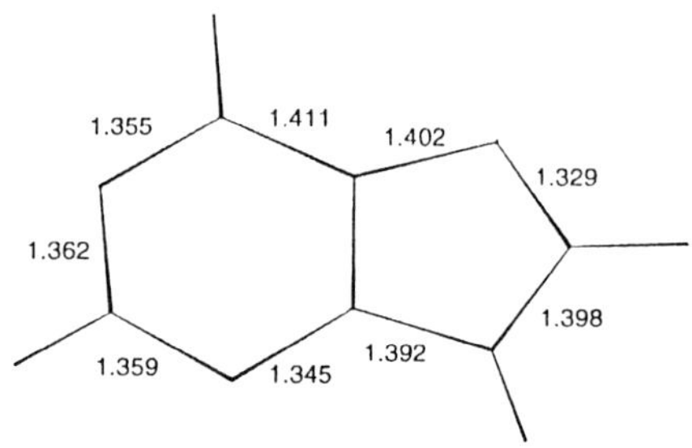

1

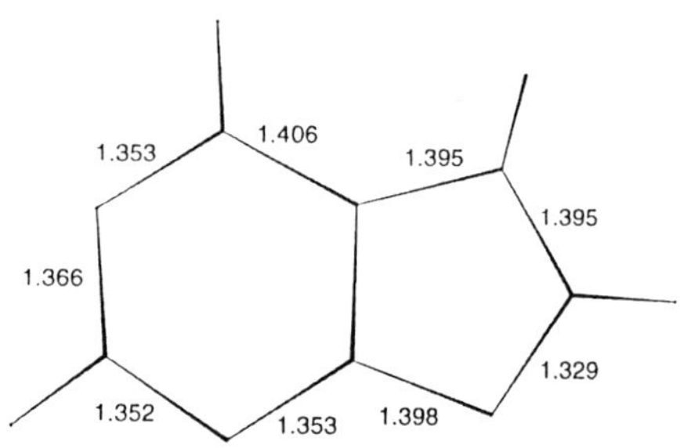

2

Figure 2: Quantum mechanically optimized structures of the N9-H (**1**) and N7-H (**2**) tautomeric forms of purine. All the interatomic distances are in Å.

Agreement Between Experimental and Calculated Values

The linear parameters of the correlation between the experimental and calculated principal values of the chemical shifts using different geometrical models are given in Table 3. The correlations have been produced only for the N7-H tautomeric form, because the data for the N9-H tautomeric form is incomplete and exhibit larger experimental errors due the overlap of the resonance lines in the sample containing both tautomers.

All of the models exhibit reasonably good agreement between experimental and calculated results. The standard deviations for the nitrogen shifts are approximately three times larger than those for the carbon shifts. This may be expected since the chemical shift range for nitrogen is more than three times larger than that for carbon. In both cases, the HB calculations exhibit the least amount of scatter from the experimental data. The calculations with the X-ray structures always produce the largest deviations from the experimental results, most likely due to lack of precision in the location of the protons.

Orientation of the Principal Values

Calculations of the chemical shifts allows for the determination of the orientation of the principal components of the chemical shift tensors. The calculations in purine agree with the well established results found in previous studies of aromatic compounds (*6, 24*). For the carbon tensors the most shielded component, δ_{33} lies perpendicular to the molecular plane, while, δ_{11} and δ_{22} are in the molecular plane. For the protonated carbons δ_{11} is oriented close to the radial direction, and δ_{22} is close to the tangential direction; for the quaternary carbons δ_{11} is approximately along the central bond of the molecule. For the nitrogen nuclei the orientation of the chemical shifts is similar to that of the carbons, but the direction of δ_{11} and δ_{22} depends on the hybridization of the atom. For non-protonoted nitrogens, δ_{11} is the tangential component and δ_{22} is the radial component, while the reverse is true for the protonated atoms. These orientations do not depend on the size of the aromatic ring.

Intermolecular effects on the principal values

Comparison of the calculated results given in columns three (N7-H/HB model) and six (N7-H/Hopt model) of Tables I and II, provides a measure of the importance of the intermolecular interactions on the principal components of the chemical shift tensors. The difference between these two calculations arise due to the surrounding molecular fragments used in the HB model, since both calculations use the X-ray geometry with optimized X-H distances for the central molecule. For both nitrogen and carbon chemical shifts, most of the effects of the intermolecular interactions are small. The only exceptions are the principal values in N9, which is acting as the proton acceptor in the HB model. This is the only non-protonated nitrogen with a close hydrogen bond contact in the plane of the molecule (see Figure 1). For this nitrogen δ_{11} and δ_{22} exhibit large shielding effects of 35 ppm and 14 ppm, while δ_{33}

Table III: Correlation parameters between experimental and calculated ^{15}N and ^{13}C chemical shift principal values in purine.[a]

	Model	Slope	Intercept	Std. Dev.
^{15}N	HB	1.05	10.5	23.9
	OPT	1.10	32.8	26.6
	Hopt	1.09	18.2	28.9
	X-ray	1.12	21.9	32.2
^{13}C	HB	0.88	12.5	7.8
	OPT	0.89	15.3	9.7
	Hopt	0.87	12.1	10.3
	X-ray	0.91	5.2	10.5

[a] RMS and intercept in ppm. Models described in the text.

shows a de-shielding of 15 ppm. Note that δ_{11} is in the plane of the molecule and close to the perpendicular direction of the HB interaction. The shifts induced by the HB are similar to those observed in other nucleic acid bases shifting the principal components of the non-protonated nitrogens towards the values observed in protonated nitrogens.

From the observations in purine and other nucleic bases (*4*) it appears that the components perpendicular to the direction of the HB bond are the most affected by the interactions. Unfortunately, it is difficult to secure quantitative information from observations in actual compounds because the direction of the HB interaction is not aligned with any of the principal components of the chemical shift tensors.

A better understanding can be gained from calculations in model systems in which the direction of the HB interactions can be aligned with one of the principal components by enforcing the symmetry of the model. A convenient model for these studies is the pyridine-methanol model given in Figure 3. Calculations were performed, as described in the Computational Procedures section, for N-H distances between 1.6 Å and 2.4 Å. The remaining geometrical parameters of the models were fully optimized and the results of these calculations are presented in Figure 4.

From Figure 4 it is apparent that the components perpendicular to the N-H direction, δ_{11} (in the plane along the tangential direction) and δ_{33} (perpendicular to the pyridine plane), are the ones affected most by the HB interaction. The tangential component, δ_{11}, changes by more than 25 ppm for the range of N-H distances studied here. The shielding in this direction increases with the strength of the HB, i.e. decreasing N-H distances. This can be explained by the HB stabilization of the n-π orbital, which makes this component less paramagnetic (*6*). Note that, as previously described, the tangential component in the protonated nitrogens is much more shielded, becoming δ_{22} (*6*). The component perpendicular to the pyridine plane exhibits the opposite behavior, but its range is smaller, ~20 ppm. The shielding along this direction is dominated by the less polarizable σ electrons and therefore is less affected than the tangential component, δ_{11}, which is dominated by the more polarizable π electrons. The component along the N-H direction, δ_{22} is very insensitive to the strength of the HB interaction, and changes only 4 ppm for the range of N-H distances studied here.

It is apparent from the calculations in the model pyridine-methanol system, that various principal components of the chemical shift tensor are affected differently by the HB. Consequentely, the principal values perpendicular to the direction of the interaction will exhibit a greater effect than the shielding components along the direction of the interaction. This general argument is in agreement with the qualitative observations in purine. A full 3D study of the interaction in the pyridine-methanol complex is required to quantitatively interpret the results in purine.

Conclusions

The NMR results confirm the X-ray determinations that the N7-H tautomer is the most stable in the solid and the DFT calculations agree with all previous evidence that the N9-H tautomer is the most stable in the gas phase. The calculations show good agreement with the experimental values. When intermolecular effects are included in the calculations, they produce significant improvements in the calculated ^{15}N chemical

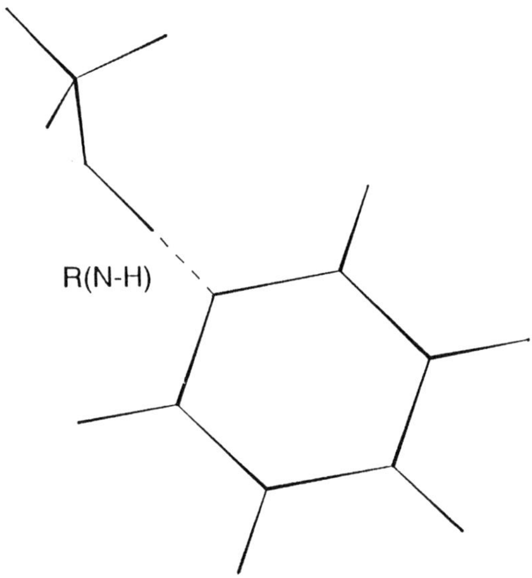

Figure 3: Pyridine-methanol dimer model used to evaluate the effect of the hydrogen bonding on the chemical shift principal components of aromatic nitrogens.

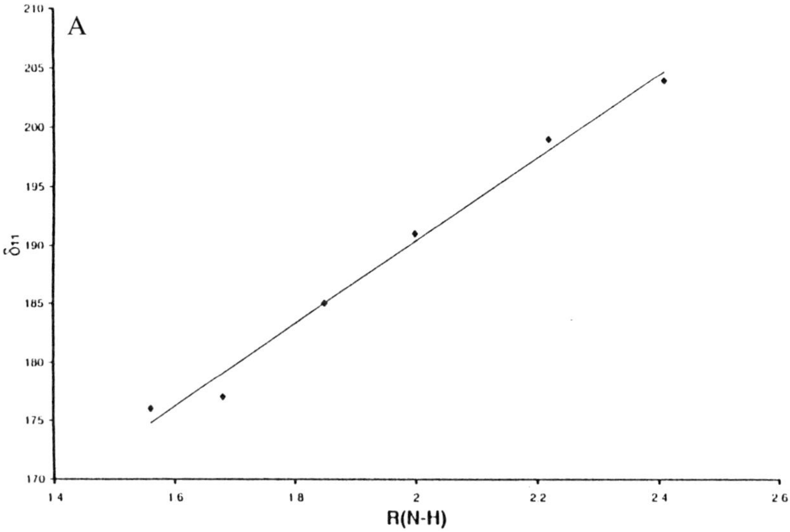

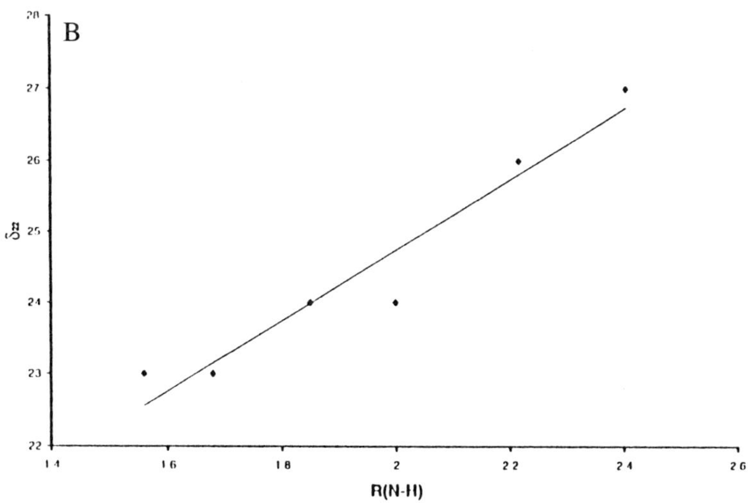

Figure 4: Dependence of the ^{15}N principal values of the chemical shift in pyridine with the N-H distance for the pyridine-methanol complex. Note the different chemical shift scales for each shift component, δ_{11}, δ_{22}, δ_{33} and δ_{iso}. The N-H distances are in Å and the horizontal scales in the figure are the same for all the shift components.

Continued on next page.

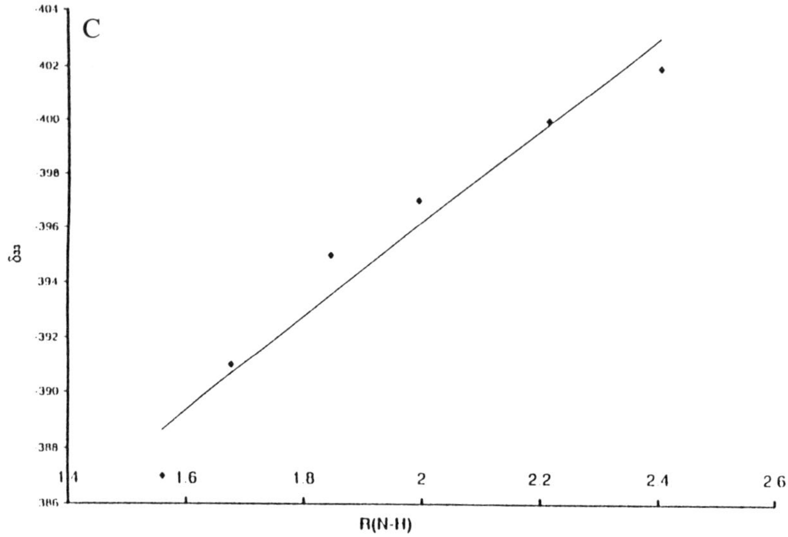

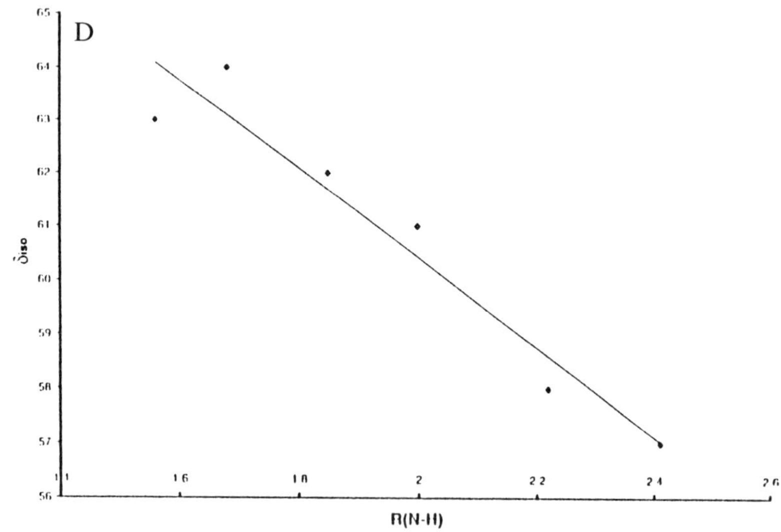

Figure 4. *Continued.*

shifts tensors. The ^{13}C chemical shifts are also well reproduced, but they are much less sensitive to the intermolecular interactions.

The ^{15}N chemical shift tensors in purine are similar to those in other nucleic acid bases for similar types of nitrogens. Comparison of the calculated values in purine for the isolated molecule and for the molecule with its nearest neighbors shows well defined trends, which could be rationalized using the results in the model pyridine-methanol complex.

Because the direction of the HB interaction (N-H direction) is not co-linear with any of the principal components of the ^{15}N chemical shifts tensor, a complete 3D model of the interaction in model systems is required to secure a quantitative model of the intermolecular effects.

Acknowledgments. Financial support was provide by NIH grant GM-08521-37 and DOE grant DE-FG02-94ER-1445. The authors also acknowledge a generous allocation of computer time at the CHPC for the use of the IBM SP system partially funded under NSF # CDA9601580 and by IBM in their SUR grant to the University of Utah. The SGI O2000 system was partially funded by the SGI Supercomputing Visualization Center.

Literature Cited.

1. Calladine, C. R.; Drew, H. R. *Understanding DNA,* Academic Press, New York, NY, 1992.
2. Hu, J. Z.; Orendt, A. M.; Alderman, D. W.; Ye, C.; Pugmire, R. J.; Grant, D. M. *Solid State NMR* **1994**, *3*, 181.
3. Hu, J. Z.; Wang, W.; Liu, F.; Solum, M. S.; Alderman, D. W.; Pugmire, R. J.; Grant, D. M. *J. Magn. Reson.* **1995**, *A113*, 210.
4. Hu, J. Z.; Facelli, J. C.; Alderman, D. W.; Pugmire, R. J.; Grant, D. M. *J. Am. Chem. Soc.* **1998**, *120, 9863.*
5. Anderson-Altmann, K. L.; Phung, C. G.; Mavromoustakos, S.; Zheng, Z.; Facelli, J. C.; Poulter, C. D.; Grant, D. M. *J. Phys. Chem.* **1995**, *99*, 10454.
6. Solum, M. S.; Altman, K.; Strohmeier, M.; Berges, D. A.; Zhang, Y.; Facelli, J. C.; Pugmire, R. J.; Grant, D. M. *J. Am. Chem. Soc.* **1997**, *119,* 9804.
7. Grant, D. M.; Liu, F.; Iuliucci, R. J.; Phung, C. G.; Facelli, J. C.; Alderman, D. W. *Acta Crystall.* **1995**, *B51*, 540.
8. Facelli, J. C.; Pugmire, R. J.; Grant, D. M. *J. Am. Chem. Soc.* **1996**, *118,* 5488.
9. Nowak, M. J. ; Rostkowska, H.; Lapiski, L.; Kwiatkowski, J. S.; Leszczyński *J. Phys. Chem.* **1994**, *98*, 2813.
10. Lin, J.; Yu, C.; Peng, S.; Akiyama I.; Li, K.; Lee, L. K.; LeBreton, P. R. *J. Am. Chem. Soc.* **1980**, *102,* 4627.
11. Watson, D. G.; Sweet, R. M.; Marsh, R. E. *Acta Crystall.* **1965**, *19*, 573.
12. Houben, L.; Schoone, K.; Smets, J.; Adamowicz, L.; Maes, G. *J. Mol. Struct.* **1997**, *410-411*, 397.
13. Gaussian 94 (Revision D.3), Frisch, M. J.; Trucks, G. W.; Schlegel, H. B.; Gill, P. M. W.; Johnson, B. G.; Robb, M. A.; Cheeseman, J. R.; Keith, T. A.; Petersson, G. A.; Montgomery, J. A.; Raghavachari, K.; Al-Laham, M. A.;

Zakrzewski, V. G.; Ortiz, J. V.; Foresman, J. B.; Cioslowski, J.; Stefanov, B. B.; Nanayakkara, A.; Challacombe, M.; Peng, C. Y.; Ayala, P. Y.; Chen, W.; Wong, M. W.; Andres, J. L.; Replogle, E. S.; Gomperts, R.; Martin, R. L.; Fox, D. J.; Binkley, J. S.; Defrees, D. J.; Baker, J.; Stewart, J. P.; Head-Gordon, M.; Gonzalez, C.; Pople, J. A. Gaussian, Inc. Pittsburgh PA, 1995.

14. Dunning, T. H.; Hay, P. J. In *Modern Theoretical Chemistry*, Schaefer III, H. F. Ed.; Plenum: New York, NY, 1976, p. 1.
15. Cheeseman, J. R.; Trucks, G. W.; Keith, T. A.; Frisch, M. J. *J. Chem. Phys.* **1996**, *104*, 5497.
16. Lee, C.; Yang, W.; Parr, R. G. *Phys. Rev.* **1988**, *B37*, 785 and Becke, A. D. *Phys. Rev.* **1988**, *A38*, 3098.
17. Wolinski, K.; Hinton, J. F.; Pulay, P.; *J. Am. Chem. Soc.* **1990**, *112,* 8251.
18. Ditchfield, R. *Mol. Phys.* **1974**, *27*, 789.
19. London, F. *J. Phys. Radium* **1937**, *8*, 397.
20. Jameson, C. J.; Mason, J. In *Multinuclear NMR*, Mason, J., Ed.; Plenum: New York, NY, 1987, p. 56.
21. Jameson, A. K.; Jameson, C. J. *Chem. Phys. Lett.* **1981**, *134,* 461.
22. Witanowski, M.; Stefaniak, L.; Webb, G. A. In *Ann. Reports on NMR Spectrosc.*, Webb, G. A., Ed.; Academic Press: London, 1993, Vol. 25; pp 1-469
23. Pugmire, R. J.; Grant, D. M.; Townsend, L. B.; Robins, R. K. *J. Am. Chem. Soc.* **1973**, *95,* 2791.
24. Facelli, J. C.; Grant, D. M. *Nature*, **1993**, *365*, 325.

Chapter 13

Effects of Hydrogen Bonding on ^{1}H Chemical Shifts

Yufeng Wei and Ann E. McDermott

Department of Chemistry, Columbia University, New York, NY 10027

The topic of "low-barrier hydrogen bonds" (LBHBs) and the question of how they are involved in enzyme function has been discussed heavily in the literature recently. Hydrogen bonds between two bases of nearly matched proton affinity often exhibit strongly perturbed bond lengths and spectroscopic parameters; it remains somewhat unclear exactly how the spectroscopic parameters reflect total energy or reactivity. In this study, we report ^{1}H NMR chemical shift data and surveys of structural preferences for the well-studied O-H···O systems, and also for less studied, but biologically important N-H···O systems, in particular the imidazole and imidazolium functionality. The ^{1}H shifts also show interesting trends in comparison with O-H···O motifs, which will require further scrutiny.

Hydrogen bonds appear to be essential in all enzyme-catalyzed reactions, although why they are essential and how they promote function is an open question. In recent years a specific hypothesis for their involvement in catalysis has emerged: so-called low-barrier hydrogen bonds (LBHB) have been proposed to lower the transition state energy for many enzymatic reactions, including those of serine protease, citrate synthase, triosephosphate isomerase(TIM), Δ^5-ketosteroid isomerase, ribonuclease A (RNase A), and mandelate racemase.(*1-5*) The transition states or intermediates, but not the ground states, putatively benefit from a 10-20 kcal/mol LBHB interaction; this proposal has been the cornerstone for a hypothesis about enzyme reactivity.(*2,4*) The LBHB is defined as a particularly short hydrogen bond that forms when the two bases competing for a single proton have nearly matched proton affinities. The short bond gives the LBHB many special spectral properties and sensitivity to isotopic substitution: highly downfield ^{1}H chemical shifts (16-22 ppm), small or inverted isotope effects in infrared vibrational frequencies ($\nu_{AH}/\nu_{AD}\approx1.0$), and low isotope fractionation factors (0.5-1.0) are observed.(*6*) Extensive studies and debate have appeared in the literature since the idea of LBHB was first proposed.(*7-11*) A key question in the debate is the energetics of a short hydrogen bond (or a LBHB) as contrasted with more typical hydrogen bonds between two groups of different proton affinity.

Several studies have shown that a pK_a match provides only moderate energy (~5 kcal/mol) in solution.(*8,12-15*) *Ab initio* calculations also confirmed this conclusion.(*16*) A linear correlation between ΔG^{HB} and ΔpK_a was proposed in both experimental and computational studies where the dimensionless Brønsted slope β is defined as the slope relating $\log K^{HB}$ and ΔpK_a. The experimental value for β was reported to be ~0.7 in DMSO and 0.05-0.14 in water.(*8,15,16*) In contrast, gas phase studies showed 30-40 kcal/mol energy gains when the pK_a values are matched (implying $\beta >> 1$).(*6*) We have focused on similar questions regarding the solid or crystalline state, because we propose that the crystalline database is a relevant database for enzyme active sites and enzyme-substrate interactions. In particular, we are addressing the question of the bond lengths and proton chemical shifts when hydrogen bonds in the crystalline states involve matched vs. mis-matched pK_a's, and we also discuss the question of their energies.

NMR spectroscopy is probably the most popular spectroscopic tool used to study hydrogen bonding. The ^{1}H chemical shift is a well-established indicator of hydrogen bonding; NMR signatures for hydrogen bonding were established very soon after the invention of NMR. The ^{1}H shifts have also been used extensively by enzymologists. Strongly deshielded signals (ca. 16-20 ppm) and anomalous isotope effects and fractionation factors are observed in species with very short hydrogen bonds, for both small molecules(*17-22*) and enzymes [reviews;(Mildvan, A. S.; Harris, T. K.; Abeygunawardana, C. *Methods Enzymol.* **1998**, *in press.*) serine protease;(*5,23-26*) aspartate aminotransferase;(*27,28*) triosephosphate isomerase (TIM);(*29*) ketosteroid isomerase;(*30,31*) AKB-ligase(*32*)]. Many of these measurements for small crystalline compounds were performed using solid state NMR methods, such as Combined Rotation And Multiple Pulse Spectroscopy (CRAMPS).(*33-37*) Previous databases focused mainly on O-H···O systems; we have established an additional database of high resolution solid state ^{1}H NMR chemical shifts of the imidazole-carboxylate hydrogen bonding system, which is much more relevant for enzymatic studies. In an effort to understand our NMR shifts, we have also summarized the hydrogen bonding structural trends of O-H···O and N-H···O systems by surveying the Cambridge Structural Database (CSD).(*38-40*)

Results and Discussion

^{1}H NMR chemical shifts in solid state. Harris, *et. al.*, reported many ^{1}H chemical shifts of acidic protons in carboxylic acids;(*20*) McDermott and Ridenour also reported some ^{1}H chemical shifts in amino acids and small peptides.(*21*) In Figure 1a, we plot the ^{1}H chemical shifts from literature(*20,21*) against the corresponding hydrogen bonding O···O distances from crystal structures from the CSD. We have organized the data according to the pK_a difference between the partners. It is clear that the pK_a matched and mis-matched species are distinct in terms of their chemical shifts and hydrogen bonding distances. All the matched species (COO-H···COO$^-$) are short in distance (2.4 - 2.6 Å) and deshielded in chemical shift (14 - 21 ppm). A good correlation can be found between ^{1}H chemical shifts and hydrogen bond distances (R^2 = 0.763). In contrast, the mis-matched species (COO-H···COOH) have longer distances (2.6 - 2.7 Å), and more shielded chemical shifts (12 - 14 ppm). The correlation between ^{1}H chemical shifts and hydrogen bond distance in the mis-matched range is weak (R^2 = 0.191). The outlying datum in the mis-matched group is for maleic acid, which adopts a short intramolecular hydrogen bond (2.502 Å), and a deshielded ^{1}H chemical shift (16.6 ppm). Despite this outlying point, we conclude that pK_a matched species exhibit special properties.

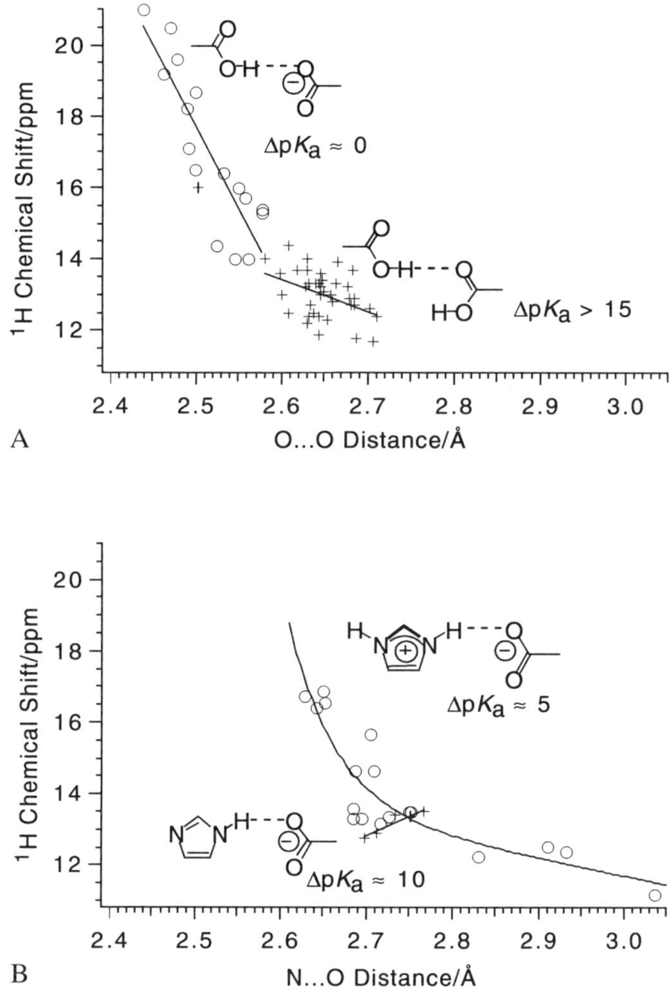

Figure 1. Solid state ^{1}H NMR chemical shifts of carboxylic acids and imidazoles are plotted against the hydrogen bonding distance: a) data for carboxylic acids (data from reference (*20,21*)), circles are compounds with COO—H ··· COO⁻ motif, $\Delta pK_a \approx 0$, and crosses are compounds with COO—H ··· COOH motif, $\Delta pK_a > 15$; b) data for imidazoles (data from Table I), circles present cationic imidazoles (Im^+—H- ··· COO^-, $\Delta pK_a \approx 3$), and crosses are neutral imidazoles (Im ··· COO^-, $\Delta pK_a \approx 10$).

Table I. Solid State ^{1}H NMR Chemical Shifts of Imidazoles and Hydrogen Bonding Distances in Crystals

	Compound	^{1}H Chemical Shifts δ_H/ppm	N···O Å	CSD Refcode
Neutral species	L-Carnosine	12.8	2.698	BALHIS01
	DL-histidine (monoclinic)	13.5	2.767	DLHIST
	L-His-L-Leu	12.9	2.713	JUKMOR
	L-histidine (monoclinic)	13.4	2.733	LHISTD02
	L-histidine (orthorhombic)	13.3	2.752	LHISTD13
Cationic species	DL-histidine hydrochloride dihydrate	13.3	2.726	DCHIST
	L-histidine perchlorate	12.4 11.2	2.933 3.037	GASKAM
	L-histidine hydrochloride monohydrate	12.3 16.4	2.830 2.642	HISTCM12
	L-histidine L-aspartate	14.7	2.687	LHLASP01
	L-histidine acetate (monoclinic)	13.3 13.3	2.685 2.694	POPGUW
	L-histidine hydrogen oxalate	16.7 12.5	2.627 2.910	RARXOX
	L-His-Gly hydrochloride	16.9	2.650	RAVMAC
	Gly-L-His hydrochloride	13.6	2.685	TEJGAO
	DL-histidine glycolate	13.5 13.5	2.749 2.752	TEJVUZ
	L-histidine glycolate	16.6 15.7	2.651 2.704	TEJWAG
	L-histidine squarate	14.7 13.2	2.709 2.716	TIWXEC

In this paper we report for the first time our studies of hydrogen bonding in the imidazole-carboxylate system, a common motif in enzyme active sites. The ^{1}H solid state NMR data of polymorphic crystalline histidines were taken on a Chemagnetic CMX-400 spectrometer using the BR-24 sequence,(*34,36*) with pulse length 2.2 μs, and magic angle spinning (MAS) speed at 2.0 kHz. The results are listed in Table I along with the hydrogen bonding N···O distances according to their ionization state. The reason we emphasize the ionization states is that the neutral imidazole species have quite different pK_a values (p$K_a \approx 14$) from the cationic ones (p$K_a \approx 7$). In comparison with the carboxylic acids, which are the hydrogen bond acceptors in this system, the neutral imidazole species are far mis-matched in proton affinity (Δp$K_a \approx$ 9-11), while the cationic species are nearly matched (Δp$K_a \approx$ 2-4). In Figure 1b, ^{1}H chemical shifts of imidazoles were plotted against the hydrogen bonding N···O distance. The cationic species behave normally: the ^{1}H chemical shifts become more deshielded when hydrogen bonding distances shrink. It is worth noting in comparison with these data that the ^{1}H shifts for the $H^{N\delta}$ in cationic imidazole of His-57 at the active site of α-chymotrypsin exhibit a 18-19 ppm signal: 18.0 ppm for free enzyme at pH 4.1;(*23*) 18.7 ppm in *N*-acetyl-L-Leu-L-Phe trifluoromethyl ketone inhibited complex at pH>7.6;(*23*) 18.9 ppm in *N*-acetyl-L-Leu-DL-Phe trifluoromethyl ketone inhibited complex at pH 7.0;(*26*) 18.6 ppm in *N*-acetyl-DL-Phe trifluoromethyl ketone inhibited complex at pH 7.0(*26*). Thus we tentatively conclude that under some circumstances the imidazolium carboxylate pair in serine proteases acquires a rather short N···O bond length of about 2.6Å. We are unaware of any other instance of imidazole proton shifts in enzymes at such deshielded values, other than the serine protease family.

For neutral species the plot does not resemble the known hydrogen bonding trend at all. As the hydrogen bond stretches, the ^{1}H chemical shifts become more deshielded, and thus the plot exhibits a positive slope. The reason for the unusual behavior of the neutral species is still unclear, but the fact that the pK_a is very mis-matched is probably relevant. Ultimately, we want to rationalize these chemical shift trends in terms of structure and hydrogen bonding, using *ab initio* calculations. This is an ambitious goal for the moment. Our first step towards this will be to achieve a qualitative understanding of the hydrogen bonding trends in matched and mis-matched systems by carefully surveying structural trends. These are reviewed in the following.

Hydrogen bonding geometry in crystals. A survey of the N-H···O and O-H···O hydrogen bonding systems of small crystalline molecules in the CSD has been performed. The O-H···O motif is a classic case, and many studies have focused on it. Fewer studies have focused on the biologically relevant N-H···O motif, and so we have selected that motif for further study. Because of the diversity of hydrogen bonding definitions, we chose to use the simplest one: two groups are said to be involved in a hydrogen bond when the distance between the two heteroatoms is smaller then the sum of their van der Waals radii. When we mention the proton affinity or pK_a values for two functional groups, we are referring to the relative typical values in aqueous solutions (despite the fact that the measurements we report are in the crystalline state). We ignore the question of whether the absolute proton affinity is important (i.e. strong vs. weak bases), and simply analyze structure and spectroscopy as a function of difference in affinity of the two sites.

We began by establishing the simple fact that the pK_a matching condition strongly affects the hydrogen bonding geometry. In the O-H···O systems, for example, the distribution of O···O distances in carboxyl-carboxyl (COO-H···COOH) and carboxylate-carboxyl ($COO-H\cdots COO^-$) pairs differs dramatically (Figure 2). Here, carboxyl-carboxyl is a far mis-matched case ($\Delta pK_a > 15$), and carboxyl-carboxylate is matched ($\Delta pK_a \approx 0$). As one can see in the histograms, which include both X-ray and neutron diffraction structures, the center of the distance distribution for all O-H···O cases is 2.8 Å; in the carboxyl-carboxyl system, the center moves to 2.65 Å; when the pK_a values become closely matched in the case of carboxylate-carboxyl, the most probable O···O distance shortens to 2.5 Å. Apparently it does not matter if the hydrogen bonding partners are *syn* or *anti*, the hydrogen bonding distances are all in the same region. An analogous trend relating O···O distance to ΔpK_a can be seen for water species. The distribution for asymmetric and pK_a mis-matched $HO-H\cdots H_2O$ is centered at 2.8 Å, while $H_2O^+-H\cdots H_2O$ species, with $\Delta pK_a \approx 0$, is centered at 2.45 Å, and includes the shortest O···O distances found in any O-H···O system.

We looked at several chemical motifs to test whether the O-H covalent bond length is perturbed by interactions involving two pK_a matched bases. Figure 3 displays the "stretching" of O-H bond lengths in O-H···O hydrogen bonding systems. We focused on structures involving two motifs, carboxyl and water. Only neutron diffraction data are included in these plots, because the hydrogen position in X-ray structures is poorly defined. The master plot of all O-H···O hydrogen bonds is well-known and well documented.(*6*) As O···O distances shorten, O-H bond lengths are stretched. When we look at the individual systems in detail, we can isolate motifs that are "matched" or "mis-matched" in pK_a. The symmetric systems or "matched" cases (with $\Delta pK_a \approx 0$), included carboxyl-carboxylate ($COO-H\cdots COO^-$) and hydroxium-

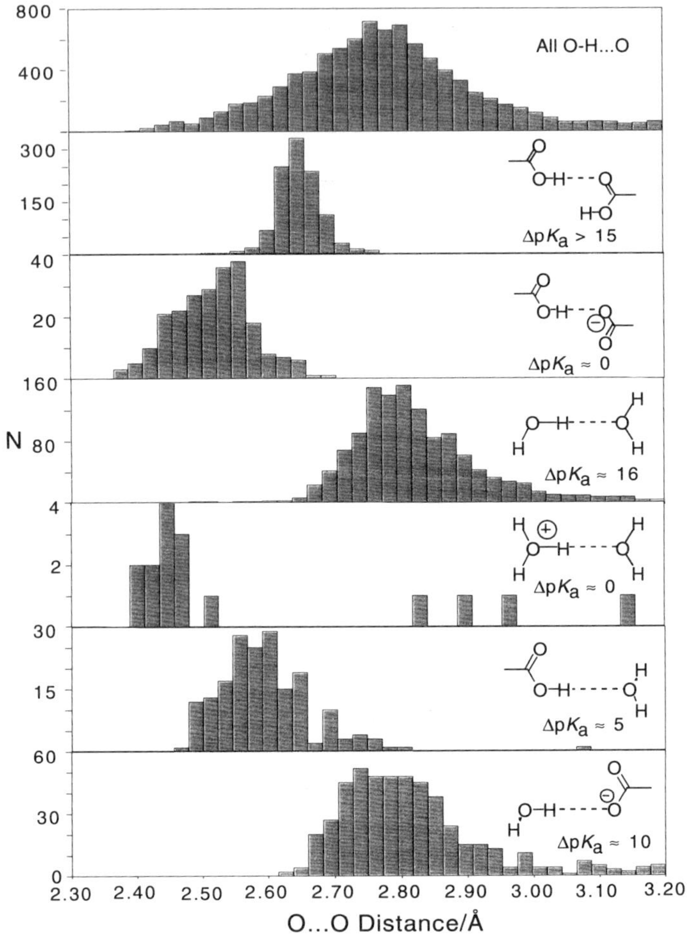

Figure 2. O ⋯ O hydrogen bond length histograms indicate a difference in the intermolecular potential for various chemical species. Closely matched (symmetric) hydrogen bonds clearly adopt shorter O ⋯ O distances. Both X-ray and neutron diffraction data are from the CSD. From top to bottom: all O—H ⋯ O containing species; carboxyl-carboxyl, $\Delta pK_a > 15$; carboxyl-carboxylate, $\Delta pK_a \approx 0$; water-water, $\Delta pK_a \approx 16$; hydroxium-water, $\Delta pK_a \approx 0$; carboxyl-water, $\Delta pK_a \approx 5$; water-carboxylate, $\Delta pK_a \approx 10$. The pK_a values used here are based on typical aqueous values.

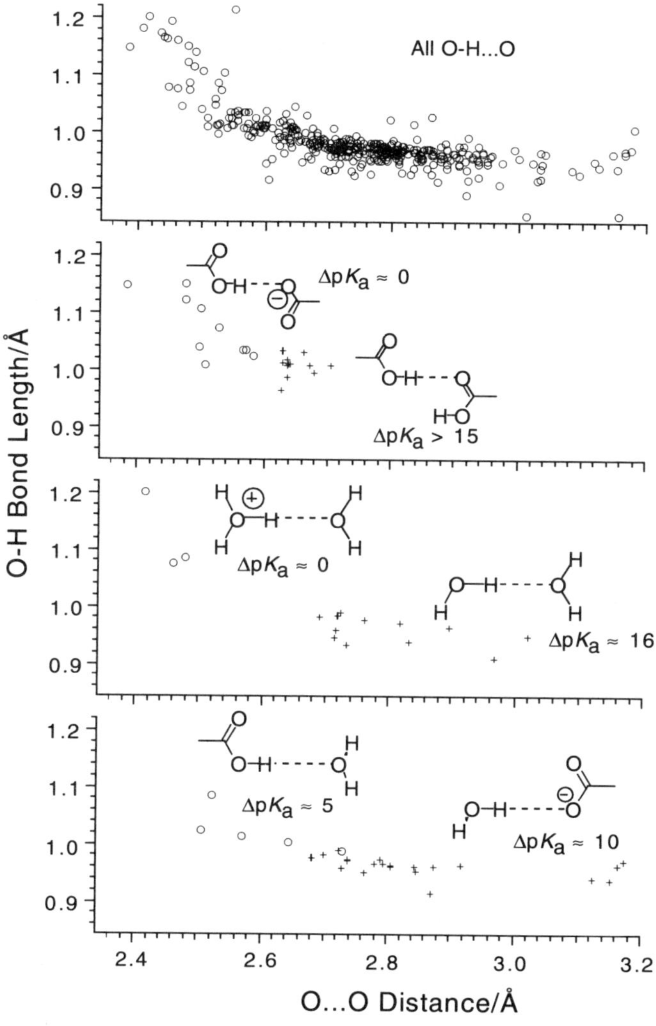

Figure 3. O—H bond lengths in O—H···O hydrogen bonds. The phenomenon of long O—H bonds appdars to be primarily for the cases in which $\Delta pK_a \approx 0$. Neutron diffraction data are from the CSD. From top to bottom: all O—H···O; carboxyl-carboxylate, $\Delta pK_a \approx 0$ (circles), and carboxyl-carboxyl, $\Delta pK_a > 15$ (crosses); hydroxium-water, $\Delta pK_a \approx 0$ (circles), and water-water, $\Delta pK_a \approx 16$ (crosses); carboxyl-water, $\Delta pK_a \approx 5$ (circles), and water-carboxylate, $\Delta pK_a \approx 10$ (crosses). The pK_a values used here are based on typical aqueous values.

water (H_2O^+-H···H_2O). The O···O distances are shortest in matched cases, from 2.4 - 2.6 Å, and O-H bond lengths are significantly stretched, 1.0 - 1.2 Å. The structurally similar, but mis-matched systems (asymmetric systems), including carboxyl-carboxyl (COO-H···COOH) and water-water (HO-H···H_2O), have $\Delta pK_a > 15$. The O···O distances are long (>2.6 Å), and O-H bond lengths are in normal range (0.9 - 1.0 Å). In contrast to matched cases, the relation between O-H bond length and O···O distance is poor for the mis-matched cases.

It could be argued that the O···O distances and O-H bond length are not perturbed so much by pK_a match, but by the total charge of the system, because the shorter O···O distances and longer O-H bond lengths appear in charged species as described above. To distinguish the effects of pK_a match from total charge, we selected two motifs: one is carboxyl-water (COOH···H_2O), a neutral but near matched pair ($\Delta pK_a \approx 5$); the other is water-carboxylate (H_2O···COO^-), charged and more mis-matched pair ($\Delta pK_a \approx 10$). The central distribution of O···O distances for the neutral, nearly pK_a matched carboxyl-water motif is short (2.6 Å), compared with the charged, pK_a mis-matched water-carboxylate system (2.8 Å), as seen in Figure 2. In the plot correlating O-H and O···O distances (Figure 3), the carboxyl-water system shows longer O-H bond lengths and shorter O···O distances than the water-carboxylate system. Apparently, it is the pK_a match, not the total charge, that correlates with the perturbation of O···O distances and O-H bond lengths.

N-H···O hydrogen bonding is more common than O-H···O bonding in enzyme active sites. Figure 4 shows the histograms of the distribution of N···O distance in several N-H···O hydrogen bonding systems, including both X-ray and neutron diffraction structures. The pK_a values of the imidazolium-carboxylate (Im^+-H···COO^-) motifs are the most closely matched, with $\Delta pK_a \approx 3$. The center of the histogram of N···O distances is 2.7 Å. The next two cases, ammonium-carboxylate (R_3N^+-H···COO^-, R=H or C) and amide-amide, with ΔpK_a values about 5 and 15 respectively, exhibit N···O histograms centered at 2.8 and 2.9 Å, respectively. The last case, the amide-water system, is generally not a good hydrogen bonding system, with $\Delta pK_a \approx 17$. The distribution of this system is pretty broad, compared with other systems, evenly ranging from 2.7-3.0 Å, with a small peak at 2.85 Å. The master plot of all N-H···O systems, is centered at 2.9 Å. It is clear that when the pK_a matching condition achieved, the hydrogen bonding distance between the two heteroatoms (N···O) is shorter.

The stretching of the N-H covalent bond in N-H···O systems, involving N-H bond lengths of 0.95-1.10 Å, is not very significant (Figure 5), in comparison with the O-H···O system, 0.85-1.20 Å (Figure 3) or in comparison with experimental error. The reason for the small variation in N-H bond lengths is probably that there are no or few exactly-matched structures reported in the CSD. We noted a difference between near-matched species (R_3N^+-H···COO^-, $\Delta pK_a \approx 5$) and mis-matched species (neutral aromatic heterocycles···COO^-, $\Delta pK_a \approx 10$, and amide-amide, $\Delta pK_a \approx 15$). The data for near-matched species are mainly in the upper-left region in the master plot, and those for the mis-matched species make up the middle to right part of the master plot. The pK_a matching condition could still be an important factor in determining the length of N-H···O hydrogen bonds in crystalline state, but these data do not present a compelling cases. Future NMR studies will focus on N-H bond length for pK_a matched systems.

Energetics of hydrogen bonding in crystals. As an indication of the formation energy of a hydrogen bond, we counted the percentage of structures with

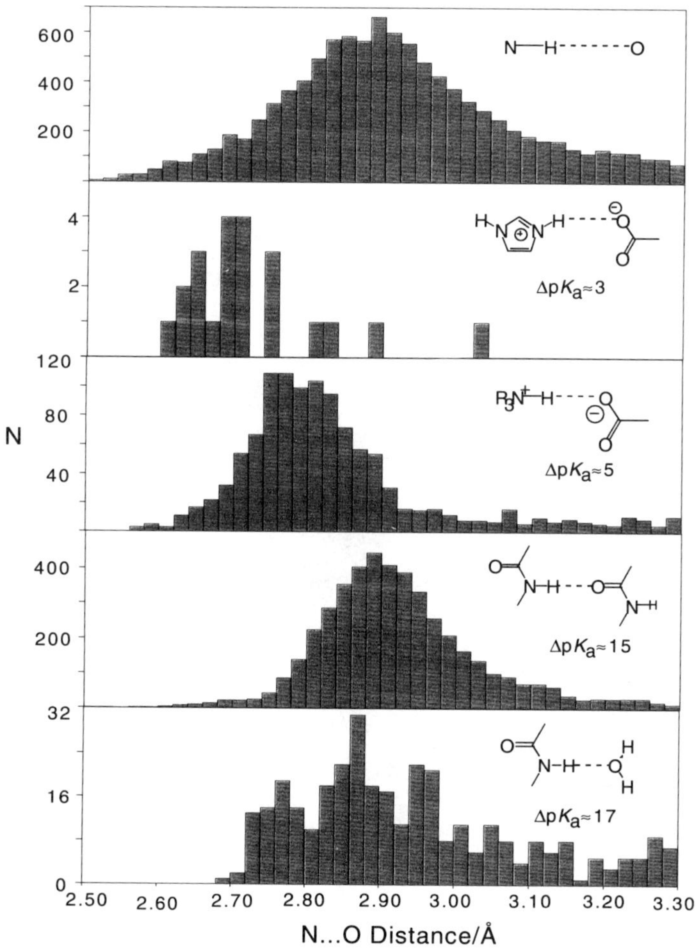

Figure 4. Histograms of N ··· O hydrogen bonding distances. A systematic dependence of N ··· O distance on ΔpK_a appears in these data. Both X-ray and neutron diffraction data are from the CSD. From top to bottom: all N—H ··· O; imidazolium-carboxylate, $\Delta pK_a \approx 3$; ammonium-cargoxylate, $\Delta pK_a \approx 5$; amide-amide, $\Delta pK_a \approx 15$; amide-water, $\Delta pK_a \approx 17$. The pK_a values used here are based on typical aqueous values.

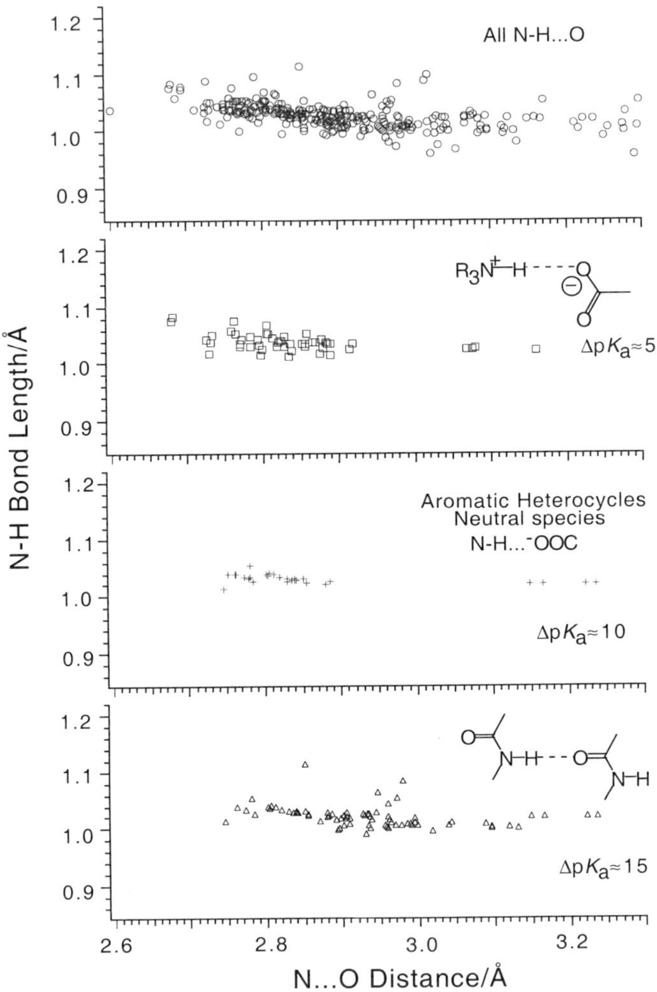

Figure 5. N−H bond lengths in N−H ··· O hydrogen bonding systems. The phenomenon of N−H bond "stretching" is only weakly represented (if at all), probably because there are only few nearly matched systems that have been characterized by neutron diffraction. Neutron diffraction data are from the CSD. From top to bottom: all N−H ··· O; ammonium-carboxylate, $\Delta pK_a \approx 5$; neutral heterocycle-carboxylate, $\Delta pK_a \approx 10$; amide-amide, $\Delta pK_a \approx 15$. The pK_a values used here are based on typical aqueous values.

and without hydrogen bonding motifs in the CSD when the specific hydrogen bonding motifs were presence in the structure. The statistical data are tabulated in Table II. For the matched pK_a cases, like H_2O^+-H···H_2O and COO-H···COO^-, the percentage of structures with hydrogen bonding was 80% to 100%; for near matched cases (R_3N^+-H···COO^-), the percentage was about 70-80%; while for the mis-matched cases (amide···amide, COOH···COOH and H_2O···H_2O) the percentage of structures with hydrogen bonding was only 20-40% (concerning neutron diffraction data for amide-amide motifs, a larger number of hydrogen bonds (70%) may be due to a lack of other hydrogen bonding possibilities). The percentage of "missed opportunities" for hydrogen bonding becomes larger when ΔpK_a increases, which we loosely interpret to mean that a hydrogen bond with pK_a match is energetically more favored than one with mis-matched pK_a, but not dramatically so. To extract an enthalpy from such a study would be ridiculous, but the enthalpy appears not to be dominant over other crystal packing considerations.

Scheme 1

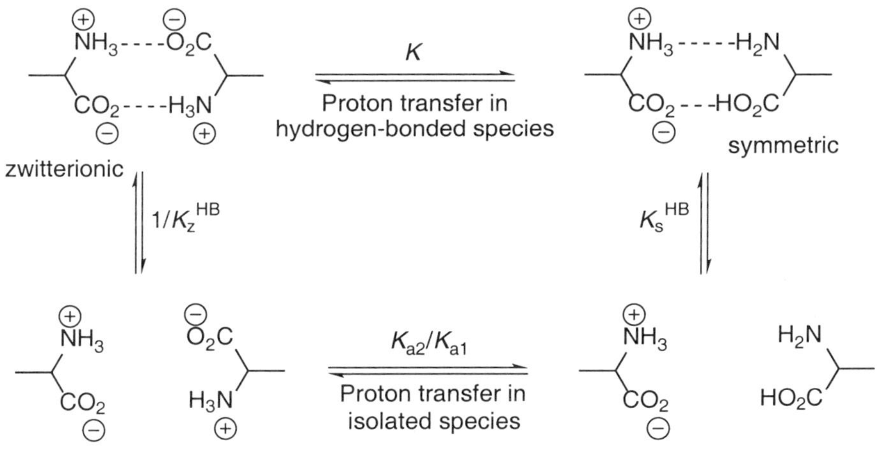

We add a few comments on the linear relationship between the ΔG^{HB} and ΔpK_a(*8,15,16*) in the crystalline state. The following argument is based on the observation that formation of LBHB does not drive uphill proton transfers in the crystalline state under any known circumstances (and has not been reported in solution). Based on this observation, using a thermodynamic cycle, we conclude that $\beta << 0.5$. Consider an amino acid which undergoes an equilibrium as illustrated in a thermodynamic cycle in Scheme 1, where the equilibrium constant for the reaction can be written as

$$K = \left(\frac{1}{K_z^{HB}}\right)^2 \cdot \frac{K_{a2}}{K_{a1}} \cdot \left(K_s^{HB}\right)^2 \qquad (1)$$

Here K describes a proton transfer in the hydrogen bonded zwitterion pair to generate two symmetric hydrogen bonds or LBHBs. Equation (1) can be rewritten as

$$\log K = 2(\log K_s^{HB} - \log K_z^{HB}) - \Delta pK_a \qquad (2)$$

Table II Hydrogen Bonding Preferences: Symmetric vs. Asymmetric

	Both X-ray and Neutron Structures				Neutron Structures Only			
	Total Number	Number with HB	Number without HB	Percentage of HB	Total Number	Number with HB	Number without HB	Percentage of HB
H_3O^+-H_2O $\Delta pK_a \approx 0$	15	13	2	86.7%	3	3	0	100%
COO^--COOH $\Delta pK_a \approx 0$	363	278	85	76.6%	15	12	3	80%
R_3NH^+-COO^- $\Delta pK_a \approx 5$	1007	690	317	68.5%	34	29	5	85.3%
Amide-Amide $\Delta pK_a \approx 15$	4964	2851	2113	42.6%	57	40	17	70.2%
COOH-COOH $\Delta pK_a > 15$	2402	861	1541	35.8%	57	13	44	22.8%
H_2O-H_2O $\Delta pK_a \approx 17$	2492	678	1814	27.2%	44	10	34	22.7%

where $\Delta pK_a = pK_{a2} - pK_{a1}$. Assuming a linear free energy correlation, we have $\log K_s^{HB} - \log K_z^{HB} = \beta \Delta pK_a$. Substituting this into equation (2), we get

$$\log K = (2\beta - 1)\Delta pK_a \tag{3}$$

It is well-known that amino acids are in the zwitterionic state in essentially every crystalline solid, not in a mixed neutral/zwitterionic form. Therefore the equilibrium strongly favors the reverse direction, i.e. $\log K << 0$. Thus, we conclude

$$\beta << 0.5 \tag{4}$$

This argument probably also applies to most solvent environments, as such mixed neutral/zwitterionic dimers are unknown: generally the species are either charged or neutral, but not in mixed dimers. We present this thermodynamic cycle because the resulting estimate for β is small compared some solution estimates.(*8,15,16*) If these crystalline and solution systems were realistic models for the transition state and ground state of a reaction, our argument would put a limit on the role of LBHB in enzymatic catalysis. However, our observation may have validity only for ground states, and studies of reactions and transition states may require a more sophisticated analysis.

Conclusions

In recent studies of enzymatic catalysis, deshielded ^{1}H chemical shifts were frequently used as an indication of the presence of "low barrier hydrogen bonds", which are symmetric (matched in pK_a) hydrogen bonds and/or short hydrogen bonds. Such hydrogen bonds were usually considered as energetically strong. We have discussed some important characteristics for nearly pK_a matched hydrogen bonding partners as compared with mis-matched base pairs, including short O···O or N···O distances, "stretched" O-H or N-H bond lengths, deshielded ^{1}H chemical shifts, and altered formation energy. We have concluded that the bond length and spectroscopy are strongly perturbed by pK_a match, but energetics seem to be only slightly perturbed. When the pK_a values are almost matched, then the O···O and N···O distances are obviously short and the O-H covalent bond length is stretched dramatically. The phenomenon of stretched N-H bonds is poorly documented, but we speculate that this is because there are so few neutron diffraction data on pK_a matched cases. The ^{1}H chemical shifts for O-H···O and N-H···O motifs are extensively perturbed by pK_a match. For both N-H···O and O-H···O systems, different slopes with hydrogen bonding distances can be observed for different ΔpK_a values (3-5 vs. 10). The interpretation of ^{1}H chemical shifts for N-H···O systems clearly differs from those for O-H···O systems, and imidazole differs from imidazolium, and this fact should be considered in the context of NMR studies of enzymes.

We propose that crystals provide a useful model for the active sites of enzymes, to supplement data from the solution state, because enzyme active sites are ordered, and chemically complex. But what happens in enzyme active sites regarding the transition state is of course unclear to us, and issues of reactivity in hydrogen bond networks still await further investigation.

Acknowledgments

This work was supported by National Institute of Health (NIH RO1 GM 49964). We are also grateful to Prof. Angel C. de Dios for informative advice and discussions, and Dr. Zhengtian Gu and Xiangjin Song for suggestions.

Supplementary Materials

The original ^{1}H solid state NMR spectra (2 pages) are available.

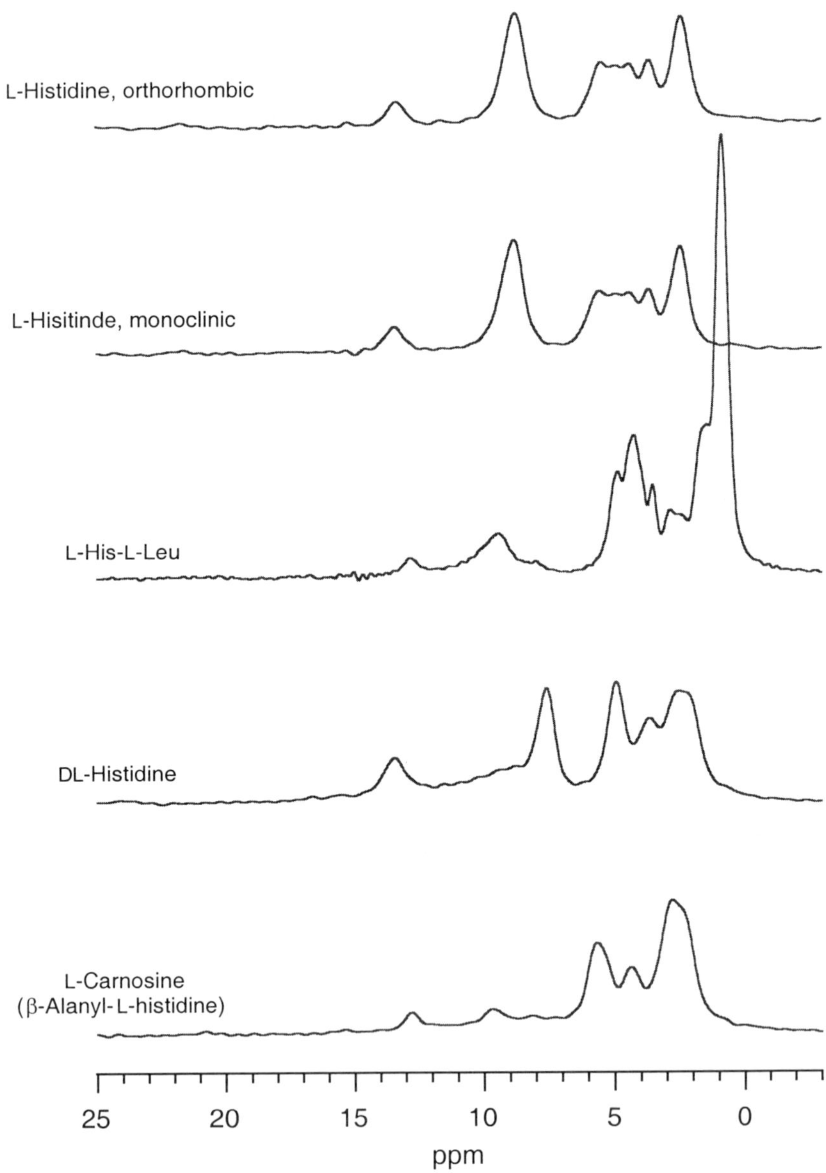

^{1}H NMR spectra of neutral histidines and histidine-containing peptides

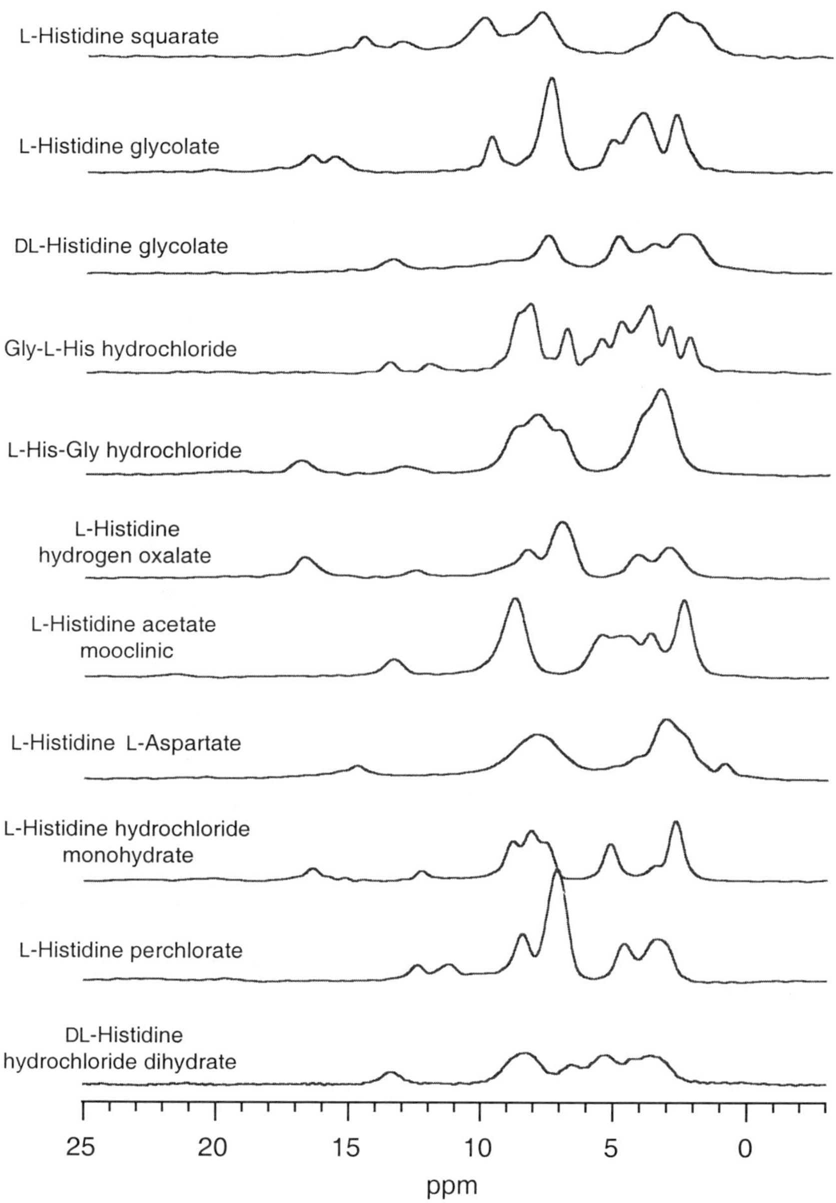

^{1}H NMR spectra for cationic histidines and histidine-containing peptides

Literature Cited

1)Cleland, W. W. *Biochemistry* **1992**, *31*, 317-319.
2)Gerlt, J. A.; Gassman, P. G. *Biochemistry* **1993**, *32*, 11943-11952.
3)Gerlt, J. A.; Gassman, P. G. *J. Am. Chem. Soc.* **1993**, *115*, 11552-11568.
4)Cleland, W. W.; Kreevoy, M. M. *Science* **1994**, *264*, 1887-1890.
5)Frey, P. A.; Whitt, S. A.; Tobin, J. B. *Science* **1994**, *264*, 1927-1930.
6)Hibbert, F.; Emsley, J. *Adv. Phys. Org. Chem.* **1990**, *26*, 255-379.
7)Tobin, J. B.; Whitt, S. A.; Cassidy, C. S.; Frey, P. A. *Biochemistry* **1995**, *34*, 6919-6924.
8)Shan, S.-o.; Loh, S.; Herschlag, D. *Science* **1996**, *272*, 97-101.
9)Ash, E. L.; Sudmeier, J. L.; De Fabo, E. C.; Bachovchin, W. W. *Science* **1997**, *278*, 1128-1132.
10)Guthrie, J. P. *Chem. Biol.* **1996**, *3*, 163-170.
11)Gerlt, J. A.; Kreevoy, M. M.; Cleland, W. W.; Frey, P. A. *Chem. Biol.* **1997**, *4*, 259-267.
12)Schwartz, B.; Drueckhammer, D. G. *J. Am. Chem. Soc.* **1995**, *117*, 11902-11905.
13)Shan, S.-o.; Herschlag, D. *J. Am. Chem. Soc.* **1996**, *118*, 5515-5518.
14)Kato, Y.; Toledo, L. M.; Rebek, J. *J. Am. Chem. Soc.* **1996**, *118*, 8575-8579.
15)Shan, S.-o.; Herschlag, D. *Proc. Natl. Acad. Sci. USA* **1996**, *93*, 14474-14479.
16)Scheiner, S.; Kar, T. *J. Am. Chem. Soc.* **1995**, *117*, 6970-6975.
17)Berglund, B.; Carson, D. G.; Vaughan, R. W. *J. Chem. Phys.* **1980**, *72*, 824-827.
18)Berglund, B.; Vaughan, R. W. *J. Chem. Phys.* **1980**, *73*, 2037-2043.
19)Jeffrey, G. A.; Yeon, Y. *Acta Crystallogr.* **1986**, *B42*, 410-413.
20)Harris, R. K.; Jackson, P.; Merwin, L. H.; Say, B. J.; Hägele, G. *J. Chem. Soc., Faraday Trans. 1* **1988**, *84*, 3649-3672.
21)McDermott, A.; Ridenour, C. F. In *Encyclopedia of Nuclear Magnetic Resonance*; Grant, D. M. and Harris, R. K., Ed.; John Wiley and Sons: New York, 1996; Vol. 6; pp 3820-3825.
22)Smirnov, S. N.; Golubev, N. S.; Denisov, G. S.; Benedict, H.; Schah-Mohammedi, P.; Limbach, H.-H. *J. Am. Chem. Soc.* **1996**, *118*, 4094-4101.
23)Liang, T.-C.; Abeles, R. H. *Biochemistry* **1987**, *26*, 7603-7608.
24)Zhong, S.; Haghjoo, K.; Kettner, C.; Jordan, F. *J. Am. Chem. Soc.* **1995**, *117*, 7048-7055.
25)Markley, J. L.; Westler, W. M. *Biochemistry* **1996**, *35*, 11092-11097.
26)Cassidy, C. S.; Lin, J.; Frey, P. A. *Biochemistry* **1997**, *36*, 4576-4584.
27)Kintanar, A.; Metzler, C. M.; Metzler, D. E.; Scott, R. D. *J. Biol. Chem.* **1991**, *266*, 17222-17229.
28)Mollova, E. T.; Metzler, D. E.; Kintanar, A.; Kagamiyama, H.; Hayashi, H.; Hirotsu, K.; Miyahara, I. *Biochemistry* **1997**, *36*, 615-625.
29)Harris, T. K.; Abeygunawardana, C.; Mildvan, A. S. *Biochemistry* **1997**, *36*, 14661-14675.
30)Zhao, Q.; Abeygunawardana, C.; Talalay, P.; Mildvan, A. S. *Proc. Natl. Acad. Sci. USA* **1996**, *93*, 8220-8224.
31)Zhao, Q.; Abeygunawardana, C.; Gittis, A. G.; Mildvan, A. S. *Biochemistry* **1997**, *36*, 14616-14626.
32)Tong, H.; Davis, L. *Biochemistry* **1995**, *34*, 3362-3367.
33)Rhim, W.-K.; Elleman, D. D. *J. Chem. Phys.* **1973**, *59*, 3740-3749.
34)Burum, D. P.; Rhim, W. K. *J. Chem. Phys.* **1979**, *71*, 944-956.
35)Ryan, L. M.; Taylor, R. E.; Paff, A. J.; Gerstein, B. C. *J. Chem. Phys.* **1980**, *72*, 508-515.
36)Burum, D. P.; Linder, M.; Ernst, R. R. *J. Magn. Reson.* **1981**, *44*, 173-188.

37)Bronnimann, C. E.; Hawkins, B. L.; Zhang, M.; Maciel, G. E. *Anal. Chem.* **1988**, *60*, 1743-1750.
38)Allen, F. H.; Bellard, S.; Brice, M. D.; Cartwright, B. A.; Doubleday, A.; Higgs, H.; Hummelink, T.; Hummelink-Peters, B. G.; Kennard, O.; Motherwell, W. D. S.; Rodgers, J. R.; Watson, D. G. *Acta Crystallogr.* **1979**, *B35*, 2331-2339.
39)Allen, F. H.; Kennard, O.; Taylor, R. *Acc. Chem. Res.* **1983**, *16*, 146-153.
40)Allen, F. H.; Kennard, O. *Chemical Design Automation News* **1993**, *8*, 1 & 31-37.

Chapter 14

An Empirical Analysis of Proton Chemical Shifts in Nucleic Acids

Annick Dejaegere[1,2], Richard A. Bryce[3], and David A. Case[4]

[1]Groupe RMN-UPR 9003, Ecole Supérieure de Biotechnologie de Strasbourg, 67400 Illkirch, France
[2]Laboratoire de Chimie Biophysique, ISIS-UPRESA-7006 CNRS, rue B. Pascal, 67000 Strasbourg, France
[3]Department of Chemistry, University of Manchester, Manchester M13 9PL, United Kingdom
[4]Department of Molecular Biology, The Scripps Research Institute, La Jolla, CA 92037

Abstract. We present an empirical analysis of chemical shifts for non-exchangeable protons in eleven B-form DNA duplex structures and nine other DNA oligonucleotides whose conformations have been determined from solution NMR analysis or from X-ray crystallography. The candidate structures are used to estimate the conformation-dependent part of the shift, that is, the difference between the observed shift and a reference shift found in sugars or bases. Factors influencing the choice of the appropriate reference shifts are discussed. The results indicate that a substantial portion of the observed chemical shift dispersion can be explained by ring-current theories, with smaller estimated contributions coming from electrostatic effects. The shifts of non-exchangeable base protons and of H1' are predicted better than the remaining sugar protons, but even for these latter resonances many features of the structural shifts are reproduced by a simple theory. Prospects for using proton chemical shifts in the analysis of nucleic acid structures are discussed.

1. Introduction.

Advances in NMR instrumentation and methodology have now made it possible to determine site-specific proton chemical shift assignments for a large number of proteins and nucleic acids *(1,2)*. It has been known for some time that in proteins the "structural" chemical shifts (the differences between the resonance positions in a protein and in a "random coil" polypeptide *(3-5)*,) carry useful structural information. We have previously used a database of protein structures to compare shifts calculated from simple empirical models to those observed in solution *(6)*. Here we demonstrate that a similar analysis appears promising for nucleic acids as well. Our conclusions are similar to those recently reported by Wijmenga *et al.* *(7)*, extending them by considering different structures and by an explicit consideration of reference shifts for model compounds that are fragments of nucleic acids.

Empirical analyses of the type considered here decompose contributions to proton chemical shifts into local and non-local contributions *(8)*:

$$\delta_{tot} = \delta_d(local) + \delta_p(local) + \delta_{rc} + \delta_m + \delta_{el} + \delta_s \quad (1)$$

The first two terms represent local "diamagnetic" and "paramagnetic" contributions; here we will assume that these terms can be approximated by the observed shifts of reference compounds. In principle, these terms should also contain contributions to the shift dispersion arising from differences in local bond lengths and angles (compared to the reference structures); such effects are generally smaller for proton shifts than for heavier elements, and are not included in the present model. The final four terms in Eq. (1) represent contributions from more distant parts of the molecule deriving from aromatic ring currents, other magnetic anisotropies, electrostatic and solvent effects. Although this decomposition is neither complete nor unique, it represents well a considerable amount of empirical knowledge about chemical shift propensities *(9)*. In this paper, we consider reference compounds that can provide estimates of $\delta_d(local) + \delta_p(local)$, and compute values for δ_{rc} and δ_{el} based on NMR or X-ray structures. In our current model, there are no magnetic anisotropy terms δ_m (other than ring-current shifts); for peptides, bond anisotropy contributions appear to be small but non-negligible *(10)*, and future studies will explore these terms. For the protons bonded to carbon that are considered here, differential solvation effects are expected to be small, so that our current model does not attempt to estimate δ_s. Hydrogen bonding interactions would generally be included in the "solvent" contribution, and are also expected to be of minor importance for protons bonded to carbon.

In spite of the limitations of this model, the resulting shifts capture many aspects of the observed chemical shift dispersion, but have some key limitations that are discussed in the final section. We believe that this model captures enough of the key physics involved in shift dispersion in DNA duplexes to provide a useful foundation for analyses of new structures and for the development of improved theories.

2. Methods

2.1. Database of observed shifts

The nucleic acids structures used to generate the chemical shift database are summarized in Table 1. The goal in selecting these was to examine DNA in a variety of environments, where a significant number of assigned shifts were available along with a reasonable estimate of structural parameters. This yields a database with 20 structures and a total of 1631 proton shifts; in this study, we consider only protons bonded to carbon, since shifts for protons bonded to nitrogen and oxygen are influenced strongly by hydrogen bonding and solvent effects that are not considered here *(9)*.

Table 1. Data used in the calculations

PDB ID	*Sequence*	*Reference*
	B-form duplexes:	
132d	GCCGTTAACGGC	*(11)*
142d	AGCTTGCCTTGAG	*(12,13)*
1d18	GTACGTAC	*(14)*
1d19	CATGCATG	*(14)*
1d20	TCTATCACCGCGGTGATAGA	*(15)*
1d42 (M5)[a]	GTATATAC	*(16)*
1d68	GCGTATACGC	*(17)*
1d70 (M3)[a]	GTATAATG	*(18)*
1bna[b]	CGCGAATTCGCG	*(19,20)*
b500[c]	GTCTGACGTCAGAC	*(21)*
gcg10[c]	GCGTTAACGC	*(22)*
	Triple helices	
135d	TCCTCCT-AGGAGGA-TGGTGGT	*(23)*
149d (M3)[a]	CCTATTC-GAATAGG-CTTGTCC	*(24)*
177d (M4)[a,d]	CCTGTTC-GAACAGG-CTTTTCC	*(25)*
	Quadruple helices	
139d (M2)[a]	TTGGGGT	*(26)*
186d (M6)[a]	$(TTGGGG)_4$	*(27)*
148d (M11)[a]	GGTTGGTGTGGTTGG	*(28)*
	Mismatched duplexes	
175d	GCGAATGAGC	*(29)*
1d69	ATGAGCGAATA	*(30)*
	Hairpin	
1ac7 (M1)[a]	ATCCTA-GTTA-TAGGAT	*(31)*

a. Where several structures have been deposited, the model number used is given in parenthesis i.e. M3 means model 3.

b. Crystal structure *(20)* ; NMR assignments from *(19)*

c. Structures and assignments kindly provided by the authors.

d. Contains protonated cytosines; which are treated as regular cytosines by the shift model.

The structures were either taken from the Brookhaven Databank, or made available by the authors. All the structures were from solution NMR data excepted *1bna*, where a crystal structure was used. When more than one model is deposited for the same structure, the model giving the best agreement between calculated and experimental shifts was used. The rms error obtained by averaging the shifts from all deposited structures is always within 0.02 ppm of the rms error obtained by using one structure.

2.2. Choice of reference compounds

The question of identifying model compounds to provide reference shifts for nucleic acids has been a vexing one. For proteins, short, disordered peptides can be used, whose shifts are reasonably independent of the exact sequence or solution conditions (at least for protons bonded to carbons) *(3-5)*. This is much less true for nucleic acids *(1)*, presumably because the conformations and electrostatic

---

Table 2. Potential reference shifts for sugar protons

	DNA					RNA		
	[1]	*[2]*	*[3]*	*[4]*	*[5]*	*[6]*	*[7]*	*[8]*
H1'	**5.23**	5.21	5.19	5.14	5.41	4.88	5.06	5.18
H2'	**2.24**	2.23	2.28	2.27	2.14	4.02	4.03	4.52
H2"	**2.17**	2.16	2.47	1.60	1.97	-	-	-
H3'	**4.36**	4.33	4.72	4.66	4.07	4.14	4.47	4.22
H4'	**4.00**	3.97	4.17	3.46	4.13	4.00	3.90	4.36
H5'	**3.72**	3.69	4.02	3.66	3.53	3.78	3.67	3.46
H5"	**3.60**	3.57	3.98	3.84	3.74	3.58	3.95	3.71

[1] methyl β-D deoxy ribofuranoside; in D_2O, referenced to 5% TMS in $CDCl_3$ *(32)*.

[2] methyl β-D deoxy ribofuranoside, in D_2O, at 292K *(33)*.

[3] Reference shifts extracted from a DNA database *(7)*. The H1' listed is for ADE; H1' values for other bases are GUA:5.25; CYT: 5.48, THY: 5.80.

[4] methyl β-D deoxy ribofuranoside in North (C3' endo, P=15°) conformation, ab-initio calculations (D1 set), referenced to calculated values for TMS *(34)*.

[5] As in [4], for the South (C2' endo, P=165°) conformation.

[6] β-ribofuranoside, relative to sodium 3(trimethylsilyl)-1-propanesulfonate *(35)*.

[7] As in [4], for the North (C3' endo, P=30°) conformation, and a ribose sugar.

[8] As in [7], for the South (C2' endo, P=180°) conformation.

contributions from phosphates are variable in short oligonucleotide fragments. In their recent survey of nucleic acid shifts, Wijmenga *et al.* avoided the problem entirely by using the average value (for each type of proton) found in a database of well-structured nucleic acids *(7)*. This allows one to study the conformational variability of shifts at a particular position, but does not provide information about the performance of empirical models in comparing shifts at different chemical positions. Hence, a more stringent test of computational models is possible if suitable reference compounds can be chosen, since then both the mean shifts and the deviations about the mean must be correctly predicted by the theory.

Table 2 gives some possible reference shifts for ribose and deoxyribose sugars. The experimental results for a model sugar (columns [1] and [2]) are generally upfield (have smaller shifts) compared to the average values extracted from a DNA database (column [3]); we show below that this systematic shift is predicted by our current computational model. This implies that the intrinsic "local" reference shift for sugars is probably not the average from the database, but is likely to be closer to results seen for sugars by themselves. This is supported by the large basis-set density functional calculations *(34)* reported in columns [4] and [5]: the observed sugar shifts are generally between the computed values for the N and S conformers (*i.e.* with pseudorotation angles near 0° and 180°), consistent with the notion (supported by coupling constant analysis) that these isolated sugars are a mixture of N and S conformers in solution *(33)*. As a practical matter, we have chosen the values in column [1] (highlighted in bold) for our reference shifts for the present work. This is effectively a reference model with mixed sugar pucker. It might be possible to obtain a model for analysis of DNA that is more representative of the (predominant) S pucker conformer through some combination of theoretical calculations and empirical analysis; attempts to establish such values are currently in progress.

One obvious problem with our choice of model sugar is that C1' is connected to a methoxy group in place of the nitrogenous base that would be found in nucleic acids. This clearly introduces a bias at the H1' position that needs to be corrected for. Based on results discussed below, we have added 0.32 ppm from the reference shift for H1' in Table 1 for pyrimidines, and have subtracted 0.20 ppm for purines. This difference of 0.52 ppm in H1' shift between purines and pyrimidines is close to the value of 0.44 ppm extracted from an empirical DNA database *(7)*.

Table 3 considers potential reference shifts for base protons. Here there is considerable variability, but still a rough correspondence between shifts for the isolated bases in acid or base solution (column [1]) and average values from a DNA database (column [5]). The exception is the H8 resonance of guanine, which in basic aqueous solution is about 0.4 ppm upfield from its value in nucleotides or DNA structures. We do not yet understand the origin of this behavior. For the present calculations, we have used the isolated bases as references (shown in bold in Table 3), but have increased the reference shift of guanine H8 by 0.34 ppm.

Table 3. Potential reference shifts for base protons

	[1]	*[2]*	*[3]*	*[4]*	*[5]*	*[6]*	*[7]*
H8 A	**8.60**	8.50	8.55	8.34	8.50	8.58	8.31
H8 G	**7.69**	8.18	8.20	8.01	8.11		8.00
H2 A	**8.68**	8.12	8.12	8.15	8.60	8.18	8.22
H5 C	**6.20**	6.11	6.14	6.08	6.21		6.05
H5 U	5.54	5.94	5.97	5.89		5.97	5.95
H6 C	**7.80**	8.04	8.13	7.93	7.70		7.65
H6 U	7.50	8.05	8.11	7.89		8.10	7.69
H6 T	**7.45**		7.89		7.64		
H7 T	**1.90**		1.92		2.00		

[1] isolated bases: cytosine in acidic D_2O, thymine, guanine, adenine in basic D_2O, uracil in DMSO *(36)*.

[2] 5'-deoxy ribonucleotides (ref. to TSP) in 0.1 M H_2O soln at 20±2 C *(37)*.

[3] 5' ribonucleotides, same conditions as [1] *(37)*.

[4] 3'-ribonucleotide (ref. to TSP) in 0.1 M D_2O soln at 291 K. H5 and H6 CYT are also reported as 6.062, 7.874 (Li+ vs. H+ salt) *(38)*.

[5] Fitted random coil for DNA *(7)*.

[6] 5' ribonucleotides in D_2O at 25C, ref. to DSS *(39)*.

[7] Erythronucleosides, (relative to internal HOD at 4.8 ppm) *(40)*.

2.3. Ring current calculations

The conformation-dependent contribution to chemical shifts that is best understood is the ring-current contribution associated with conjugated groups. Various functional forms have been tested and calibrated against observed data *(41-43)*. The basic ideas may be found in textbooks *(8)* and reviews *(41)*. The general form of ring current contributions is:

$$\delta_{rc} = iBG(\mathbf{r}) \tag{2}$$

where **r** is the vector from the observed proton to the aromatic ring, G(**r**) is a geometric factor and i and B are constants. It is conventional to incorporate into B those constants that would yield the expected contribution from a benzene ring, and to use i (the "ring-current intensity" factor) to represent the ratio of the intensity expected for the ring in question relative to that of a benzene ring. For example, in the Haigh-Mallion theory, which is one of the simplest to implement, the geometric factor is

$$G(\mathbf{r}) = \sum_{ij} s_{ij} \left\{ \frac{1}{r_i^3} + \frac{1}{r_j^3} \right\} \tag{3}$$

Here r_i and r_j are the distances from ring atoms *i* and *j* to the proton and s_{ij} is the area of the triangle formed by atoms *i* and *j* and the proton projected onto the plane of the aromatic ring. The sum is over the bonds in the ring. In this case, $B = 5.455 \times 10^{-6}$ Å. Other estimates of the ring current contribution can be formulated in a similar fashion.

We have used density functional calculations to calibrate intensity factors for the DNA and RNA bases *(43)*. While these are not necessarily optimal values, they are used here to provide an initial idea of the performance of ring-current models in explaining DNA shift patterns. A more extensive test of alternative values is in progress.

2.4. Electrostatic contributions

The ring current effects described above are "direct", in the sense that currents induced in the nucleic acid bases give rise to local magnetic fields that contribute directly to the observed shielding at probe nuclei. A significant but more indirect contribution to chemical shifts also arises from distant polar groups, which polarize the electron cloud around the probe nucleus and thereby increase or decrease the local shielding by electrons. The most significant term for a proton is expected to be proportional to the projection of the local electric field onto the X-H bond vector, where X is the atom connected to H *(44)*. The isotropic shift due to polarization effects is generally written in terms of an *A* coefficient:

$$\delta_{pol} = A(\mathbf{E} \cdot \hat{\mathbf{r}}) \qquad (4)$$

where **E** is the electric field, $\hat{\mathbf{r}}$ is a unit vector along the bond direction, and *A* is a proportionality constant specific to the X-H bond. Basically, fields that push electrons away from the H atom towards X (which have a positive projection on the X–H bond in the convention used here) will reduce the electron density near the H nucleus, tending to deshield it.

Many years ago, Buckingham suggested that an appropriate value for *A* for a C-H bond would be 2×10^{-12} e.s.u.$^{-1}$ *(44)*, or 9.6 ppm-Å^2/e. Modern quantum mechanical methods can now be used to estimate the derivative of the proton shielding with respect to an external electric field *(9)*. These calculations suggest larger values of *A*, close to 20 ppm-Å^2/e. Fields in condensed phases should be smaller than those in the gas-phase, and it is common to represent this (in a approximate fashion) by using an estimated gas-phase value for **E** along with a reduced value for *A (6)*. Empirical fits to a database of protein shifts yielded a value of 5.8 ppm-Å^2/e, and we use this value here. Fields were calculated using the assumed structures and partial charges from the Cornell *et al.* empirical force field *(45)*. This is clearly only an approximate model, but is equivalent to what we have used earlier for proteins *(6)*.

3. Results

The basic results for the complete DNA database are shown in Fig. 1, which compares calculated and observed values of the structural shift, *i.e.* the values of $\delta_{obs} - \delta_{ref}$, where the reference shifts are given in boldface type in Tables 2 and 3. Aside from the adjustments to the H1' and guanine H8 reference shifts discussed above, there are no adjustable parameters in this model. The calculated results show good general agreement with experiment: in Fig. 1 the calculated sugar shifts have been shifted upwards by 2 ppm so that they may be more easily distinguished from the base protons. In each case the dashed line corresponds to $y = x$. It is clear that the general trend that sugar protons in these oligonucleotides are downfield from their reference positions (*i.e.* that they have positive structural shifts), and that base protons are upfield of their positions in the isolated bases, are well reproduced by this initial model.

More details of the fits are given in Table 4, which shows the number of protons of each type, the observed and computed mean structural shifts, and the linear correlation coefficient r between calculated and observed values for each type of proton. It is clear that base protons are better accounted for than are sugar protons, both in terms of the mean shifts and in terms of the correlation between calculated and observed results. Probable reasons for this behavior are discussed below.

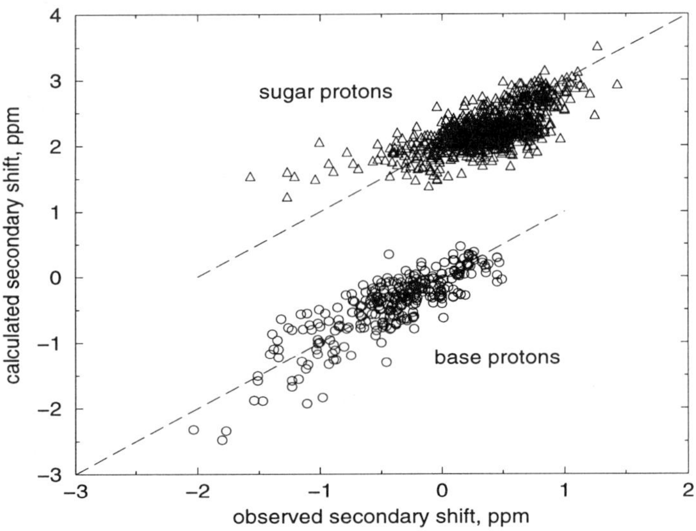

Fig. 1. Calculated vs. observed shifts.

Table 4. Mean structural shifts

proton	#	*observed*	*calculated*	*r*
H1'	216	0.68	0.64	0.54
H2'	215	0.14	0.09	0.73
H2"	216	0.34	0.14	0.66
H3'	212	0.49	0.20	0.34
H4'	206	0.26	0.13	0.35
H5'	120	0.49	0.18	0.25
H5"	106	0.33	0.16	0.11
H2	21	-1.30	-1.28	0.86
H5	38	-0.80	-0.75	0.86
H6	99	-0.26	-0.35	0.57
H7	55	-0.39	-0.45	0.71
H8	115	-0.12	-0.10	0.80

Finally, Table 5 shows statistics grouped by structure. For all of the structures, a root-mean-square difference between calculated and observed structural shifts of 0.25 to 0.35 ppm is found. Also shown are the slope and intercept of the best-fit line and the linear correlation coefficient for various portions of the total database. Statistics for all structures are not significantly different than those for the B-form DNA duplexes alone. Again, the best results are obtained for base protons and for the H1' position on the sugar.

4. Discussion.

Although a large number of NMR studies of nucleic acids have been carried out over the years, and general chemical shift trends are known *(1,2)*, attempts to understand this behavior have lagged behind similar work for proteins. In part, this is because proteins have been more frequently studied by NMR than have nucleic acids, but is also in large measure due to the relative paucity of good structural information on oligonucleotides. X-ray crystal structures of proteins are very often excellent models for protein behavior in solution, but this is less true for nucleic acids, since changes in water and salt conditions on going from solution to crystalline form are often associated with significant structural changes. Solution structure determinations by NMR have been hindered by a general lack of long-range constraints that help define global configurations for proteins. In spite of these difficulties, there is now a large enough body of reasonably secure structural information about nucleic acids to allow an empirical examination of the sort reported here. The present paper deals solely with DNA; RNA structures are becoming increasingly available and will be considered elsewhere.

As we discussed in the introduction, there are many potential contributions to chemical shift dispersion in nucleic acids. Here we have considered two that are likely to be the most important: ring current contributions from the aromatic bases,

Table 5. Statistics grouped by structure

Name	*rms error*	*slope*	*intercept*	*r*
132d	0.28			
142d	0.27			
1d18	0.25			
1d19	0.31			
1d20	0.24			
1d42	0.30			
1d68	0.26			
1d70	0.26			
1bna	0.26			
b500	0.26			
gcg10	0.27			
135d	0.24			
149d	0.20			
177d	0.30			
139d	0.28			
186d	0.35			
148d	0.34			
175d	0.28			
1d69	0.30			
1ac7	0.28			
All Struct.	0.28	0.736	-0.07	0.84
B-DNA	0.27	0.772	-0.07	0.86
base & H1'	0.25	0.926	-0.03	0.92
base & H1' BDNA	0.23	0.988	0.00	0.94
base only	0.26	0.881	-0.06	0.84
base only B-DNA	0.25	1.01	0.00	0.87
sugar only	0.29	0.572	0.00	0.67
sugar only B-DNA	0.28	0.522	0.03	0.62

and electrostatic interactions from partial charges in the sugars and bases. The results show that this is a reasonable choice, giving a model that incorporates in a qualitative way many of the features seen in experimental shifts, in particular the relative disposition of sugar and base proton positions compared to isolated fragments. There are significant remaining errors, however, especially for sugar protons other than H1'. As with proteins *(6)*, the three main sources of likely error are *(a)* errors in our estimates of ring current and electrostatic effects; *(b)* other physical contributions to chemical shift dispersion; and *(c)* differences between the structural models we have used and actual structures in solution. Further work will be required to sort out these possibilities.

We have noted that results for sugar proton shifts are substantially poorer than those for the bases. There are several possible reasons for this that point the way to further research. First, ring current contributions are probably the best-understood part of empirical relations like those of Eq. (1), and these are largest for base protons. Second, nucleic acid sugars have conformational flexibility (primarily in their pseudorotation angle) that is ignored in the present analysis; the data in columns [4], [5], [7] and [8] of Table 2 suggest that this may be an important effect to consider. Finally, sugar resonances are likely to be more strongly influenced than base protons by electrostatic perturbation by charges on the phosphates and counterions in solution, and these effects are probably poorly represented in the current model. Work to ameliorate these deficiencies is in progress.

The present results can usefully be compared to a similar study by Wijemnga *et al. (7)*, which used the ring-current and susceptibility model developed by Giessner-Prettre and co-workers *(42)* to evaluate a structural database similar to the one used here. Wijmenga *et al.* matched the computed average shift for each type of proton to that in the database of observations, essentially looking only at the structural variability within each class. They obtained an overall root-mean-square error of 0.17 ppm. If we use the same approach (*i.e.* set the reference shifts to the mean in our observed database) we obtain a nearly identical statistical results (data not shown). This implies that the ring-current model used here, which is based on density functional electronic structure calculations *(43)*, is not statistically much different for this sort of problem than the earlier models, which were based on semi-empirical calculations *(42)*. Assigning reference shifts from fragments increases the overall error to 0.25 to 0.28 ppm (Table 5) but does allow discrimination between protons at different chemical positions. It is to be hoped that both the present work and that of Wijmenga *et al.* will serve as a base-line for more extensive and sophisticated future examinations.

5. Acknowledgments.

This research was supported by NIH grant GM45811. We thank Doree Sitkoff for advice and help. A.D. thanks Prof. J.-F. Lefevre and Prof. M. Karplus for their support of this work.

6. References.

1. Wüthrich, K., *NMR of Proteins and Nucleic Acids,* Wiley, New York (1986).
2. Hilbers, C.W.; Wijmenga, S.S., in *Encyclopedia of Nuclear Magnetic Resonance*, ed. D.M. Grant; R.K. Harris, John Wiley, London (1996). pp. 3346-3359.
3. Bundi, A.; Wüthrich, K. *Biopolymers* **1979,** *18,* 285-297.
4. Merutka, G.; Dyson, H.J.; Wright, P.E. *J. Biomol. NMR* **1995,** *5,* 14-24.
5. Wishart, D.S.; Bigam, C.G.; Holm, A.; Hodges, R.S.; Sykes, B.D. *J. Biomol. NMR* **1995,** *5,* 67-81.
6. Ösapay, K.; Case, D.A. *J. Am. Chem. Soc.* **1991,** *113,* 9436-9444.

7. Wijmenga, S.S.; Kruithof, M.; Hilbers, C.W. *J. Biomol. NMR* **1997,** *10,* 337-350.

8. Harris, R.K., *Nuclear Magnetic Resonance Spectroscopy, A Physicochemical View,* Longman Scientific & Technical, Essex, England (1986).

9. Sitkoff, D.; Case, D.A. *Prog. NMR Spectr.* **1998,** *32,* 165-190.

10. Sitkoff, D.; Case, D.A. *J. Am. Chem. Soc.* **1997,** *119,* 12262-12273.

11. Kim, S.-G.; Reid, B.R. *Biochemistry* **1992,** *31,* 12103-12116.

12. Mujeeb, A.; Kerwin, S.M.; Egan, W.; Kenyon, G.L.; James, T.L. *Biochemistry* **1992,** *31,* 9325-9338.

13. Mujeeb, A.; Kerwin, S.M.; Kenyon, G.L.; James, T.L. *Biochemistry* **1993,** *32,* 13419-13431.

14. Baleja, J.D.; Germann, R.W.; van de Sande, J.H.; Sykes, B.D. *J. Mol. Biol.* **1990,** *215,* 411-428.

15. Baleja, J.D.; Pon, R.T.; Sykes, B.D. *Biochemistry* **1990,** *29,* 4828-4839.

16. Schmitz, U.; Pearlman, D.A.; James, T.L. *J. Mol. Biol.* **1991,** *221,* 271-292.

17. Cheng, J.-W.; Chou, S.-H.; Salazar, M.; Reid, B.R. *J. Mol. Biol.* **1992,** *228,* 118-137.

18. Schmitz, U.; Sethson, I.; Egan, W.M.; James, T.L. *J. Mol. Biol.* **1992,** *227,* 510-531.

19. Hare, D.R.; Wemmer, D.E.; Chou, S.H.; Drobny, G.; Reid, B.R. *J. Mol. Biol.* **1983,** *171,* 319-336.

20. Dickerson, R.E.; Drew, H.R. *J. Mol. Biol.* **1981,** *149,* 761-786.

21. Nishantha, I.; Gorenstein, D.G., personal communication 1995.

22. Smith, J.; Chazin, W.C.; Case, D.A., in preparation 1998.

23. Radhakrishnam, I.; Patel, D.J. *Structure* **1993,** *1,* 135-152.

24. Radhakrishnam, I.; Patel, D.J. *Structure* **1994,** *2,* 17-32.

25. Radhakrishnam, I.; Patel, D.J. *J. Mol. Biol.* **1994,** *241,* 600-619.

26. Wang, Y.; Patel, D.J. *J. Mol. Biol.* **1993,** *234,* 1171-1183.

27. Wang, Y.; Patel, D.J. *Structure* **1994,** *2,* 1141-1156.

28. Schultze, P.; Macaya, R.F.; Feigon, J. *J. Mol. Biol.* **1994,** *235,* 1532-1547.

29. Chou, Shan-Ho; Zhu, Leiming; Reid, Brian R. *J. Mol. Biol.* **Dec 2, 1994,** *244,* 259-268.

30. Chou, S.-H.; Cheng, J.-W.; Reid, B.R. *J. Mol. Biol.* **1992,** *228,* 138-155.

31. van Dongen, M.; Mooren, M.; Willems, E.; van der Marel, G.; van Boom, J.; Wijmenga, S.; Hilbers, C. *Nucl. Acids Res.* **1997,** *25,* 1537-1547.

32. Lemieux, R.U.; Anderson, L.; Conner, A.H. *Carbohyd. Res.* **1971,** *20,* 59-72.

33. Raap, J.; van Boom, J.H.; van Lieshout, H.C.; Haasnoot, C.A.G. *J. Am. Chem. Soc.* **1988,** *110,* 2736-2743.

34. Dejaegere, A.P.; Case, D.A. *J. Phys. Chem. A* **1998,** *102,* 5280-5289.

35. Serianni, A.S.; Barker, R. *J. Org. Chem.* **1984,** *49,* 3292-3300.

36. Pouchert, C.J., in *The Aldrich Library of NMR Spectra, Ed. II. Vol 2.*, Aldrich Chemical Co., Milwaukee, WI (1983). pp. 589, 593, 689, 692, 706.

37. Davies, D.B.; Danyluk, S.S. *Biochemistry* **1974,** *13,* 4417.

38. Davies, D.B.; Danyluk, S.S. *Biochemistry* **1975,** *14,* 543.

39. Ippel, J.H.; Wijmenga, S.S.; de Jong, R.; Heus, H.A.; Hilbers, C.W.; de Vroom, E.; van der Marel, G.A.; van Boom, J.H. *Magn. Reson. Chem.* **1996,** *34,* S156-S176.

40. Kline, P.C.; Serianni, A.S. *J. Org. Chem.* **1992,** *57,* 1772-1777.

41. Haigh, C.W.; Mallion, R.B. *Prog. NMR Spectr.* **1980,** *13,* 303-344.

42. Giessner-Prettre, C.; Pullman, B. *Q. Rev. Biophys.* **1987,** *20,* 113-172.

43. Case, D.A. *J. Biomol. NMR* **1995,** *6,* 341-346.

44. Buckingham, A.D. *Can. J. Chem.* **1960,** *38,* 300-307.

45. Cornell, W.D.; Cieplak, P.; Bayly, C.I.; Gould, I.R.; Merz, K.M., Jr.; Ferguson, D.M.; Spellmeyer, D.C.; Fox, T.; Caldwell, J.W.; Kollman, P.A. *J. Am. Chem. Soc.* **1995,** *117,* 5179-5197.

Chapter 15

A New Proton NMR Shielding Model for Alkenes

Ned H. Martin, Noah W. Allen, III[1], Everett K. Minga, Sal T. Ingrassia, and Justin D. Brown

Department of Chemistry, University of North Carolina at Wilmington, Wilmington, NC 28403–3297

In a strong magnetic field the nuclei of hydrogens that are covalently bonded to carbon and that are located over a carbon-carbon double bond experience a perturbed magnetic field owing in part to the magnetic anisotropy of the π bond. However, the theoretical model for predicting the shielding effect of a nearby alkene double bond is inconsistent with the observed proton NMR chemical shifts in numerous structures that have protons located over a carbon-carbon double bond. We have used the gauge including atomic orbital (GIAO) method, an ab initio subroutine in Gaussian 94, to calculate isotropic shielding values and to determine the proton NMR shielding increments for a simple model system: methane held at various positions over ethene. The proton NMR shielding increments calculated for one hydrogen of methane have been mapped as a function of position in Cartesian coordinates relative to the center of ethene. An equation of the same form has been fit to the resulting three-dimensional shielding surface at each of four distances from the face of the ethene molecule. In addition, a single mathematical equation has been developed for predicting the shielding experienced by a methane proton over the carbon-carbon double bond of ethene. In contrast to the traditionally employed shielding model, our results predict <u>deshielding</u> for protons within 3 Å above the center of a carbon-carbon double bond as is observed experimentally in several systems. The NMR shielding increments predicted by this equation are compared to observed shielding increments in several alkenes. Reasons for the differences between the traditional shielding model and our equation are discussed.

[1]Current address: Department of Chemistry, The University of North Carolina, Chapel Hill, NC, 27599.

It has been known for some time that the anisotropy of the magnetic susceptibility of certain functional groups leads to diamagnetic (upfield) shielding of nearby protons. Shielding is most dramatic in the case of aromatic compounds because of the high electron density found in aromatic rings. Shielding in the face of a benzene ring was explained by Pauling in terms of unusually large magnetic moments arising from the Larmor precession of the six π electrons in orbits containing multiple nuclei (1). This shielding effect was first estimated by Pople (2) using a point dipole calculation. McConnell (3) proposed an equation to predict the directional dependence of the NMR shielding experienced by a nucleus as a result of its proximity to an anisotropic group. A graphical representation of the sign of the shielding effect calculated by this equation gives rise to the familiar shielding cone (Figure 1) typically found in textbooks on NMR spectroscopy (4). An empirical function to predict the shielding increment of a proton held over the face of a benzene ring was developed by Johnson and Bovey (5). Schneider et al. (6) utilized a ring current model for the calculation of diamagnetic shielding in cyclophanes. Recently, Martin, Allen, Moore and Vo (7) developed an equation for predicting the NMR shielding of a proton over benzene based on ab initio calculations of a simple model system, methane over benzene.

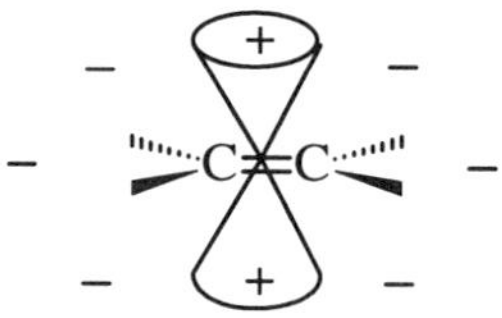

Figure 1. Traditional NMR shielding cone for a carbon-carbon double bond. (Adapted with permission from reference 9. Copyright 1998 Plenum.)

Although the magnitude of the shielding effect in alkenes is less than that observed in aromatic systems, the effect is substantial enough to be important in structural assignments based on NMR spectra. There are many molecules that contain protons held in position directly over a carbon-carbon double bond. In the traditional textbook representation (Figure 1) based on the McConnell equation (3), the currently accepted shielding model, these protons are considered to be in the shielding cone of the carbon-carbon double bond. However, a careful analysis of the NMR spectra of compounds which possess the necessary structural features to exhibit such shielding (Figure 2) reveals that in many instances the protons so oriented over a carbon-carbon double bond experience deshielding (downfield shift) instead. We have recently reported preliminary computational evidence for deshielding of protons located over a carbon-carbon double bond (8, 9). In this chapter we report a new quantitative model of the NMR (de)shielding effect for protons above the face of a carbon-carbon double bond based on ab initio calculations of a simple model system that we think better approximates the situation in real molecules than does the McConnell equation. This model accounts for the often observed deshielding of protons positioned over the center of a carbon-carbon double bond as well as for moderate shielding for protons above the plane of the double bond but farther away from the center of the bond. Our goal is to develop a mathematical model that is of practical use to chemists interested in estimating the effect on the chemical shift a proton resulting from its proximity to a carbon-carbon double bond.

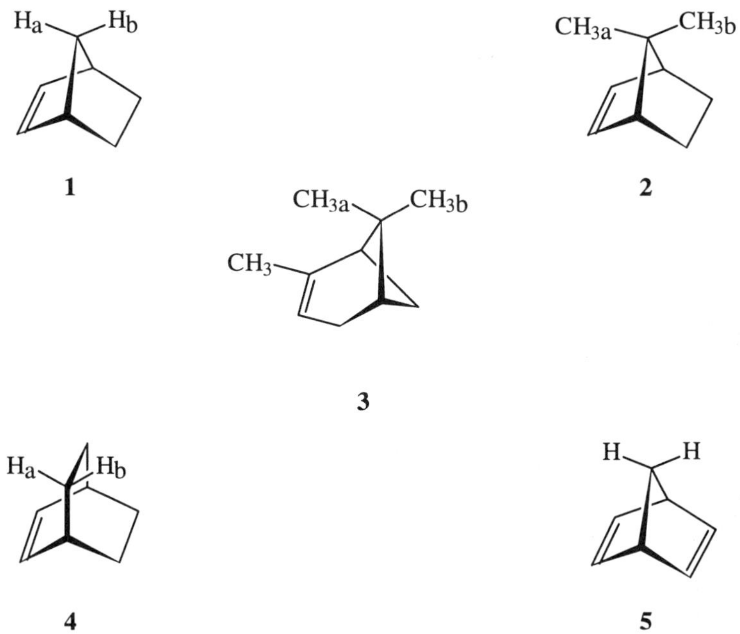

Figure 2. Examples of structures having protons over an alkene double bond.

Computational Methods

Input geometries of methane and ethene were generated using Spartan (10) molecular modeling software running on an SGI Indigo2 R4000 computer and were optimized at an ab initio level using Hartree-Fock theory, with a basis set of 6-31G(d,p) (11). The optimized structures were exported to a Brookhaven protein databank file format (.pdb) file. The two files were merged to create the geometry indicated in Figure 3, in which methane was oriented with one C-H bond normal to and directed toward the center of the double bond of ethene. The distance between the proximal hydrogen of methane and the plane of ethene was varied from 2.0 Å to 10.0 Å in seven separate input files. Each of these input files was submitted for a single point computation at Hartree-Fock theory employing a 6-31G(d,p) basis set using Gaussian 94 (12) running

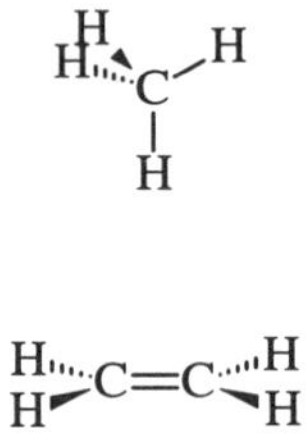

Figure 3. Initial geometry of the methane-ethene pair.

on a Cray T90. The NMR isotropic shielding values were calculated using the utility GIAO (gauge including atomic orbital), developed by Ditchfield (13) and then modified by Chesnut (14) and subsequently by Pulay (15). The isotropic shielding value of each proton on the methane molecule over ethene was transformed into the shielding increment $\Delta\sigma$ by subtracting from each data point the isotropic shielding value calculated for methane alone. A plot of the shielding increment vs. distance above ethene is shown in Figure 4.

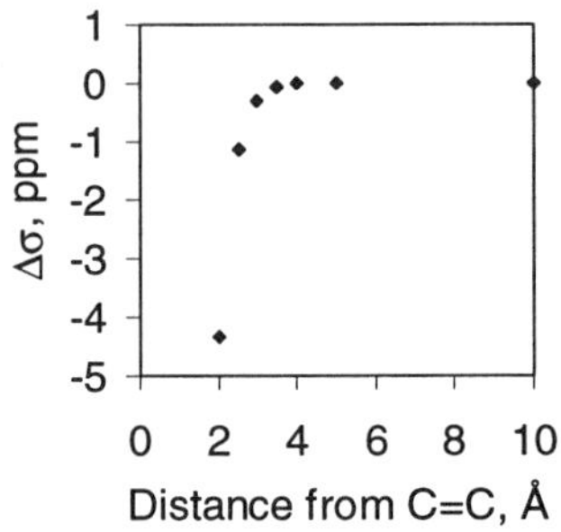

Figure 4. Graph of the GIAO-calculated shielding increment ($\Delta\sigma$) of a proton of methane (oriented as in Figure 3) vs. distance above the center of ethene.

The effect of basis set superposition error (BSSE) was estimated for this series using the counterpoise method of Boys and Bernardi (16). The isotropic shielding value of methane by itself was calculated to be 31.8276 ppm. Isotropic shielding values corrected for BSSE for the proximal proton of methane were calculated for each distance from ethene (Table I). The difference between the isotropic shielding value calculated for methane alone and the isotropic shielding values at each geometry for the proximal proton of methane obtained for the methane-ethene pair in a counterpoise calculation including the basis functions of ethene (but no electrons) is the estimated BSSE. In each calculation this is less than 0.015 ppm. Thus it appears that BSSE correction is unnecessary in this shielding calculation, and the other shielding calculations reported herein were performed without BSSE correction.

Table I. The isotropic shielding values (in ppm) calculated by GIAO for the proximal proton of methane over ethene as shown in Figure 3 with correction for basis set superposition error (BSSE).

Distance above Ethene, Å	Isotropic shielding with BSSE correction	Isotropic shielding in counterpoise calculation	Estimated BSSE
2.0	27.4943	31.8237	0.004
2.5	30.6791	31.8160	0.012
3.0	31.5278	31.8149	0.013
3.5	31.7558	31.8171	0.011
4.0	31.8109	31.8179	0.010
5.0	31.8272	31.8174	0.010
10.0	31.8218	31.8174	0.010

An input file of the molecular geometry indicated in Figure 5 was created as described above. In this geometry two protons of methane are 2.0 Å above the plane of ethene; one is directly over the center of the carbon-carbon double bond, whereas the other is in a plane normal to the carbon-carbon double bond. The other two protons of methane are more distant from the plane of ethene. After creating the merged file, multiple copies of the file were made. Coordinates of the methane portion of the input file were manipulated in these copies so as to keep ethene (in the XY plane) stationary while the methane molecule was moved over the face of the ethene molecule incrementally in the X and Y directions, keeping the Z distance above the plane of ethene constant. The symmetry of ethene was employed to limit the number of geometries to be calculated. Only one quadrant over one face of

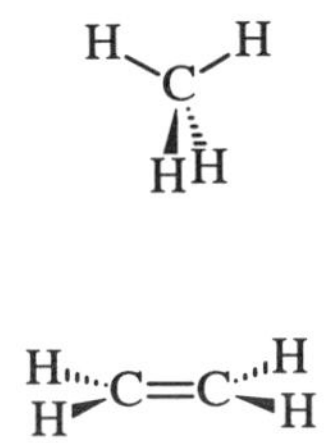

Figure 5. Geometry of the methane-ethene pair used to develop the NMR shielding increment surface.

ethene was considered. The X and Y coordinates of methane were incremented by 0.5 Å. Thirty different geometries were calculated. Similar input files were created for three other series of calculations in which methane was oriented similarly, but with the proton of interest positioned at distances (Z-axis) of 2.5 Å, 3.0 Å and 3.5 Å above ethene. Single point GIAO calculations of the isotropic shielding values were carried out using HF/6-31G(d,p) for each of the geometries. The shielding increment $\Delta\sigma$ was plotted vs. the distance (X and Y) from the center of the carbon-carbon double bond of ethene at each of four distances of separation (Z) from the plane of ethene (Figure 6). Positive values of $\Delta\sigma$ represent shielding (upfield shift); negative values represent deshielding (downfield shift).

Equations were fit to each resultant three-dimensional surface using the software TableCurve3D (16). One equation type (equation 1) was found to give a good fit to

$$\Delta\sigma = a + bX + cY + dX^2 + eY^2 + fXY \qquad (1)$$

each of the four shielding surfaces (Table II). A mathematical relationship between the constant (*a*) and coefficients (*b, c, d, e,* and *f*) in the equation for each surface and the distance (Z) above ethene was determined by least-squares fit to quadratic equations (Table III) using MSExcel (17). Substituting the appropriate quadratic equations for the constant (*a*) and each of the coefficients (*b, c, d, e,* and *f*) provided a single mathematical expression (equation 2) to predict the $\Delta\sigma$ values as a function of the displacement (X and Y) from the center of the carbon-carbon double bond and the distance (Z) above the plane of ethene.

$$\Delta\sigma = (-2.8661Z^2 + 18.378\,Z - 29.312) + (1.4039Z^2 - 8.8643Z + 13.843)X + (1.549Z^2 - 9.9345Z + 15.883)Y + (-0.0784Z^2 + 0.4602Z - 0.6438)X^2 + (-0.0795Z^2 + 0.4924Z - 0.7495)Y^2 + (-0.6311Z^2 + 4.038Z - 6.4323)XY \qquad (2)$$

Table II. The values of the constant *a* and the coefficients *b, c, d, e* and *f* in the equation $\Delta\sigma = a + bX + cY + dX^2 + eY^2 + fXY$ as a function of the distance above the plane of ethene, and the coefficient of determination (r^2) for each distance.

Distance (Å)	2.0	2.5	3.0	3.5
a	-4.1043	-1.0315	-0.2235	-0.0168
b	1.7755	0.3200	0.0216	-0.0301
c	2.2559	0.5905	0.1577	0.0413
d	-0.0407	0.0285	0.0200	0.0108
e	-0.0859	-0.0063	0.0026	0.0027
f	-0.8996	-0.2252	-0.0549	-0.0116
r^2	0.962	0.967	0.966	0.976

SOURCE: Reprinted with permission from ref. 9. Copyright 1998.

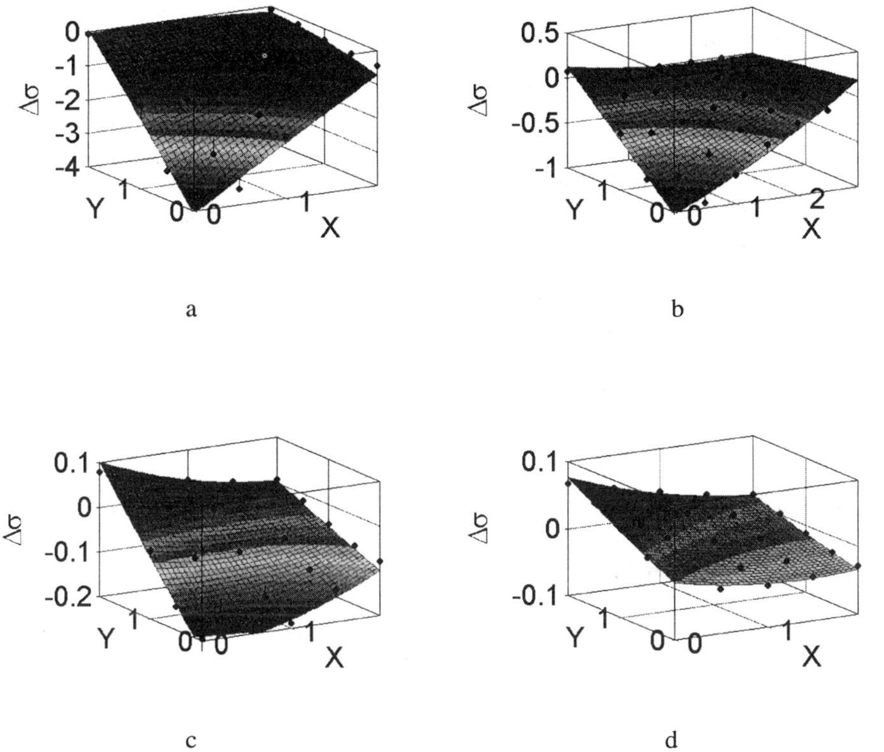

Figure 6. GIAO-calculated NMR shielding surface at (a) 2.0 Å, (b) 2.5 Å, (c) 3.0 Å and (d) 3.5 Å over ethene. (Reproduced with permission from ref 9. Copyright 1998 Plenum Press.)

Table III. Equations relating the values of a, b, c, d, e and f to the distance (Z) above the plane of ethene and their coefficient of determination (r^2).

Equation	r^2
$a = -2.8661Z^2 + 18.3780\,Z - 29.312$	0.987
$b = 1.4039Z^2 - 8.8643Z + 13.843$	0.981
$c = 1.5490Z^2 - 9.9345Z + 15.883$	0.987
$d = -0.0784Z^2 + 0.4602Z - 0.644$	0.898
$e = -0.0795Z^2 + 0.4924Z - 0.749$	0.966
$f = -0.6311Z^2 + 4.0380Z - 6.432$	0.986

SOURCE: Reprinted with permission from ref. 9. Copyright 1998.

In order to determine the effect on the calculated shielding value of the orientation of the methane C-H bond relative to ethene, two series of three calculations each were performed. In each series, methane was oriented with one proton at a fixed position above ethene such that in sequential calculations the remainder of the methane molecule had a different orientation relative to ethene. One series was performed for a methane proton position approximately above a hydrogen of ethene and another series was performed for a methane proton position directly above the midpoint of the carbon-carbon double bond of ethene. In each series the distance between the proton of methane closest to ethane and the plane of ethene was 2.5 Å. The differences among the GIAO-calculated shielding increments for each of the two positions average less than 9% of the function-predicted shielding increment for that position. We therefore conclude that the orientation of the C-H bond is of relatively minor importance, and can be ignored in the development of our model. However, it should be noted that in cases where the C-H bond crosses over the π cloud of ethene greater deviations were observed (18). Examples of molecules having this structural feature are uncommon; therefore these geometries were not considered in our model.

Calculations of the isotropic shielding by ethene independent of the presence of other atoms was accomplished by performing GIAO calculations of ethene using "ghost atoms," points in space having no charge and no mass. Input files of ethene having a square planar grid of "ghost atoms" 0.5 Å apart at several fixed distances over one quadrant of ethene were created. These input files were similarly submitted to GIAO calculations following single point HF/6-31G(d,p) calculations. This series of calculations was performed in order to provide shielding increments devoid of intermolecular shielding effects that could be compared to those calculated by the McConnell equation (3). The calculated shielding increments were plotted as described previously. A slice perpendicular to the carbon-carbon double bond through the center of the shielding surface calculated by GIAO for "ghost atoms" at a distance of 2.0 Å above ethene is shown in Figure 7. Also shown are shielding increments calculated by the McConnell equation.

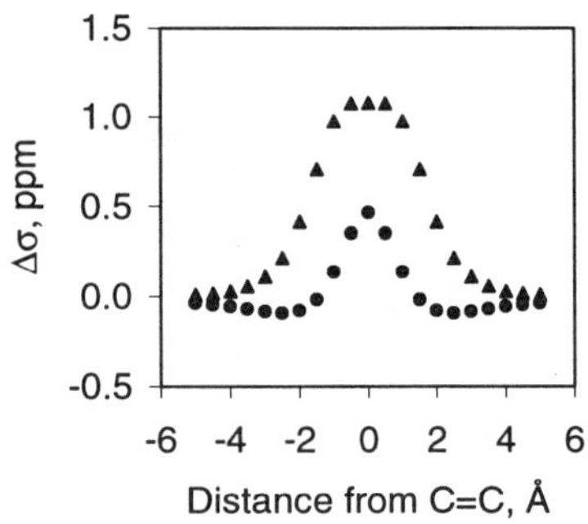

Figure 7. GIAO-calculated shielding increment (Δσ) of "ghost atoms" (triangles) and the McConnell equation-calculated shielding increment (circles) 2.0 Å above ethene vs. lateral distance from the center of the carbon-carbon double bond.

Results and Discussion

The results of the calculations of isotropic shielding of "ghost atoms" over ethene demonstrate that GIAO predicts shielding in the region above a carbon-carbon double bond to a somewhat greater extent than is predicted by the McConnell equation. However, the NMR shielding observed in molecules that have protons over a carbon-carbon double bond is not well described by a "ghost atom" calculation (or by the McConnell equation). A realistic model must include intramolecular chemical shielding effects, which can be approximated by the intermolecular chemical shielding effects obtained using the simple molecule methane at various juxtapositions over ethene. In this way the resultant of the various shielding effects can be modeled more accurately, and more realistic net shielding effects can be calculated.

The applicability of equation 2 to other molecular systems was tested by predicting the shielding increments of protons in several rigid structures (Figure 2) and comparing the results to experimentally observed shielding increments (Table IV). Note that for all five structures this function predicts deshielding (a downfield, or paramagnetic shift) for the protons over the carbon-carbon double bond. This is in sharp contrast to the predicted shielding (upfield, or diamagnetic shift) according to the traditional shielding cone model based on the McConnell equation (3). In these examples, the shielding increment is defined as the difference in the chemical shifts of protons in similar environments except for their relationship to a carbon-carbon double bond, and is given by the formula $\Delta\sigma = \delta H_b - \delta H_a$, where proton a is located over the carbon-carbon double bond.

Table IV. Shielding increments ($\Delta\sigma$, in ppm) of selected protons over a carbon-carbon double bond on structures in Figure 2 calculated by GIAO, predicted by equation 2, and experimentally observed. See text for definition of $\Delta\sigma$.

Structure	GIAO (calculated)	Equation 2 (predicted)	Experimental (observed)
1	-0.33	-1.23	-0.24
2	-0.06	-0.33	-0.09
3	-0.08	-0.44	-0.44
4	0.15	-0.36	0.27
5	-0.66	-1.47	-0.80

SOURCE: Adapted with permission from reference 9. Copyright 1998 Plenum.

Structures in which a methyl group is held over an alkene double bond pose a special problem. Rapid rotation of the methyl group results in the NMR chemical shift representing the average environment of the three protons. Both GIAO and the shielding equation presented herein utilize a specific geometry (single conformation). Therefore two conformers were used in an attempt to estimate an average shielding increment; the lowest energy conformer (which in all cases studied had one methyl proton directed toward the alkene double bond) and one differing from it by a 60 degree rotation. The shielding increments of the six proton environments were averaged in a 2:1 ratio (most stable rotamer: less stable rotamer) to provide an "effective shielding increment," taking into consideration predominance of the preferred conformation. These data are found along with the experimentally observed alkene shielding increments (defined here as the chemical shift difference between two methyl groups in similar environments except for their relationship to a carbon-carbon double bond) in Table IV. The experimental shielding increment is determined by the formula $\Delta\sigma = \delta CH_{3b} - \delta CH_{3a}$, where the protons labeled a are located over the carbon-carbon double bond. Thus a negative value of $\Delta\sigma$ suggests that CH_{3a} is deshielded relative to CH_{3b}.

The data in Table IV suggests that equation 2 generally predicts more deshielding than GIAO calculates directly or than is observed experimentally. For structure **4** (Figure 2), the equation predicts deshielding although both the direct GIAO calculation and the experimental observation show slight shielding. There are several reasons why the predicted shielding increment may differ from the observed increment. Equation 2 is based on a simple model system, one orientation of methane over ethene, and it is unreasonable to expect it to predict shielding or deshielding effects accurately in all systems. Furthermore, the model was developed from data for protons from 2.0 to 3.5 Å above ethene, yet each of the structures in Figure 2 have at least one proton closer than 2.0 Å from the plane of the carbon-carbon double bond. It is also apparent from our work that the interaction of the orbitals involved and their geometric orientation relative to one another can play an important role in determining the net shielding effect (18). Additionally, the structures (**1** and **5**) which show the greatest deviation from the equation-predicted shielding increment are those in which the angle of the deshielded C-H bond is one that is geometrically (sterically) impossible to model with a methane-ethene pair. The bond angle strain associated with most of the bicyclic test structures is another potential source of deviation from the model system results. Each of these factors would be expected to influence the shielding experienced by a proton, yet none is included in this simple model.

The shielding increment for each methane proton position used in generating the shielding surfaces was also calculated using the shielding function (equation 2). The root mean square (rms) deviation between the chemical shift increment calculated by equation 2 and that calculated by GIAO is reported in Table V; this is a measure of the fit of the shielding function to the GIAO calculated shielding values, and provides an indication of the precision of the predictions made using this equation.

Deshielding is predicted by equation 2 for protons closer than 3 Å above the center of a carbon-carbon double bond (X and Y values less than 1 Å). Similar deshielding observed for protons held close over an aromatic ring has been attributed to high-order compression effects, which generate deshielding by deformation of the electron

Table V. The rms deviation of the $\Delta\sigma$ values calculated by equation 2 from the values calculated by GIAO as a function of the distance from ethene.

Distance above ethene (Å)	rms deviation (ppm)
2.0	0.17
2.5	0.07
3.0	0.06
3.5	0.02

SOURCE: Reprinted with permission from ref. 9. Copyright 1998.

cloud (6). Indeed, visualization of the highest occupied molecular orbital (HOMO) of ethene superimposed with the HOMO of an ethene-methane pair having one hydrogen of methane oriented toward the center of the carbon-carbon double bond at a distance of 2.0 Å shows perturbation of the π cloud (Figure 8). A graph of the calculated isotropic shielding values of one proton of methane in a series with similar geometry wherein the distance between the center of ethene and the proximal proton of methane is varied shows this effect dramatically (Figure 4). It should be noted that the van der Waals thickness of half the electron cloud in an aromatic ring (vdW radius) is 1.85 Å (19); the value for an alkene should be comparable. The sum of this value and the vdW radius of hydrogen (1.11 Å) is 2.96 Å, approximately the distance below which deshielding is predicted, thus supporting the suggestion (6) of van der Waals orbital compression effects as the cause of deshielding.

Figure 8. Pictorial representations of the HOMO of ethene (wiremesh) superimposed with the HOMO of ethene together with methane (solid) oriented with one proton of methane 2.0 Å above the center of the carbon-carbon double bond and directed normal to the plane of ethene. (Reproduced with permission from ref 9. Copyright 1998 Plenum Press.)

Conclusions

The (de)shielding increment $\Delta\sigma$ has been calculated for protons of methane held at various positions over the plane of ethene using GIAO in Gaussian 94. A plot of the calculated $\Delta\sigma$ against X and Y displacement from the center of the ethene molecule gives a smooth curved surface for each of four distances Z above ethene. Each of these $\Delta\sigma$ surfaces was fitted by an equation of the general form: $\boldsymbol{\Delta\sigma = a + bX + cY + dX^2 + eY^2 + fXY}$. The constant (*a*) and the coefficients (*b, c, d, e* and *f*) were each related to the distance Z above the plane of ethene by simple quadratic equations. Substitution of these equations for *a, b, c, d, e* and *f* provides a single equation for predicting the shielding increment experienced by protons held over ethene. Application of this equation to several alkenes demonstrates its predictive validity, even in cases where the position of the proton results in deshielding. Such deshielding seems to be in sharp contrast to the traditional shielding cone model, which predicts upfield shifts for protons over the center of a carbon-carbon double bond. The reason for the difference between the McConnell shielding cone model and our equation is that the McConnell model considers only the magnetic anisotropy of the double bond, whereas our equation considers that effect plus intermolecular chemical shielding effects, including polarization and orbital overlap effects. The observed deshielding is attributed in part to van der Waals orbital compression effects, which are verified by visualization of the highest occupied molecular orbital of methane over ethene.

Acknowledgments

The authors gratefully acknowledge support of this work by the donors of The Petroleum Research Fund, administered by the American Chemical Society, Cray Research, Incorporated for a Cray Fellowship (to NWA), the North Carolina Supercomputing Center, and the College of Arts and Sciences of UNCW.

Literature Cited

1. Pauling, L. *J. Chem. Phys.* **1936**, *4*, 673-677.
2. Pople, J. A. *J. Chem. Phys.* **1956**, *24*, 1111.
3. McConnell, H. M. *J. Chem. Phys.* **1957**, *27*, 226-229.
4. For example, Silverstein, R. M.; Bassler, G. C.; Morrill, T. *Spectrophotometric Identification of Organic Compounds*, 5th ed., Wiley, New York, 1991, pp 174-175.
5. Johnson, C. E., Jr.; Bovey, F. A. *J. Chem. Phys.* **1958**, *29*, 1012-14.
6. Schneider, H.-J.; Rüdinger, V.; Cuber, U. *J. Org. Chem.* **1985**, *60*, 996-999.
7. Martin, N. H.; Allen, N. W., III; Moore K. D.; Vo, L. *J. Mol. Struct. (THEOCHEM)* **1998**, *454*, 161-166.
8. Martin, N. H.; Allen, N. W., III; Minga, E. K.; Ingrassia, S. T.; Brown, J. D. *J. Am. Chem. Soc.* **1998**, *120*, 11510-11511.
9. Martin, N. H.; Allen, N. W., III; Minga, E. K.; Ingrassia, S. T.; Brown, J. D. *Struct. Chem.* **1998**, *9(6)*, 403-410.
10. Spartan version 5.0, Wavefunction, Inc., Irvine, CA, 1997.

11. Hehre, W. J.; Radom, L.; Schleyer P. v. R.; Pople, J. A. *Ab Initio Molecular Orbital Theory*, Wiley, New York, 1986.
12. Frisch, M. J.; Trucks, G. W.; Schlegel, H. B.; Gill, P. M. W.; Johnson, B. G.; Robb, M. A.; Cheeseman, J. R.; Keith, T.; Petersson, G. A.; Montgomery, J. A.; Raghavachari, K.; Al-Laham, M. A.; Zakrzewski,V. G.; Ortiz, J. V.; Foresman, J. B.; Cioslowski, J.; Stefanov, B. B.; Nanayakkara, A.; Challacombe, M.; Peng, C. Y.; Ayala, P. Y.; Chen, W.; Wong, M. W.; Andres, J. L.; Replogle, E. S.; Gomperts, R.; Martin, R. L.; Fox, D. J.; Binkley, J. S.; Defrees, D. J.; Baker, J.; Stewart, J. P.; Head-Gordon, M.; Gonzalez, C.; Pople, J. A. Gaussian 94, Revision E.2, Gaussian, Inc., Pittsburgh PA, 1995.
13. Ditchfield, R. *Mol. Phys.* **1974**, *27*, 789-807.
14. Chesnut, D. B.; Foley, C. K. *Chem. Phys. Lett.* **1985**, *118*, 316-321.
15. Wolinski, K.; Hinton, J. F.; Pulay, P. *J. Am. Chem. Soc.* **1990**, *112*, 8251-8260.
16. Boys, S. F.; Bernardi, F. *Mol. Phys.* **1970**, *19,* 553-566.
17. TableCurve3D, v1.0, AISN Software, San Rafael, CA, 1993.
18. MSExcel97, Microsoft Corporation, Redmond, WA, 1997.
19. Martin, N. H., unpublished data.
20. Pauling, L. *Nature of the Chemical Bond*, 3rd ed.; Cornell University Press, Ithaca, 1960, pp 260-261.

Chapter 16

The NMR Chemical Shift: Local Geometry Effects

Angel C. de Dios, Jennifer L. Roach, and Ann E. Walling

Department of Chemistry, Georgetown University, 37th and O Streets, NW, Washington, DC 20057

The dependence of the principal components of the nuclear magnetic resonance (NMR) chemical shift tensor of non-hydrogen nuclei in model dipeptides is investigated. It is observed that the principal axis system of the chemical shift tensors of the carbonyl carbon and the amide nitrogen are intimately linked to the amide plane. On the other hand, there is no clear relationship between the alpha carbon chemical shift tensor and the molecular framework. However, the projection of this tensor on the C-H vector reveals interesting trends that one may use in peptide secondary structure determination. Effects of hydrogen bonding on the chemical shift tensor will also be discussed. The dependence of the chemical shift on ionic distance has also been studied in Rb halides and mixed halides. Lastly, the presence of motion can have dramatic effects on the observed NMR chemical shift tensor as illustrated by a nitrosyl meso-tetraphenyl porphinato cobalt (III) complex.

The NMR chemical shift, the most prevalent parameter in NMR spectroscopy, carries a wealth of information regarding the environment and the local electronic structure in the vicinity of the nucleus under study.(*1*). For example, one normally finds a different chemical shift for the C^{α} nucleus of each alanine residue in a protein. Ideally, a thorough analysis of the NMR chemical shift can yield information regarding the structure and interactions in the vicinity of the nucleus concerned. To achieve this, a detailed understanding of how geometrical factors and intermolecular interactions influence the chemical shift is crucial. The development and validation of the methods towards this end have combined powerful and efficient *ab initio* quantum mechanical techniques, which have been extensively tested in small molecules, with a model for partitioning the contributions to the chemical shift in a system as complex as a protein and calculating each contribution as rigorously as possible.

There are two strategies important in successfully predicting chemical shifts. First, it is necessary to determine the factor, if there is one such factor, that is primarily responsible for the observed changes in chemical shift. This search becomes possible if the chemical shift is categorized into various contributions; intra- and intermolecular effects. The distinction between intra- and intermolecular effects is particularly useful in choosing an appropriate model fragment that has a size present computational capabilities can still address. Furthermore, with intramolecular factors, one can consider separately the dependence of the chemical shift on various geometrical parameters such as bond lengths, bond angles and dihedral angles. Categorization of the chemical shift into various contributions is exemplified in the rovibrational averaging of the ^{15}N shielding in NH_3 (*2*) and ^{31}P in PH_3 (*3*), following the pioneering work of Raynes and coworkers for the nuclei in H_2O (*4*) and CH_4 (*5*). From the same group, the first categorization of intermolecular effects on the NMR chemical shift was also made (*6*).

The second important strategy is choosing an appropriate model fragment. Even with present computational software and hardware, it is still necessary to limit the number of atoms in a calculation. Selecting the right model enables the use of large basis sets and higher levels of theory. In addition, a correct model should also make it possible to investigate separately the various contributions to the chemical shift. For example, seven years ago, it was not yet possible to perform shielding calculations for Xe. Using Ar as the model, experimental Xe chemical shifts in the gas phase and as adsorbed species in zeolites were successfully reproduced using *ab initio* methods (*7*).

The same methods that successfully reproduced observed chemical shift trends in small molecules have already been shown to be equally applicable to systems as complex as proteins where not only the large number of atoms impeded *ab initio* treatments, but the presence of several factors such as local structural (torsion angle, bond angle, and bond length) constraints, electrostatic interactions and hydrogen bonding served as additional challenges as well. Combining the above two strategies with the local origin methods; enhanced gauge-including atomic orbital (GIAO) method (*8-9*), individualized gauge for localized orbitals (IGLO) method (*10*), and the localized orbital local origin (LORG) method (*11*), led to the successful prediction of NMR chemical shifts in proteins via *ab initio* methods (*12*).

Most of the *ab initio* studies of NMR chemical shifts in peptides and proteins have focused on the average or isotropic value of the NMR chemical shift. The chemical shift is a tensor quantity and is, therefore, capable of providing six independent pieces of information, namely, the magnitude and direction of each of the three principal components. In general, the shielding tensor can be antisymmetric, leading to nine independent components. However, only the symmetric part of the shielding tensor is relevant to the experiments described in this work. Thus, all references to components of the shielding tensor in the remainder of this paper will concern only the symmetric part of the tensor. Any *ab initio* package that calculates shielding provides at the same time the full shielding tensor information, hence, no computation in addition to those required for deriving the isotropic value is necessary to obtain the complete NMR chemical shift tensor. Several papers on novel solid-state NMR experiments that offer the possibility of obtaining shielding tensor information in peptides have been published recently (*13-16*). Designing experiments that will take advantage of all the information available from a shielding tensor is extremely challenging. Although the magnitudes of each of the principal components can be readily extracted from a powder spectrum, the determination of the orientation of the principal axis system of a shielding tensor normally would require single crystalline samples. Thus, it may not be experimentally feasible, for example, to determine how the orientation of a shielding tensor for a given site in a polypeptide changes with secondary structure. The adequacy of present shielding methodologies in predicting not only the magnitude but also the orientation of the principal components has already been demonstrated in the case of crystalline, zwitterionic L-threonine (*17*). Reasonable agreement is achieved for all the carbon sites, even in the presence of charged (COO-, RNH_3^+) and polar groups (-OH). With this capability, it is now possible to probe theoretically the behavior of the shielding tensor, specifically, its orientation as a function of local structure and environment. Such information is impossible to obtain empirically. The shielding tensor can be influenced by either short-range or long-range factors. Unlike experiments, theoretical calculations can also focus on one factor at a time. For example, the effect of changing one of the dihedral angles can be studied separately. Hydrogen bonding can also be studied. Such studies may not be possible experimentally, especially when most of the factors are normally simultaneously changing from one system to another. Thus, theoretical work on chemical shielding tensors in peptides can be used to explore trends that may become useful in designing and verifying novel experiments which will fully utilize the information available from chemical shift measurements.

Chemical Shift Tensors in Peptides

Details of this section have already been published (*18*). The model used in calculating amide N and C' shielding tensors was an N-formyl-glycylglycine amide fragment. Hydrogen bonding was simulated by incorporating a formamide molecule. For the C^α shielding tensor, an alanyl-X-amide fragment was used where X is one of the naturally occurring amino acids. The shielding tensor is obtained using the *deMon-NMR* program (*19*). The spatial relationship of the shielding tensor of amide nitrogen sites in adjacent residues is governed by the dihedral angle ψ (see Figure 1). If the principal axis system (PAS) of the amide nitrogen shielding tensor follows the molecular framework then this spatial relationship becomes a simple function of ψ. Figure 2 displays a tensor plot for the shielding of the amide N site in an extended conformation. In this

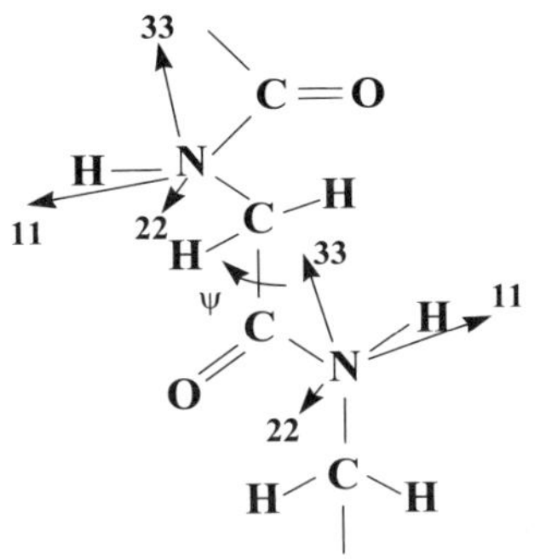

Figure 1. Model peptide fragment showing the spatial relationship between N shielding tensors of adjacent residues and the dihedral angle ψ.

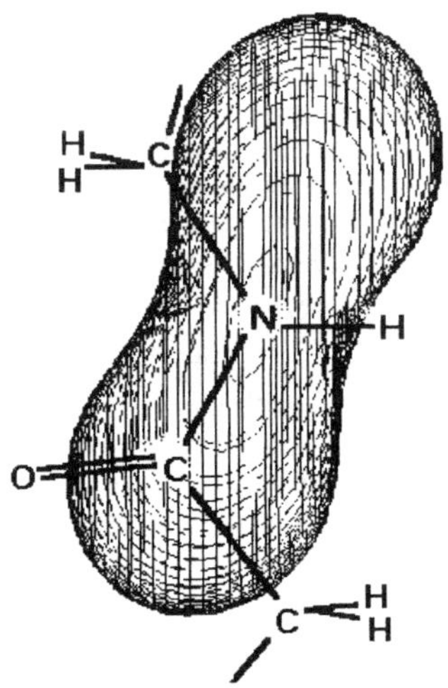

Figure 2. A tensor plot of the amide N shielding tensor for an extended conformation. (Reproduced with permission from ref. 18. Copyright 1997 American Chemical Society.)

figure, the tensor is represented by a Jorgensen-Salem plot (*20*) in which the distance of the contour from the nucleus under study (in this case, the amide N nucleus) in a given radial direction is proportional to the absolute shielding response to an external magnetic field applied in that direction. At this conformation, the least shielded component is tilted by about 19° from the N-H bond while σ_{22} lies normal to the amide plane. The σ_{11} component is very close to being perpendicular to the C'-N bond (as a result of its partial double bond character) while σ_{33} lies almost parallel to this bond. Sampling the entire *Ramachandran* surface (a surface describing the shielding at various ϕ and ψ values, where $-180° \leq \phi \leq 180°$ and $-180° \leq \psi \leq 180°$), it is found that only the least shielded component, σ_{11}, stays related to the molecular framework. In all conformations, this component lies on the amide plane and is tilted by about 18° - 22° from the N-H vector (or nearly perpendicular to the C'-N bond). The calculated magnitude and orientation of the principal components in the extended conformation are in quantitative agreement with values observed experimentally (*16, 21*). From this theoretical work, it is also observed that the other components, σ_{22} and σ_{33}, do not strictly follow the amide plane. In fact, in the helical region, σ_{22} is no longer normal to the amide plane for glycine residues. Only in extended and sheet dihedral angles does one component appear to be normal to the amide plane. One can, however, still take advantage of the fact that the least shielded component is following the N-H vector.

In the presence of hydrogen bonding as represented by a formamide molecule in the calculation, the isotropic shielding goes down by 4-5 ppm. The behavior of each principal component is dependent, however, on secondary structure. For extended and helical conformers, only σ_{11} and σ_{33} are significantly affected by hydrogen bonding. For sheet-like geometries, all three components are sensitive to hydrogen bonding, with σ_{22} showing a dependence of opposite sign. Since the overall effect of hydrogen bonding on the isotropic shielding is invariant with secondary structure, σ_{11} and σ_{33} become even more sensitive to hydrogen bonding in sheet-like conformers. Finally, hydrogen bonding does not have any significant influence in the direction of the principal components.

Figure 3 displays a tensor plot for the C' shielding in the extended conformation. Using a glycyl-glycine fragment and sampling the whole *Ramachandran* surface, the C' shielding tensor is found to follow the molecular framework. The spatial relationship between the shielding tensors of C' sites in adjacent residues is hence a simple function of the dihedral angle ϕ. The principal tensor element σ_{22} (which has a value close to that of a bare C nucleus) always lies close to the C=O bond (7° - 9°). The most shielded component always lies normal to the amide plane while σ_{11} lies on the amide plane and is perpendicular to the C=O vector. Clearly, a 2D solid-state NMR experiment that correlates two C' shielding tensors, as in the experiment proposed by Weliky et al. (*13*), is certainly ideal.

Inclusion of hydrogen bonding results in deshielding. The component most sensitive to hydrogen bonding is σ_{22} when a linear hydrogen bond is assumed for O···H-N and a 120° angle for C=O···H. In addition, it is also

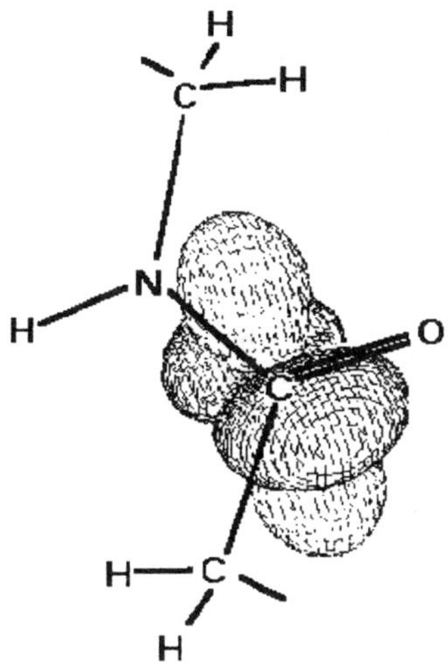

Figure 3. A tensor plot of the C′ shielding tensor for an extended conformation. (Reproduced with permission from ref. 18. Copyright 1997 American Chemical Society.).

observed that helical models are more sensitive to hydrogen bonding which perhaps explain why C' sites in helical residues are generally more deshielded compared to those in sheet regions of a protein. As in the amide N, the orientation of the principal components is also not affected by hydrogen bonding.

Of all sites in a protein, the shielding of the C^α nucleus seems to be the most tractable, being dependent only on the torsion angles ϕ, ψ and χ. Even for this site, however, the error is still within 1 ppm, which translates to about 10% of the largest observed chemical shift range for C^α in proteins for any given amino acid. Therefore, it would be useful to find a more sensitive piece of the chemical shift tensor that may have the same amount of error but a greater sensitivity to the secondary structure of the protein. The value of the shielding tensor along the C^α-H^α bond is a possible candidate. The isotropic chemical shift of helical C^α sites is found to be deshielded compared with those of sheet geometry (*12, 22*). Surprisingly, an opposite trend is seen when one looks at the projection of the C^α shielding tensor on the C^α-H^α bond. With sheet dihedral angles (ϕ = -120°, ψ = 120°), the least shielded component almost lies parallel to the C^α-H^α vector. On the other hand, with a helical conformation (ϕ = -60°, ψ = -60°), the contribution of the least shielded component to the projection of the C^α shielding tensor on the C^α-H^α bond becomes minimal. Although the difference between the isotropic chemical shifts of helical and sheet C^α is about 5-8 ppm, the difference between the projection of the C^α shielding on the C^α-H^α vector can be as large as 22 ppm. These differences become more evident when one compares the tensor plots of the C^α shielding in sheet and helical geometries as shown in Figure 4. In the sheet geometry (Figure 4A, ϕ = -120°, ψ = 120°), it can be seen that the short axis (most deshielded) of the shielding tensor coincides with the C^α-H^α bond, while for a helical residue (Figure 4B, ϕ = -60°, ψ = -60°), this is no longer true. This exciting trend has already been observed and verified in 2D and 3D solution NMR experiments (*23*).

Electrostatic Effects

There is considerable interest in measuring and interpreting the influence of electric fields on NMR chemical shifts. In real systems, the perturbing electric field can be attributed to the presence of nearby molecules or charged groups. A sound interpretation of how and why chemical shifts change with environment should lead to a better understanding of the intermolecular interactions that give rise to such perturbations. Several simplified strategies addressing the effects of long-range interactions on the NMR chemical shift are now available (*24 and references therein*). Carbonmonoxy heme proteins offer an excellent testing ground for theories regarding the effects of weak intermolecular interactions on the NMR chemical shift.

Park et al. (*25*) first noted interesting correlations between various electronic properties such as vibrational frequencies, chemical shifts and quadrupole coupling constants in various carbonmonoxy heme proteins. In this series of proteins, they established a linear relationship between the ^{13}C and ^{17}O

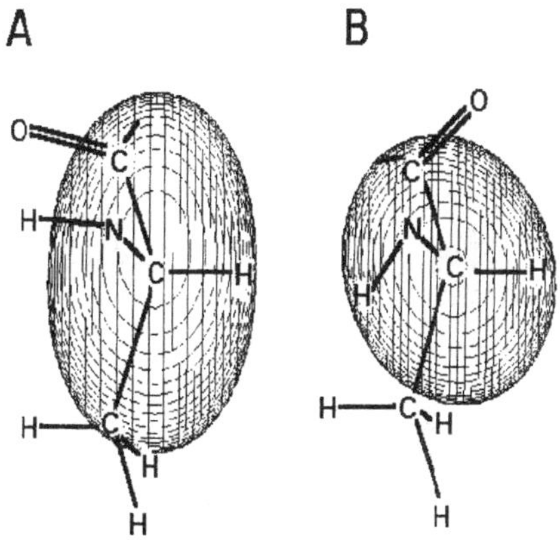

Figure 4. (A) A tensor plot of the C^{α} shielding in alanine with sheet dihedral angles ($\phi = -120°$, $\psi = 120°$). (B) same as (A), but with helical dihedral angles ($\phi = -60°$, $\psi = -60°$). (Reproduced with permission from ref. 18. Copyright 1997 American Chemical Society.)

NMR chemical shifts, and the CO vibrational stretching frequencies. Augspurger et al. (*26*) later proposed a model for these observed correlations. And in their work, the computed relationships bear the same sign as the experiment although the magnitudes seem overestimated. Using a model (see Figure 5A) consisting of Fe^{2+} octahedrally coordinated to five negative point charges (each one has a charge of -0.4*e*) and a CO molecule, the effects of uniform and non-uniform electric fields applied along the Fe-C-O axis on the chemical shifts of ^{13}C and ^{17}O and the CO stretching frequency have been reinvestigated (*27*). The correlations (as described by the slope of a linear curve relating the chemical shifts to the CO stretching frequency) obtained from this model are; ^{13}C, -0.11 ppm/cm^{-1} and ^{17}O, 0.28 ppm/cm^{-1} (Figure 5B), to be compared with the experimental values (Figure 5C); -0.07 ppm/cm^{-1} and 0.26 ppm/cm^{-1}, respectively. These calculated values were all obtained using a linear Fe-C-O geometry. This study has already been followed by a more exhaustive experimental and theoretical work supporting the fact that CO binds to Fe in a close to a linear arrangement in all conformational substates of carbonmonoxy heme proteins and model compounds (*28*). Weak electrical perturbation acting as the main factor appears to be responsible for all the observed relationships. The presence of factors other than weak electrical perturbation will not lead to such correlations between so many spectroscopic parameters. The success of the model used in this work (*27*) illustrates the adequacy of present shielding computational methodologies in extracting intermolecular effects on chemical shifts.

Hydrogen Bonding

One factor that can significantly affect the magnitude of the isotropic shielding and its principal components (as seen in amide N and carbonyl C sites) is hydrogen bonding. To study this factor in detail, theoretical and experimental ^{15}N NMR studies are ongoing (Wei, Y.; McDermott, A; Roach, J. L.; de Dios, A. C., *unpublished data*), in collaboration with McDermott at Columbia University. Specifically, the shielding tensor of one of the imidazole ring nitrogens (see Figure 6) is studied as a function of hydrogen bond distance. Shown in this figure is the calculated orientation of the PAS of the ^{15}N shielding tensor. The most shielded component is normal to the plane of the molecule while the least shielded component lies parallel to the N-H bond. The hydrogen bond partner used in the calculations is an acetate ion. A series of small histidine containing compounds (ranging from L-His through small peptides (2 - 3 residues)) which have known high-resolution structures (the hydrogen bond distance is known within 0.01 Å) is being investigated. Preliminary experimental results show that the isotropic ^{15}N NMR chemical shift changes by about 13 ppm as the hydrogen bond distance changes from 2.60 to 2.85 Å. This change is primarily due to changes in one of the components, σ_{22}. Theoretical studies which make use of a protonated imidazole molecule hydrogen bonded to an acetate ion are in agreement with experiment. In Figure 7, the measured values of the principal components of the ^{15}N shielding tensor are plotted against the hydrogen bond

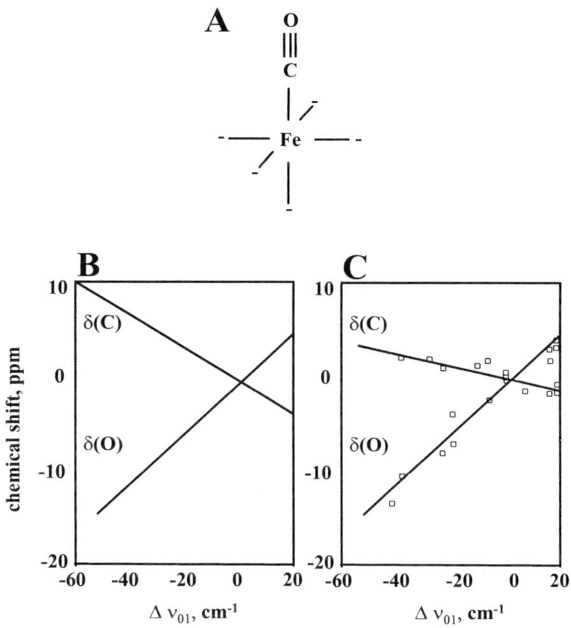

Figure 5. (A) Model fragment used in shielding and frequency calculations for $[FeCO]^{2+}$. (B) Calculated correlations between ^{13}C and ^{17}O NMR chemical shifts and CO stretching frequency. (C) as in (B), but with experimental values (*26*). ((B) Reproduced with permission from ref. 27. Copyright 1997 American Chemical Society. (C) Adapted from ref. 26.)

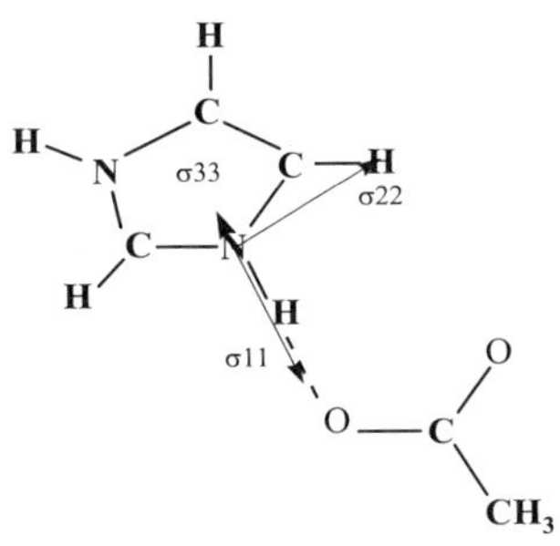

Figure 6. Model fragment used in ^{15}N shielding calculation for imidazole. Also shown is the calculated orientation of the principal axis system of the shielding tensor. (Adapted from reference 26. Copyright 1991 American Chemical Society.)

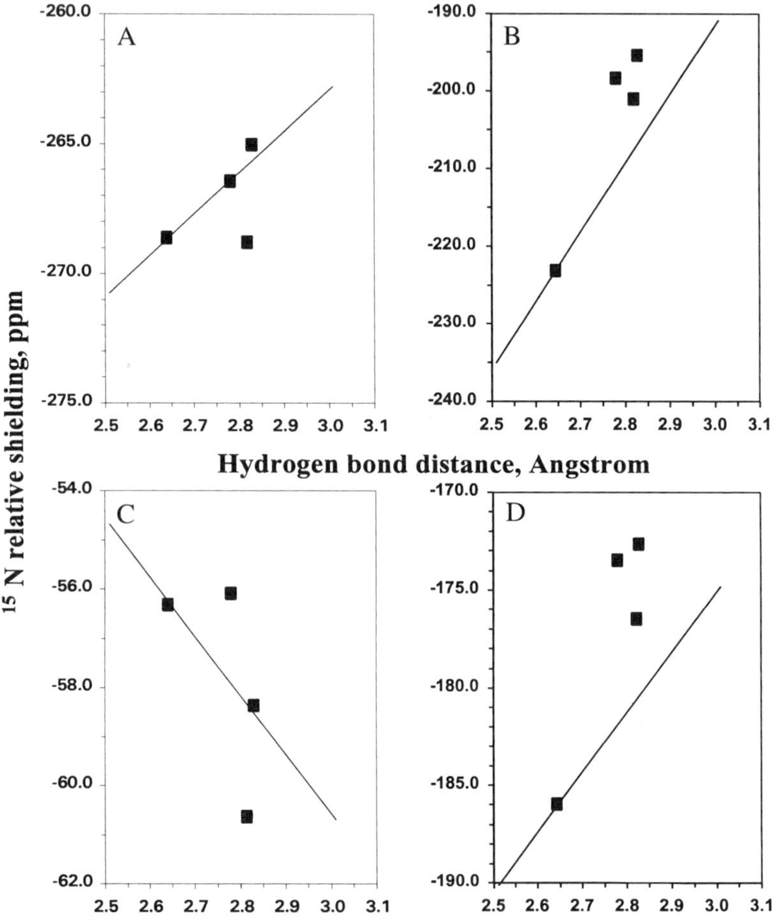

Figure 7. Experimental (■) and calculated (linear curve) ^{15}N shieldings for imidazole. (A) σ_{11}. (B) σ_{22}. (C) σ_{33}. (D) σ_{iso}.

distance. The lines drawn over these points are obtained from the *ab initio* calculations. The predicted dependence is slightly less than what is observed experimentally but as in the experiment, the middle component is also the most sensitive to hydrogen bonding (Figure 7B). This component, as observed in the calculations, lies on the plane of the molecule but is perpendicular to the N-H bond and, hence, the hydrogen bond direction.

Ionic Bonding

This section discusses work currently in progress and in collaboration with Ripmeester at the Carleton Chemistry Institute (de Dios, A. C.; Walling, A. E. Ratcliffe, C.I.; Ripmeester, J. A., *unpublished data*). Before discussing the shielding dependence on ionic distances, the adequacy of the shielding computational method (in this particular case, unlike in the previous sections, the GIAO method at the coupled Hartree-Fock level is used) is first tested using known absolute shielding values for various Rb species. Self-consistent field (SCF) optimized geometries were used for Rb_2 and $[Rb(H_2O)_6]^+$ while experimental x-ray distances were used for the Rb halides. The calculated shieldings, compared with experimental values, are presented in Table I. These absolute shieldings were obtained using the nonrelativistic diamagnetic shielding value for an isolated Rb atom (3367 ppm).

Table I. Absolute ^{87}Rb shielding values

Compound	*Theory(ppm)*	*Experiment (ppm)*
Rb_2	3239	3247 (*29*)
$[Rb(H_2O)_6]^+$	3213	3156 (*30*)
RbCl	3079	3029 (*31*)
RbBr	3073	3005 (*31*)
RbI	3051	2977 (*31*)

It can be seen that the calculated shielding values agree qualitatively with experimental values. The predicted shieldings are in correct order with Rb_2 being most shielded and RbI being least shielded. In the absence of relativistic corrections and correlation effects, the calculated shieldings at the Hartree-Fock level are already promising.

The dependence of the NMR chemical shielding on ionic distance is explored in various Rb halides. By introducing small amounts of Rb^+ (less than or equal to 5 mole percent) into crystals of other alkali and ammonium halides, lattice parameters are preserved, thus, enabling the measurement of ^{87}Rb NMR chemical shieldings as a function of the lattice parameter or interionic distance. The model used in these computations consists of a Rb atom surrounded octahedrally by six halogens. The total charge for this fragment is -5. A face-centered cubic lattice is assumed for all cases (see Figure 8). The basis set used

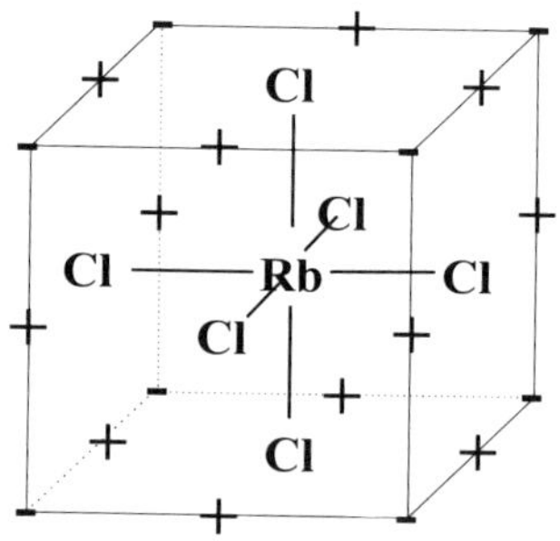

Figure 8. Model used for ^{87}Rb shielding calculations in RbCl.

for Rb was obtained from Partridge and Faegri (*32*) while the halogens were provided standard double zeta basis sets. In addition, to ease SCF convergence, twenty point charges were added to the nuclei: Twelve positive (+0.9*e*) point charges were placed at the center of each cubic edge and eight negative (-0.9*e*) point charges were placed at the remaining lattice points of the unit cell. The magnitude of these charges was chosen based on a Mulliken population analysis of an isolated rubidium halide. These charges have no significant effect on the calculated shielding values.

The experimentally determined dependence of the Rb NMR chemical shielding on ionic distance is shown in Figure 9A. Upon initial inspection, it can be seen that the halides form three parallel lines. The dependence of the ^{87}Rb shielding on the Rb-X distance is linear for all three halides. The slopes are very similar to each other (about 400 ppm/Å), indicating that the sensitivity of the shielding to interionic distance is probably a property of Rb and is independent of the identity of the halide and the other cation. The three halides form three separate but parallel lines due to an offset caused by changing the identity of the halide.

The corresponding computational results are shown in Figure 9B. Comparing this with 9A, an obvious resemblance can be seen. Like the experimental points, each halide also forms a straight line. Furthermore, the lines are also almost exactly parallel to each other indicating that the shielding dependence is mainly governed by the Rb-X distance. It should be noted that the computations here completely ignore the identity of the other cations present in the unit cell. Furthermore, a face-centered cubic lattice is assumed even for those containing Cs^+ ions. However, even with these drastic assumptions, the predictions still worked well, indicating that the present theoretical treatment has successfully identified the main factor behind the trends seen in ^{87}Rb NMR chemical shieldings in various halides.

The Presence of Motion

Details of this section will be published elsewhere (de Dios, A. C.; Roach, J. L., *submitted to J. Phys. Chem. A*). In addition to the static factors mentioned in the preceding sections, the presence of motion may also significantly contribute to the observed chemical shielding tensor. To demonstrate this aspect, the ^{15}N shielding tensor of a nitrosyl ligand bound to a *meso*-tetraphenylporphinato complex is investigated. Groombridge et al. (*33*) have measured by ^{15}N cross-polarization with magic angle spinning (CPMAS) - NMR spectroscopy the ^{15}N shielding tensor at various temperatures ranging from 200 to 298 K. The span ($\Omega = \sigma_{33} - \sigma_{11}$) is observed to change dramatically at temperatures below 200 K. At room temperature, a symmetric tensor is observed with $\Omega = 242$ ppm. Below 200 K, this value becomes 1655 ppm, which is comparable to a bent nitrosyl ligand. Incidentally, a phase transition is also observed at 206.7 K by differential scanning calorimetry (*34*). Groombridge et al. explained this observation using the disorder found by Scheidt and Hoard (*35*) after determining its x-ray

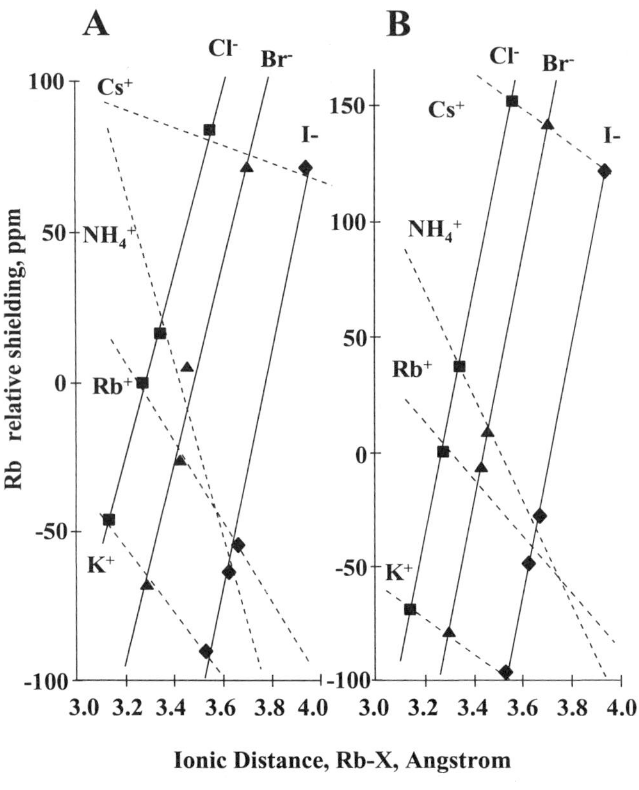

Figure 9. (A) Experimental ^{87}Rb shieldings in various ionic environments, with the following halides (■, Cl; ▲, Br; ◆, I) and in the presence of other cations (each cation drawn with a dotted line). (B), as in (A), theoretical.

structure. The x-ray structure shows a bent Co-N-O linkage (128.5°) and the presence of a fourfold axis. At room temperature, the oxygen of the nitrosyl ligand is observed to be swinging between four sites, each site lying above the region between two porphyrin nitrogen atoms. Since a four-fold axis is observed, the motion is not free swinging but a jumping between the four sites. At temperatures below 200 K, the nitrosyl ligand is no longer free to swing between such sites.

A tensor of axial symmetry usually can be taken as an indication of a linear metal-nitrosyl arrangement. Mason has compiled experimental ^{15}N shielding tensors for a variety of nitrosyl complexes, sampling both linear and bent geometries (*34*). The ^{15}N nucleus in linear nitrosyls is usually more shielded by 200-800 ppm compared to bent nitrosyls. In addition, the shielding tensor is also shown to be very sensitive to the geometry of the nitrosyl ligand and the manner it is attached to a metal atom. In general, bent nitrosyls usually show a larger span while linear nitrosyls have axial symmetry and a smaller span. The *meso*-tetraphenylporphinato complex is a very interesting system since it gives an isotropic absolute shielding of -901 ppm, close to values observed for bent nitrosyls, while displaying axial symmetry and a small span (Ω = 242 ppm). At 200 K, the isotropic absolute shielding has hardly changed (-893 ppm) but the tensor has become much wider, Ω = 1655 ppm.

The above compound is clearly an excellent system to test present computational methodologies, to see if these methods can reproduce not just the isotropic value but the tensor in the absence and presence of motion. The model employed in this work (shown in Figure 10) is similar to the one used by Jewsbury et al. (*36*) in studying heme proteins. It consists of two amidinato ligands representing the porphyrin complex. The ^{15}N shielding of the nitrosyl ligand is calculated at various geometries: Co-N-O angle = 127° with the following bond lengths, Co-N (1.833 Å, 1.927 Å, 1.964 Å) and N-O (1.01 Å, 1.10 Å, 1.15 Å) plus Co-N-O angle varying from 170° to 120° at Co-N = 1.927 Å and N-O = 1.10 Å. The shielding was also calculated for an optimized geometry (where Co-N-O angle = 124°, Co-N = 1.83 Å, N-O = 1.19 Å, with the Co atom 0.1 Å above the plane of the amidinato ligands). Lastly, the effect on the calculated ^{15}N shielding tensor of placing the Co atom above the plane of the amidinato ligands is also investigated. An all-electron basis set for Co is used (*37*) while a triple-zeta basis is assigned to all the nitrogen atoms and the oxygen atom of the nitrosyl ligand. The other atoms are given a standard double zeta basis set. The heavy atoms of the amidinato ligand are placed at the coordinates of the atoms they represent in the porphyrin complex as found in the x-ray structure (*35*). The C-H and N-H bonds employed are identical to those of Jewsbury at al. (*36*).

The results of the calculations are presented in Table II. The ^{15}N shielding of the nitrosyl ligand is extremely sensitive to the lengths of the Co-N and N-O bonds. This should not be surprising as it is observed from gas phase NMR, in NNO, the central nitrogen has a shielding derivative (averaged between the N-N and the N-O bonds), of -1030 ppm/Å (*38*). The shielding derivative with

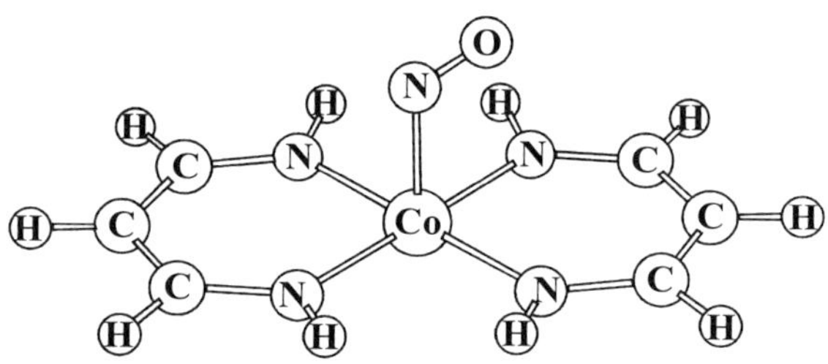

Figure 10. Model used for calculating ^{15}N shielding in Co(NO)(TPP).

Table II. Calculated ^{15}N shieldings in Co(NO)(TPP)

< *Co-N-O* (*degrees*)	*R Co-N* (*Å*)	*R N-O* (*Å*)	σ_{iso} (*ppm*)	σ_{11} (*ppm*)	σ_{22} (*ppm*)	σ_{33} (*ppm*)	Ω^{a} (*ppm*)
127	1.83	1.01	-440	-1194	-181	55	1249
127	1.83	1.10	-585	-1514	-264	22	1536
127	1.83	1.15	-662	-1667	-315	-5	1662
127	1.93	1.01	-486	-1324	-181	43	1367
127	1.93	1.10	-647	-1687	-267	14	1701
127	1.93	1.15	-731	-1860	-319	-13	1847
127	1.96	1.01	-506	-1377	-182	43	1420
127	1.96	1.10	-672	-1758	-268	10	1768
127	1.96	1.15	-759	-1939	-321	-17	1922
120	1.93	1.10	-648	-1638	-299	-8	1630
140	1.93	1.10	-631	-1821	-151	79	1900
150	1.93	1.10	-514	-1928	29	176	2104
160	1.93	1.10	-396	-1947	372	387	2334
170	1.93	1.10	171	-1482	843	1153	2635
127[b]	1.83	1.10	-580	-1505	-263	28	1533
124[c]	1.83	1.19	-704	-1737	-353	-24	1713
Experiment[d]			-893	-1853	-626	-198	1655

[a] $\Omega = \sigma_{33} - \sigma_{11}$.

[b] In this geometry, the cobalt atom is placed 0.1 Å above the plane of the amidinato ligands.

[c] Optimized geometry.

[d] From ref. (*33*).

respect to the N-O distance is very large and negative: -1760 ppm/Å, while for the Co-N distance, it is -660 ppm/Å. The two derivatives are not independent from each other. Each one becomes more negative as the other bond length increases. For example, when the Co-N distance is changed to 1.964 Å, the first derivative of the shielding with respect to the N-O distance becomes -1810 ppm/Å. The magnitude of each of the three principal components are also displayed in Table II. It can be seen that the least shielded component is most sensitive to changes in both bond lengths. This component is found to lie along the N-O bond. The isotropic shielding is not that sensitive to changes in the Co-N-O angle in the vicinity of 127°. By transforming this angle from 120° to 140°, the isotropic shielding changes by only 17 ppm. The span, on the other hand, appears to be a little bit more sensitive to changes in the Co-N-O angle. Within the same range of Co-N-O angles, the span has increased from 1630 to 1900 ppm. Changing this angle from 120° to 127° causes a 71 ppm change in the span. The span is also very sensitive to bond lengths, its first derivative with respect to N-O is large and positive: 3500 ppm/Å, and with respect to Co-N, also large and positive: 1800 ppm/Å.

The calculated orientation of the PAS of the ^{15}N shielding of the nitrosyl ligand in Co(NO)(TPP) agrees with the predicted orientation made by Groombridge et al. (*33*) based on symmetry arguments. The least shielded component, σ_{11}, lies along the N-O bond and the most shielded component lies normal to the Co-N-O plane. From the localized molecular orbital contributions to the shielding, the bulk of the shielding comes from an orbital shared between Co and the nitrosyl N atoms. For the model where Co-N = 1.927 Å, N-O = 1.10 Å, and <Co-N-O = 127°, this orbital accounts for about -600 ppm and almost all of it are in σ_{11} (-1700 ppm of σ_{11} is attributable to the Co-N bond). Comparing the experimental values of the principal components with the calculated values at this geometry, one finds that the computed numbers are more positive (around 200 ppm for σ_{11} and σ_{33}, which is outside the experimental uncertainty of 60 ppm). Since the errors in computing σ_{11} and σ_{33} are comparable in magnitude, the resulting span is very close to experiment ($\Omega_{calculated}$ = 1701 ppm compared to $\Omega_{experimental}$ = 1655 ppm).

With rapid hopping through the four sites, the tensor now becomes (These are values for Co-N = 1.927 Å, N-O = 1.10 Å, and <Co-N-O = 127°): $\sigma_{||}$ = -781 ppm (to be compared with the experimental value, -1062 ppm (*33*)) and $\sigma_{\perp}$ = -580 ppm (experimental value, -820 ppm). Thus, the calculated motion-averaged span is 201 ppm to be compared with the experimental value at 298 K, 242±30 ppm (*33*).

These studies have shown one example of a shielding tensor that is very sensitive to local geometry. It is very sensitive to the two bond lengths (Co-N and N-O) and the bond angle (< Co-N-O). Inspecting the results in Table II for the geometries where the Co atom has been placed above the plane of the amidinato ligands leads to the conclusion that the change in the shielding tensor can be easily explained by the shortened Co-N bond length. Based on this results, one can predict that displacing the nitrosyl N nucleus from the fourfold symmetry axis can also be explained by taking into account the three parameters, Co-N and N-O bond lengths and the Co-N-O bond angle.

Acknowledgments. This work was supported in part by a New Faculty Award from the Camille and Henry Dreyfus Foundation. Acknowledgement is made to the donors of the Petroleum Research Fund, administered by the American Chemical Society, for partial support of this research. We also thank Prof. D.R. Salahub for providing us a free copy of the *deMon* code, Prof. A. McDermott and Mr. Y. Wei. for sharing with us unpublished data on ^{15}N shielding in imidazole, Prof. J. A. Ripmeester and Dr. C. A. Ratcliffe for sharing with us unpublished ^{87}Rb NMR data.

Literature Cited

1. de Dios, A. C.; Jameson, C. J. In *Annual Reports on NMR Spectroscopy;* Webb, G. A., Ed., Academic Press: London, 1994, Vol. 29.; Chapter 1, pp 1-69.

2. Jameson, C. J.; de Dios, A.C.; Jameson, A. K. *J. Chem. Phys.* **1991,** *95,* 1069.
3. Jameson, C. J.; de Dios, A.C.; Jameson, A. K. *J. Chem. Phys.* **1991,** *95,* 9042.
4. Fowler, P. W.; Riley, G.; Raynes, W. T. *Mol. Phys.* **1981,** *42,* 1463.
5. Lazzeretti, P.; Zanasi, R.; Sadlej, A.J.; Raynes, W.T. *Mol. Phys.* **1987,** *62,* 605.
6. Raynes, W.T.; Buckingham, A.D.; Bernstein, H.J. *J. Chem. Phys.* **1962,** *36,* 3481.
7. Jameson, C. J.; de Dios, A.C. *J. Chem. Phys.* **1992,** *97,* 417.
8. Ditchfield, R. *Mol. Phys.* **1974,** *27,* 789.
9. Wolinski, K.; Hinton, J.; Pulay, P. *J. Am. Chem. Soc.* **1990,** *112,* 8251.
10. Schindler, M; Kutzelnigg, W. *J. Chem. Phys.* **1982,** *76,* 1919.
11. Hansen, A. E.; Bouman, T. D. *J. Chem. Phys.* **1985,** *82,* 5035.
12. de Dios, A.C.; Pearson, J.G.; Oldfield, E. *Science* **1993,** *260,* 1491.
13. Weliky, D. P.; Dabbagh, G.; Tycko, R. *J. Magn. Reson. A* **1993,** *104,* 10.
14. Tomita, Y.; O'Connor, E. J.; McDermott, A. *J. Am. Chem. Soc.* **1994,** *116,* 8766.
15. Schmidt-Rohr, K. *J. Am. Chem. Soc.* **1996,** *118,* 7601.
16. Lee, D.-K.; Ramamoorthy, A. *J. Magn. Reson.* **1998,** *133,* 204.
17. de Dios, A. C.; Laws, D. D.; Oldfield, E. *J. Am. Chem. Soc.* **1994,** *116,* 7784.
18. Walling, A. E.; Pargas, R. E.; de Dios, A. C. *J. Phys. Chem. A* **1997,** *101,* 7299.
19. Malkin, V. G.; Malkina, O. L.; Casida, M. E.; Salahub, D. R. *J. Am. Chem. Soc.* **1994,** *116,* 5898.
20. Hansen, A. E.; Bouman, T. D. *J. Chem. Phys.* **1989,** *91,* 3552.
21. Wu, C. H.; Ramamoorthy, A.; Gierasch, L. M.; Opella, S.J. *J. Am. Chem. Soc.* **1995,** *117,* 6148.
22. Spera, S.; Bax, A. *J. Am. Chem. Soc.* **1991,** *113,* 5490.
23. Tjandra, N.; Bax, A. *J. Am. Chem. Soc.* **1997,** *119,* 9576.
24. de Dios, A. C.; Oldfield, E. *Chem. Phys. Lett.* **1993,** *205,* 108.
25. Park, K. D.; Guo, K.; Adebodun, F.; Chiu, M. L.; Sligar, S. G.; Oldfield, E. *Biochemistry* **1991,** *30,* 2333.
26. Augspurger, J. D.; Dykstra, C. E.; Oldfield, E. *J. Am. Chem. Soc.* **1991,** *113,* 2447.
27. de Dios, A. C.; Earle, E. M. *J. Phys. Chem. A* **1997,** *101,* 8132.
28. McMahon, M. T.; de Dios, A. C.; Godbout, N.; Salzmann, R.; Laws, D.D.; Le, H.; Havlin, R.H.; Oldfield, E. *J. Am. Chem. Soc.* **1998,** *120,* 4784.
29. Huber, R.; Knapp, M.; König, H.; Reinhard, H.; Weber, H.G. *Z. Physik A - Atoms and Nuclei* **1980,** *296,* 95.
30. Lutz, O.; Nolle, A. *Z. Naturforsch A* **1972,** *27,* 1577.
31. Gauß, W.; Günther, S.; Haase, A. R.; Kerber, M.; Kessler, D.; Kronenbitter, J.; Krüger, H.; Lutz, O.; Nolle, A.; Schrade, P.; Schülle, M.; Siegloch, G.E. *Z. Naturforsch A* **1978,** 33, 934.
32. Partridge, H.; Faegri, K. *NASA Technical Memo* **1992,** *103918.*
33. Groombridge, C. J.; Larkworthy, L. F.; Mason, J. *Inorg. Chem.* **1993,** *32,* 379.
34. Mason, J. In *Nuclear Magnetic Shieldings and Molecular Structure;* Tossell, J. A., Ed.; NATO ASI Series C; Kluwer Academic Publishers: Dordrecht, 1993, Vol. 386.; pp 449-471.
35. Scheidt, W. R.; Hoard, J. L. *J. Am. Chem. Soc.* **1973,** *95,* 8281.
36. Jewsbury, P.; Yamamoto, S.; Minato, T.; Saito, M.; Kitagawa, T. *J. Am. Chem. Soc.* **1994,** *116,* 11586.
37. Partridge, H. *NASA Technical Memo* **1987,** *89449.*
38. Jameson, C. J.; Osten, H.-J. *J. Chem. Phys.* **1984,** *81,* 2556.

Chapter 17

Correlations between Transition-Metal NMR Chemical Shifts and Reactivities

M. Bühl

Organisch-Chemisches Institut, Universität Zürich, Winterthurerstrasse 190, CH-8057 Zürich, Switzerland

Modern methods based on density-functional theory (DFT) can describe relative activation barriers of organometallic reactions, i.e. relative reactivities, as well as the transition-metal NMR chemical shifts of the reactant complexes involved. It is thus possible to reproduce or rationalize observed correlations between these properties or to predict new ones. NMR/reactivity correlations that could be reproduced theoretically ("intrinsic correlations") are summarized. Newly predicted NMR/ reactivity correlations are discussed for the ethylene polymerization with $V(=O{\cdots}X)R_3$ or $V(=Y)R_3$ catalysts. When X or Y are varied (X = AlH_3, Li^+, SbF_5, H^+; Y = NH, O, S, Se), both the olefin insertion barrier and the ^{51}V chemical shift are computed to change in a parallel way, albeit with a different sensitivity for both types of compounds. ^{51}V NMR spectroscopy could thus be a valuable tool for a screening of potential polymerization catalysts.

Catalytic speedup of many reactions is indispensable for their practical application in synthesis. Homogeneous catalysis with transition-metal complexes continues to receive much attention in that respect (*1*), and many research efforts are devoted to the "tailoring" of ligands for special catalytic purposes. Rational design of such catalysts, however, usually requires detailed knowledge of the reaction mechanism(s) involved, which is often very difficult to elucidate experimentally. The modern tools of computational chemistry, on the other hand, may provide such important information concerning the structures and energetics that characterize the catalytic cycle. This is true in particular because the currently available variants of density-functional theory (DFT) perform quite well for the description of these properties (*2*) and allow the treatment of ever larger, increasingly realistic, systems.

One of the reasons for the difficult experimental characterization of catalytic mechanisms is that the intermediates involved are usually so short-lived that the single most powerful analytical tool, NMR spectroscopy, cannot be applied. Fourteen years ago it was discovered that this technique may be useful for a more direct estimation (or prediction) of substituent effects on the rate of a catalytical reaction: in the pyridine synthesis catalyzed by $[Co(C_5H_4X)(COD)]$ (COD = 1,5-cyclooctadien), the catalytic activity and the ^{59}Co chemical shifts were found to be correlated for various substituents X (*3*). Since then, several systems were identified empirically for which transition-metal chemical shifts can be correlated with rate constants of stochiometric reactions (for instance, equations 1 and 2) (*4,5*), and another such relationship was reported for the catalytic activity in CO_2 hydrogenation (equation 3) (*6*). Also, stability constants of transition-metal complexes sometimes parallel their δ(metal) values (*7*).

$$Rh(C_5H_4X)(CO)_2\ (\mathbf{1}) + PR_3 \longrightarrow Rh(C_5H_4X)(PR_3)(CO) \qquad (1)$$

$$Fe(C_5H_5)(CO)_2R\ (\mathbf{2}) + PR_3 \longrightarrow Fe(C_5H_5)(PR_3)(CO)(COR) \qquad (2)$$

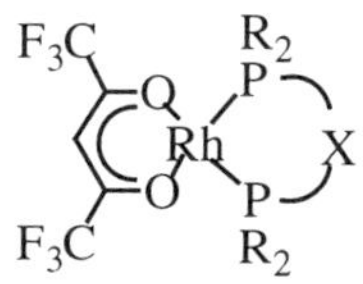

(**3**)

$$CO_2 + H_2 \longrightarrow HCOOH \qquad (3)$$

There is certainly no general, universal connection between reactivity and chemical shift, but apparently substituents can affect either one in a similar, parallel way. Once such a correlation is established it can be of great practical use because reactivities or catalytic activities of newly prepared compounds can be estimated from their NMR spectra alone.

With the advent of appropriate DFT-based methods, NMR properties of transition-metal compounds have now become amenable to theoretical computations (*8*). Suitable density functionals have been identified (*9*) which permit calculations of transition-metal chemical shifts with reasonable accuracy, typically within a few percent of the respective shift ranges. Thus, it is now possible to investigate possible NMR/reactivity correlations for transition-metal complexes from first principles; several such studies have already been undertaken (*10,11,12*).

The first part of the present paper summarizes attempts to reproduce or rationalize NMR/reactivity correlations known empirically. The second part is devoted to the search for prospective candidates for new such correlations, with the emphasis on catalytic reactions.

Reproducing NMR/Reactivity Correlations

The first successful reproduction of an empirical NMR/reactivity correlation was achieved for the substitution reaction at a rhodium center, equation 1. For several substituents X at the cyclopentadienyl ring, the logarithm of the observed rate constant had been found to correlate linearly with the corresponding ^{103}Rh chemical shift of the reactant complex *(4a)* (see Figure 1a for a schematic representation). For a model reaction (involving PH_3 instead of PPh_3), the suggested associative, "ring-slippage" mechanism (equation 4) was confirmed at the BP86/ECP1 level (see Computational Details at the end) and formation of the η^3-intermediate **4** was indicated to be the rate-determining step *(11)*.

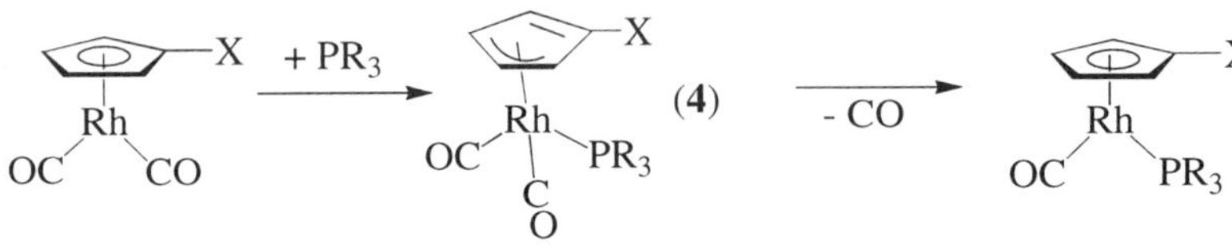

(4)

When the substituents X at the Cp ring are varied (from X = H to Cl, NMe_2, and NO_2), the barrier for formation of **4** is successively reduced. Both the trend and the magnitude of this variation are consistent with the experimental k_{obs} values. Except for a minor detail, the observed, concomitant deshielding of the ^{103}Rh nucleus is qualitatively reproduced at the SOS-DFPT-IGLO/II level. Thus, the overall NMR/reactivity relation in this system is well reproduced with DFT-based methods (Figure 1).

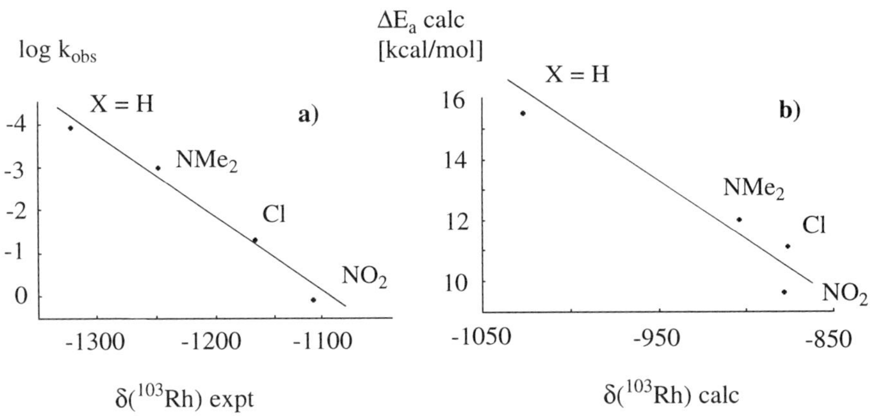

Figure 1. a) Empirical correlation between rate constants (log k_{obs}) and $\delta(^{103}Rh)$ in $Rh(C_5H_4X)(CO)_2$ complexes (Adapted from ref. 4a). b) Correlation between ΔE_a and $\delta(^{103}Rh)$ values computed for the same compounds (From the data given in ref. 11).

The first theoretical rationalization of an NMR/reactivity correlation was offered for an insertion reaction involving iron complexes **2**, equation 2. With increasing bulkiness of the alkyl ligands at iron, PPh_3-induced insertion of CO into the Fe-C(alkyl) bond proceeds more readily. At the same time, the ^{57}Fe nucleus becomes more deshielded (*4b*). The relative ease of insertion was suspected to be related to variations in the Fe-C(alkyl) bond strength. Indeed, in the series **2a** (R = Me), **2b** (R = Bu), and **2c** (R = *i*-Pr), a significant elongation of the Fe-C bond and a notable decrease of the corresponding bond dissociation eneergy (BDE) was computed at the BP86/AE1 level (*10*), entirely consistent with the observed increase of reactivity in this sequence.

Despite serious shortcomings of the SOS-DFPT-IGLO method for a larger range of ^{57}Fe chemical shifts, the trend in the $\delta(^{57}Fe)$ values between **2a**, **2b**, and **2c** was well reproduced (*10*). The same qualitative trend is computed at the GIAO-B3LYP level (Table I), even if the individual chemical shifts appear to be overestimated at that level. The latter functional is preferable because of its good performance over the whole ^{57}Fe chemical-shift range (*9,13*).

It is probably safe to assume that the insertion barriers into the Fe-C(alkyl) bonds will decrease with the BDEs (even though the range of k_{obs} values, which cover two orders of magnitude, suggest a narrower span of actual insertion barriers than obtained for the BDEs, which vary by 11 kcal/mol). Nevertheless, the data in Table I constitute a plausible rationalization of the observed NMR/reactivity correlation.

Table I. Observed and Computed Kinetic Parameters and $\delta(^{57}Fe)$ Values for $Fe(C_5H_5)(CO)_2R$ Complexes

R	BDE[a,b] (UBP86)	k_{obs}[c]	$\delta(^{57}Fe)$[d] SOS-DFPT[b]	B3LYP[e]	Exp.[c]
Me	51.9	$2.1\cdot10^{-2}$	286	808	684
Bu	46.9	$1.6\cdot10^{-1}$	345	912 [g]	716
i-Pr	40.8	1.6	401	984	796

[a][kcal/mol], AE1 basis. [b]From referecne (*10*), basis II. [c][s^{-1}], from reference (*4b*). [d][ppm], relative to $Fe(CO)_5$. [e]From reference (*9*). [g]This work.

For CO_2 hydrogenation with rhodium-based catalysts **3** bearing chelating phosphine ligands (equation 3), the observed correlation between catalytic activities and ^{103}Rh chemical shifts (*6a*) appeared to be mediated by the steric requirements of the ligands, in particular by the bite-angle. First, this parameter itself (as measured in the solid state) was found to correlate with the $\delta(^{103}Rh)$ values (*6a*). Second, an ab initio study of the catalytic cycle has indicated that dissociation of the product from the rhodium-bis(phosphine) complex is the rate-determining step (*14*); the "ease" of this dissociation was related to the so-called "accessible surface" at the metal fragment (*6b*), which in turn depends on the steric demand of the ligands, and is quite sensitive to the bite angle.

When the explicit dependence of $\delta(^{103}Rh)$ on the bite angle was computed for model compound $Rh(acac)(PH_3)_2$ (**5**), however, only minor variations were found, much smaller than the range observed experimentally (see Figure 2a) (*15*). What other structural features could be responsible for the observed trend in $\delta(^{103}Rh)$? It had been noted earlier that not only the bite angles but also the Rh-P distances *r* in complexes **3** show substantial variations (up to nearly 0.05 Å), and in fact a roughly linear correlation of the two had been found (*6b*). Consequently, $\delta(^{103}Rh)$ should also correlate with *r*. This empirical correlation is depicted in Figure 2b, together with the DFT results for **5** as a function of *r*. Indeed, both data sets show a similar variation of $\delta(^{103}Rh)$.

Complexes of type **3** with four-membered chelate rings (not included in the data of Figure 2) are a special case: the observed, substantial deshielding of the ^{103}Rh nucleus (around ca. 1100 ppm) cannot be explained by the Rh-P bond lengths (*15*). According to the model calculations for **5**, it is the tilt of the lone pairs at phosphorus away from the Rh-P axis that is responsible for the deshielding in this case.

The results summarized in Figure 2 imply that the observed correlation between $\delta(^{103}Rh)$ and catalytic activities for complexes **3** may be indirect rather than intrinsic. The activity is attributed to steric effects related to the bite angle which in turn affects the chemical shifts only indirectly via concomitant changes in the Rh-P bond lengths.

Predicting NMR/Reactivity Correlations

In the preceding chapter it has been shown that the DFT methods currently available can be used to reproduce relative trends in both reactivities and transition-metal NMR chemical shifts. Thus, NMR/reactivity correlations can be modeled theoretically, at least when relative reactivities are reflected in relative energies on the potential energy surfaces (activation barriers, BDEs). It should in principle also be possible to predict new such correlations. This is done in the following, with the emphasis on olefin polymerization with vanadium-based catalysts.

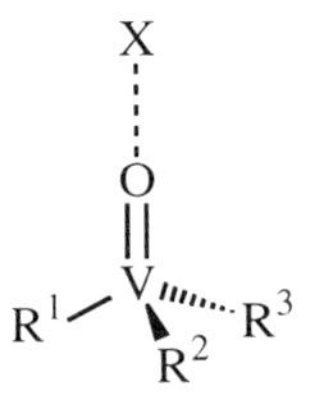

6 X = $Al(CH_2SiMe_3)_3$
$R^1, R^2, R^3 = CH_2SiMe_3, OSiPh_3$

7 X = AlH_3
a $R^1=R^2=R^3$ = Me
b $R^1=R^2$ = Me, R^3 = OMe

Lewis-acid complexes **6** are formed upon addition of the aluminium alkyl to oxovanadium(V) species and are mildly active catalysts for homogeneous ethylene polymerization (*16*). System **6** appeared to be a promising candidate for an

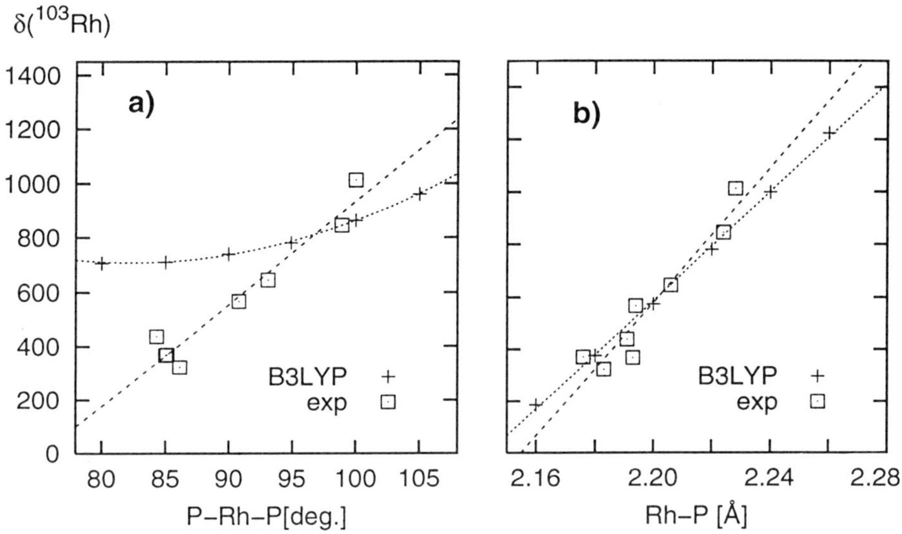

Figure 2. a) Dependence of $\delta(^{103}Rh)$ in bis(phosphine complexes) on the P-Rh-P bite angle; experimental for chelates **3** (Adapted from ref. 15), computed for model compound **6** (GIAO-B3LYP). b) The same for the Rh-P distance *r*.

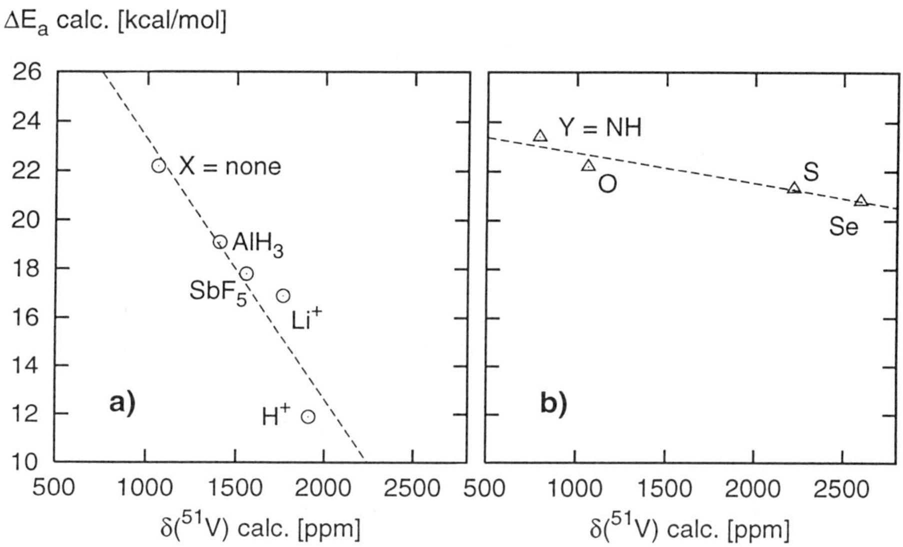

Figure 3. a) Predicted correlation between the ethylene-insertion barrier ΔE_a and $\delta(^{51}V)$ for $V(=O\cdots X)Me_3$ complexes (Adapted from ref. 12b). b) The same for $V(=Y)Me_3$ complexes. (Figure 3a is adapted with permission from reference 12b. Copyright 1998 Wiley-VCH.)

NMR/reactivity relation because the available experimental evidence had already suggested such a possibility when the substituents R at vanadium are varied; with an increasing number of oxygen donors, the catalytic activity decreases and the ^{51}V nucleus becomes more shielded (*16*).

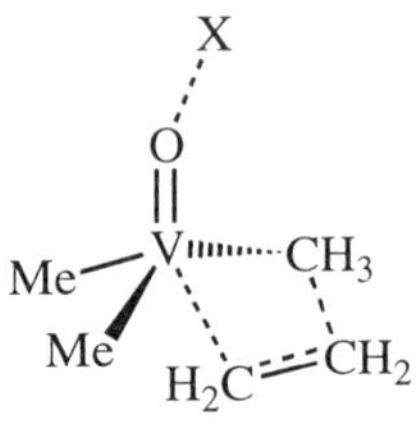

7a(TS)

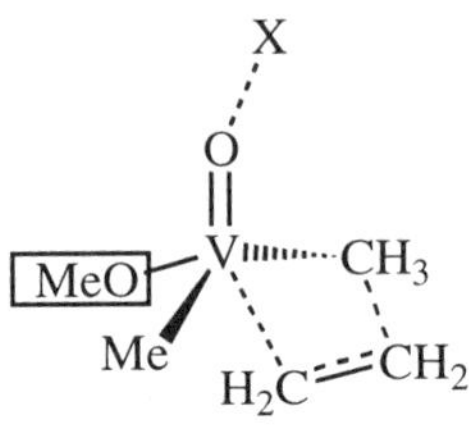

7b(TS)

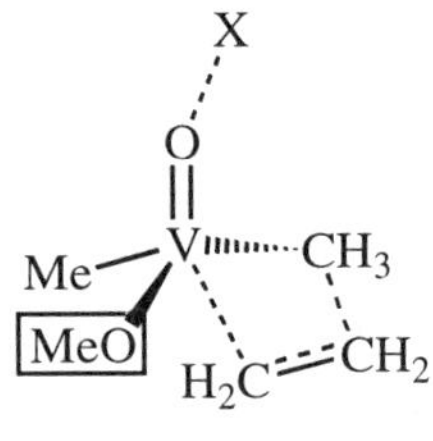

7b(TSa)

First, a plausible reaction mechanism was established theoretically for model complex **7a**, which involves insertion of ethylene into a V-C bond via **7a(TS)** as key step (*12a*). This transition state leads to the propyl derivative which can continue chain growth. When one methyl group in **7a** is replaced by a methoxy group, there are two such transition states, **7b(TS)** and **7b(TSa)**, leading to the same product. The difference between the two is that the methoxy group is cis to the leaving methyl group in **7b(TS)** and trans in **7b(TSa)**. Quite surprisingly, there is a large energetic difference between both; **7b(TSa)** is computed to be higher in energy than **7b(TS)** by more than 6 kcal·mol^{-1}. The methoxy group thus exerts a pronounced stereoelectronic effect in the transition state (*12a*).

Since this stereoelectronic effect is absent in the reactant minimum, it is unlikely that a general, intrinsic correlation between ^{51}V chemical shifts and reactivities could exist when the substituents at vanadium are varied. Accordingly, a sharp increase in the barrier is computed when the second methoxy group is introduced (which must now be trans to the leaving methyl in the transition state), affording different trends for the barriers and the $\delta(^{51}V)$ values upon successive methoxy substitution (*12a*).

In search for better candidates for an NMR/reactivity calculation, the role of the Lewis-acid cocatalyst was studied subsequently. In the experimental systems **6**, the alkyl aluminium additive is necessary for the catalytic activity (*16*). According to DFT computations for model **7a**, the Lewis acid lowers the insertion barrier by stablizing the increasing negative charge at the oxo substituent in the transition state **7a(TS)** (*12b*). Consequently, stronger Lewis acids X should lower the barrier even more. Indeed, successively lower barriers were computed in the series X = none, AlH_3, Li^+, SbF_5, and H^+. In the same series, the ^{51}V nuclei of the reactant Lewis-acid complexes become increasingly deshielded (*12b*). Thus, both properties are roughly correlated, as illustrated by the plot in Figure 3a. As with the rhodium-based system **1** discussed above, the complexes with the most deshielded metal center should be the most reactive ones.

The results summarized in Figure 3a were the first theoretical prediction of an NMR/reactivity correlation. Transferred to the experimental system **6**, these data indicate that the polymerization activity towards ethylene could be increased by Lewis acids stronger than the alkylaluminium reagents employed so far. Most significantly, a screening of suitable cocatalysts should be possible by ^{51}V NMR spectroscopy: the Lewis-acid complexes with the most deshielded $\delta(^{51}V)$ resonance should also be the most active catalysts.

Finally one can modify the precatalyst in **6** by substituting the terminal oxo ligand with other heteroatoms. Compounds **8 - 10** are recent examples for such imido- (*17*), thio- and seleno (*18*) derivatives, respectively.

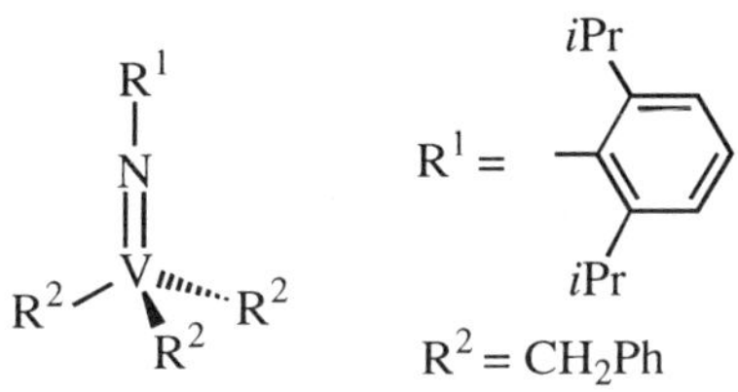

8

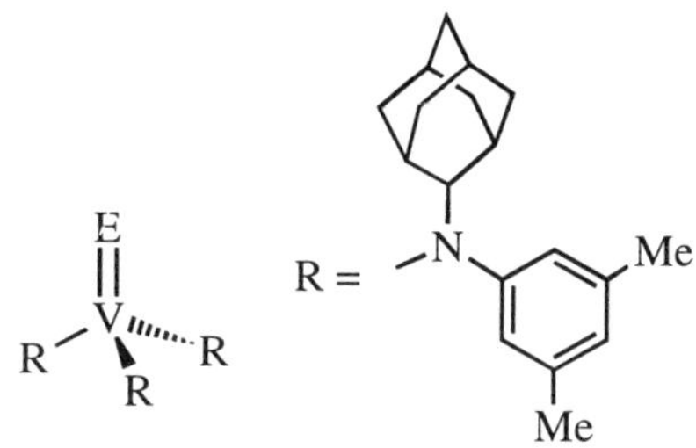

E = S (**9**), Se (**10**)

Can the NMR/reactivity correlation of Figure 3a be extended to model species $V(=Y)Me_3$ (Y = NH, S, Se)? Indeed, large variations are computed for the $\delta(^{51}V)$ values in this series, and enormous shifts to high frequency are predicted for the higher chalcogenides, up to ca. 2600 ppm for the seleno derivative (Table II). [Similarly deshielded ^{51}V resonances are known for selenovanadates, e.g. for VSe_4^{3-}, δ = 2570 (*19*)] The computed barriers for ethylene insertion are only moderately sensitive to this hetero substitution. The trend in the ΔE_a values follows that in the ^{51}V chemical shifts, and as with the Lewis acids discussed above, more deshielded resonances are connected with lower barriers (see Figure 3b for a graphical representation).

Table II: Predicted ^{51}V Chemical Shifts and Ethylene Insertion Barriers for $V(=Y)Me_3$

Y	$\delta(^{51}V)^a$	$\Delta E_a{}^b$
NH	791	23.4
O[c]	1062	22.2
S	2218	21.3
Se	2597	20.8

[a]In ppm relative to $VOCl_3$, GIAO-B3LYP/AE1 level.

[b][kcal/mol], BP86/I + ZPE level, [c]From reference (*12b*).

However, the two NMR/reactivity correlations depicted in Figure 3 are quantitatively different: the slopes of the two linear regression lines differ by one order of magnitude (ca. 1 and 11 cal·ppm^{-1} in Figure 3a and 3b, respectively). Thus, compounds of type **9** or **10** (with alkyl instead of amino rests) would probably hold little promise as active catalysts, despite the expected huge deshielding of the metal center.

The same might be expected for imidovanadium(V) species, based on the higher barrier and larger shielding of $V(NH)Me_3$ relative to the $VOMe_3$ parent (Table II). A higher shielding of imidovanadium specied with respect to their oxo counterparts is also known experimentally. cf. $VO(CH_2SiMe_3)_3$, $\delta(^{51}V)$ = 1205 (*16*) vs. V(N*t*Bu)-$(CH_2SiMe_3)_3$, δ = 877 (*20*). The latter value is in good accord with that obtained for model $V(NH)Me_3$ (Table II), and the substance should be inactive as catalyst for ethylene polymerization.

Arylimido derivatives, on the other hand, are deshielded with respect to alkyl imido complexes, cf. $V(NTol)(CH_2SiMe_3)_3$ (Tol = tolyl), δ = 1048 (*21*). Provided the NMR/reactivity correlations hold also when the substituents at the imido nitrogen are varied, one might speculate that suitably derivatized aryl rests (for instance, by introducing electron-withdrawing groups) could produce more deshielded ^{51}V resonances and, at the same time, more active catalysts for ethylene polymerization. Further experimental and theoretical studies in that direction could be rewarding.

Conclusions

The rate of elementary reactions of certain transition-metal complexes, such as insertions or substitutions, can be controlled by the substituents at the metal center. In favorable cases, usually in families of closely related systems, these substituents can affect the reactivities and the chemical shifts of the transition metal nuclei in a similar, parallel fashion, resulting in an apparent correlation of both properties. Modern DFT methods can reproduce these findings, provided that changes in rate constants are reflected in corresponding trends in activation barriers or BDEs on the potential energy surface.

When an elementary reaction that shows such an NMR/reactivity correlation is the rate-determining step of a catalytic cycle, the overall activities of the corresponding catalysts can be related to their chemical shifts. Despite the great potential use, only few such examples are yet known empirically. With the approach detailed in the present paper, theoretical searches for new such correlations are now possible from first principles. Even though NMR/reactivity relations have been predicted only for model compounds, as with the alkylvanadium(V) species discussed, there is a good chance that similar relationships would be observable for the corresponding real systems. Once such a correlation is established, potential catalysts could be readily screened via NMR spectroscopy. It is well possible that in this way active catalysts may be identified which were not or could not be considered in the theoretical computations.

Computational Details

The same methods and basis sets were used as in the previous studies (*9-12*): geometries were optimized at gradient-corrected DFT levels using the Gaussian 94 program (*22*), employing exchange and correlation functionals according to Becke (*23*) and Perdew (*24*) (designated BP86), and a fine integration grid, together with the following basis sets: Wachters' all-electron basis augmented with two diffuse p and one diffuse d set for Fe and V (*25*), a quasi-relativistic effective core potential and the corresponding [6s5p3d] valence basis set for Rh (*26*), the 641G(d) basis for Se (*27*), and the standard 6-31G* basis (*28*) for all other atoms (designated AE1 or ECP1).

Magnetic shieldings have been computed with the GIAO (gauge-including atomic orbitals)-DFT method as implemented (*29*) in Gaussian 94, employing Becke's three-parameter hybrid exchange functional (*30*) in conjunction with the correlation functional of Lee, Yang, and Parr (*31*) (denoted B3LYP) and the following basis sets: Wachters' basis contracted to [8s7p4d] and augmented with two diffuse p and one diffuse d functions on Fe, well-tempered [16s10p9d] basis (contracted from the 22s14p12d set and augmented with two d-shells of the well-tempered series) for Rh (*32*), and the recommended IGLO-basis II on the ligands (*33*) (designated II); AE1 basis was used for the vanadium species. Chemical shifts are reported relative to the usual standards, $Fe(CO)_5$, $VOCl_3$, and relative to a standard frequency of Ξ = 3.16 MHz for $\delta(^{103}Rh)$ [see references (*9,12a*) for the corresponding shielding values].

Acknowledgments

Continuous support by Prof. W. Thiel is greatfully acknowledged, as is a Liebig fellowship from the Fonds der chemischen Industrie. I thank Prof. W. v. Philipsborn for his interest and for a preprint of reference (*5*). Calculations were performed on a Silicon Graphics PowerChallenge (Organisch-chemisches Institut, Universität Zürich) and on IBM RS6000 workstations (C4 cluster, ETH Zürich), as well as on a NEC-SX4 (CSCS, Manno, Switzerland).

Literature Cited

1. e.g. *Applied Homogeneous Catalysis with Organometallic Compounds*; Cornils, B.; Herrmann, W. A., Eds.; VCH Publishers: Weinheim, New York, 1996; Vol. 1 and 2.
2. Review: Ziegler, T., *Can. J. Chem.* **1995**, 73, 743.
3. a) Bönnemann, H.; Brijoux, W.; Brinkmann, R.; Meurers, W.; Mynott, R.; von Philipsborn, W. ; Egolf, T., *Journal of Organometallic Chemistry* **1984**, *272*, 231. b) Bönnemann, H. *Angew. Chem. Int. Ed. Engl.* **1985**, *24*, 248.
4. a) Koller, M. ; Philipsborn, W. v., *Organometallics* **1992**, *11*, 467. b) Meier, E. J.; Kozminski, W.; Linden, A.; Lustenberger, P. ; Philipsborn, W. v., *Organometallics* **1996**, *15*, 2469.
5. Review: Philipsborn, W. v., *Chem. Soc. Rev.*, in press.
6. a) Fornica, R.; Görls, H.; Seeman, B. ; Leitner, W., *J. C. S. Chem. Comm.* **1995**, 1479. b) Angermund, K.; Baumann, W.; Dinjus, E.; Fornika, R.; Görls, H.; Kessler, M.; Krüger, C.; Leitner, W.; Lutz, F., *Chem. Eur. J.* **1997**, *3*, 755.
7. Öhrström, L., *Comments Inorg. Chem.* **1996**, *18*, 305.

8. Recent reviews: a) Schreckenbach, G., Ziegler, T. *Theor. Chem. Acc.* **1998**, *2*, 71. b) Bühl, M.; Kaupp, M.; Malkin, V. G.; Malkina, O. L. *J. Comput. Chem.*, in press.
9. Bühl, M. *Chem. Phys. Lett.* **1997**, 267, 251.
10. Bühl, M.; Malkina, O. L.; Malkin, V. G. *Helv. Cim. Acta* **1996**, *79*, 742.
11. Bühl, M. *Organometallics*, **1997**, *16*, 261.
12. a) Bühl, M.; Hamprecht, F. A. *J. Comput. Chem.* **1998**, *119*, 113. b) Bühl, M. *Angew. Chem. Int. Ed.***1998**, *37*, 142.
13. Godbout, N.; Havlin, R.; Salzmann, R.; Debrunner, P. G.; Oldfield, E. *J. Phys. Chem. A* **1998**, *102*, 2342.
14. F. Hutschka, A. Dedieu, W. Leitner, *Angew. Chem.* **1995**, *107*, 1905.
15. W. Leitner, M. Bühl, R. Fornika, C. Six, W. Baumann, E. Dinjus, M. Kessler, C. Krüger, A. Rufinska, submitted for publication.
16. Feher, F. J.; Blanski, R. L. *Organometallics* **1993**, *12*, 958.
17. Murphy, V. J.; Turner, H. *Organometallics* **1997**, *16*, 2495.
18. Ruppa, K. B. P.; Desmangles, N.; Gambarotta, S.; Yap G.; Rheingold, A. L. *Inorg. Chem.* **1997**, *36*, 1194.
19. Horwath, O. W. *Prog. NMR Spectrosc.* **1990**, *22*, 453.
20. Preuss, F.; Becker, H. *Z. Naturforsch.* **1986**, *41b*, 185.
21. Devore, D. D.; Lichtenhahn, J. D.; Takusagawa, F.; Maatta, E. A. *J. Am. Chem. Soc.* **1987**, *109*, 7408.
22. Frisch, M. J.; Trucks, G. W.; Schlegel, H. B.; Gill, P. M. W.; Johnson, B. G.; Robb, M. A.; Cheeseman, J. R.; Keith, T.; Petersson, G. A.; Montgomery, J. A.; Raghavachari, K.; Al-Laham, M. A.; Zakrzewski, V. G.; Ortiz, J. V.; Foresman, J. B.; Peng, C. Y.; Ayala, P. Y.; Chen, W.; Wong, M. W.; Andres, J. L.; Replogle, E. S.; Gomperts, R.; Martin, R. L.; Fox, D. J.; Binkley, J. S.; DeFrees, D. J.; Baker, J.; Stewart, J. J. P.; Head-Gordon, M.; Gonzales, C. ; Pople, J. A., Gaussian 94, Pittsburgh PA, 1995.
23. Becke, A. D. *Phys. Rev. A* **1988**, *38*, 3098.
24. Perdew, J. P. *Phys. Rev. B* **1986**, 33, 8822; *ibid.* **1986**, 34, 7406.
25. a) A. J. H. Wachters, *J. Chem. Phys.* **1970**, *52*, 1033; b) P. J. Hay, *J. Chem. Phys.*, **1977**, *66*, 4377.
26. Andrae, D.; Häußermann, U.; Dolg, M.; Stoll, H. ; Preuß, H., *Theor. Chim. Acta* **1990**, *77*, 123.
27. a) Dunning, T. H. *J. Chem. Phys.* **1977**, *66*, 1382. b) Binning, R. C.; Curtiss, L. A. *J. Comput. Chem.* **1990**, *11*, 1206.
28. Hehre, W. J.; Radom, L.; Schleyer, P. v. R. ; Pople, J. A., *Ab initio Molecular Orbital Theory*, Wiley: New York, 1986.
29. Cheeseman, J. R.; Trucks, G. W.; Keith, T. A.; Frisch, M. J. *J. Chem. Phys.* **1996**, *104*, 5497.
30. Becke, A. D., *J. Chem. Phys.* **1993**, *98*, 5648.
31. Lee, C.; Yang, W. ; Parr, R. G., *Phys. Rev. B* **1988**, *37*, 785.
32. Huzinaga, S. ; Klobukowski, M., *J. Mol. Struct.* **1988**, *167*, 1.
33. Kutzelnigg, W.; Fleischer, U. ; Schindler, M., in *NMR Basic Principles and Progress;* Springer-Verlag: Berlin, 1990, Vol. 23, pp 165-262.

Chapter 18

Calculated Chemical Shielding Tensors as an Aid to Elucidating the Method of Attachment of Alkoxysilanes to Magnesium Chloride

E. A. Moore[1] and N. J. Clayden[2]

[1]Department of Chemistry, The Open University, Milton Keynes MK7 6AA, United Kingdom
[2]School of Chemical Sciences, University of East Anglia, Norwich NR4 7TJ, United Kingdom

Ab initio calculations of ^{29}Si and ^{13}C NMR chemical shielding tensors have been used as an aid in the interpretation of experimental results aimed at elucidating the role of alkoxyalkylsilanes as external electron donors in Ziegler-Natta catalysis of propene. The results support a model in which dialkoxydialkylsilanes are loosely attached to two adjacent magnesium ions on the surface of magnesium chloride, but trialkoxyalkylsilanes are attached primarily by only one magnesium ion.

Nuclear magnetic resonance spectroscopy is a powerful and widely-used tool for investigating structure. Its utility is enhanced by the use of computational chemistry to aid in the interpretation. In this work we present an example of the use of calculated nuclear magnetic resonance parameters to help elucidate the role of alkoxyalkylsilanes in Ziegler-Natta catalysis.

The detail of the structure of the polymerisation centre present in suppported Ziegler-Natta catalysts for α-olefin polymerisation has been the subject of much research effort (e.g.*1-12*) The catalyst consists of a solid catalyst $MgCl_2/TiCl_4$/electron donor and a co-catalyst, an aluminium alkyl complexed with an electron donor. Proposed mechanisms for the polymerisation involve a titanium species attached to magnesium chloride with the olefin coordinated to titanium. The detail of the site at which the titanium species is attached is an important area of study in understanding the mechanism of catalysis and several recent papers (*10-12*) have investigated the surface structure of magnesium chloride and the attachment of $TiCl_4$, in particular the interaction of titanium species with the 100 and 110 planes of α and β- magnesium chloride.

The calculations reported here are concerned with the use of alkoxyalkylsilanes as external electron donors. When used as external electron donors, alkoxyalkylsilanes improve the stereoregularity of polypropene, reducing the level of atactic polymer present(*13*). The mechanism by which this occurs has been proposed as the stabilisation of magnesium chloride crystallites and inactivation of non-stereo-specific polymerisation sites. Nuclear magnetic resonance studies of alkoxyalkylsilanes in the presence of components of the catalyst have been carried out (*1, 14-16*) in order to investigate how these electron donors interact. We have used ab initio calculations of nuclear magnetic resonance chemical tensors to aid in the interpretation of these results.

Method

^{29}Si and ^{13}C nuclear magnetic resonance chemical shielding tensors were calculated using ab initio methods at the Hartree-Fock level. Early runs and those involving point charges used the GIAO method in Gaussian94 (*17*). These runs were performed on the EPSRC Convex C3800 vector supercomputing facility at the University of London and on the EPSRC superscalar Digital 8400 facility Columbus at the Atlas Centre of the Rutherford Appleton Laboratories. These facilities provided access to greater computer resource than was initially available at the author's institution. Later runs used GAMESS (*18*) to produce the wavefunctions and the molecular properties package RPAC (*19*) to obtain the chemical shielding. GAMESS and RPAC runs were performed on on a DEC alpha 600 cluster. Point charge runs used Gaussian94 because such runs are routine with this program but not with GAMESS. All runs in a set studying the effect of a particular change were performed on one machine and with the same program.

The basis sets employed were Pople's 6-311G (*20*) with d polarization functions on silicon, carbon and oxygen and a p polarization function on hydrogen where practicable. For the larger calculations it was necessary to use fewer basis functions. The strategy adopted was to retain a large basis set on silicon and smaller basis sets on the other atoms. Pople's 3-21G set (*21*) was used for magnesium and chlorine. Molecular modeling employed the Spartan program suite with the magnesium chloride structure generated by Cerius ^{2} .

Dimethoxydimethylsilane

It has been reported (*1*) that when dialkoxydialkylsilanes were milled with magnesium chloride, the ^{29}Si chemical shift was found to move by about 20 ppm downfield relative to that for the free silyl ether. There is also an increase in the anisotropic part of the chemical shielding tensor. The dialkoxydialkylsilanes were milled under an inert atmosphere with precautions taken to ensure the absence of water so that hydrolysis of the silane did not occur. The chemical shielding tensor was described by the parameters δ ($=\sigma_{33}-\sigma(\text{iso})$)and η ($=\sigma_{22}-\sigma_{11})/\delta$). Values of from -32 to -72 were observed for δ and from 0 to 0.11 for η when the silyl ethers were milled with magnesium chloride. ^{13}C nuclear magnetic resonance spectroscopy by contrast shows little difference in the ^{13}C chemical shift on milling with magnesium chloride.

Geometry Changes Modelling of the magnesium chloride surfaces in the presence of $TiCl_4$ has indicated strong binding to the 100 and 110 surfaces(*10, 12*). The 100 and 110 planes of magnesium chloride contain rows of magnesium ions with the distance between adjacent ions being 3.6 Å in the bulk solid. This distance would be expected to be less at the surface. A recent calculation (*12*) predicts a Cl-Cl distance of 3.3Å and the Mg-Mg distance would be similar. On the 111 surface, it is possible to locate groups of four magnesium ions in a diamond with a short diagonal of 3.6Å and a long diagonal of 6.2Å. Comparison of the optimized conformation of dimethoxydimethylsilane and the lattice spacing in $MgCl_2$ suggests that if the silyl ether does attach itself to the surface through both oxygens, then in doing so, it will probably be distorted. A widening of the OSiO angle would enable the distance between the oxygen atoms to more nearly match the distance between adjacent magnesium ions and the silyl ether to be anchored via the interaction of both oxygens with magnesium ions. It is also possible that two oxygen atoms coordinate to one magnesium and that this could lead to a decrease in OSiO bond angle. A previous preliminary molecular modeling result (Cox, P.; ICI internal report, unpublished) indicated extensive modifications around the geometry, in particular an increase in the OSiO angle. Ab initio optimization of $SiH_2(OH)_2$ in the presence of two Mg^+ ions (*22*) produced an increase in the SiO bond distance of 0.1Å and a decrease in the OSiO angle of 9°. (However the positions of the magnesium ions in these calculations were not fixed and the Mg-Mg distance increased to 6.17Å with the oxygen ions located between the magnesiums.) As a further guide to the expected geometry changes and mode of attachment to magnesium chloride, we optimized dimethoxydimethylsilane close to the surface of magnesium chloride. A slab of magnesium chloride was generated using Cerius^2 and then transferred to Spartan where the silane was optimized with the magnesium and chlorine positions fixed. A molecular mechanics optimization was performed using the SYBYL force field. An optimized position was attained in which the oxygen atoms of both methoxy groups settled between magnesium ions, the OSiO angle was 109°, the SiO bond distance 1.8Å and the CO bond distance 1.41Å. Optimization of dimethoxydimethylsilane itself at the 6-311G level gave Si-O = 1.67Å, C-O = 1.41Å and the OSiO angle as 111° and at the 6-311G* level Si-O = 1.64Å, C-O = 1.39Å and the OSiO angle as 111°. In our previous paper (*22*), calculations on a model compound, dihydroxysilane, indicated that the changes in silicon chemical shift were due to interaction with the magnesium chloride surface and not simply due to geometry changes in the molecule. In particular, the changes produced by varying the OSiO bond angle were much smaller than the observed changes in NMR chemical shift. By using dimethoxydimethylsilane as our model compound we are now able to investigate geometry changes involving the carbon atoms bonded to the oxygens. Rotating the methyl groups about the SiO bond produced very small changes in the shielding, up to 2 ppm for silicon and 1 ppm for carbon as shown in the top part of Table I.

If partial breaking of the C-O bonds and partial formation of an Mg-O bond were occurring when the oxygen interacted with the surface then a lengthening of the C-O bond would be expected. Increasing the CO bond distance led to a large decrease in ^{13}C shielding but a smaller change in ^{29}Si shielding. For an increase of 0.165Å in the CO bond distance, the isotropic ^{13}C shielding of the methoxy carbons decreased by 44 ppm and the anisotropy, δ, doubled. The ^{29}Si shielding decreased by 9 ppm . Such a large change in ^{13}C shielding accompanied by a smaller change in ^{29}Si shielding is not observed.

Since geometry changes alone do not account for the observed changes in shielding, we then considered interaction of the alkoxyalkylsilane with the magnesium chloride surface. The most likely way in which alkoxysilanes are attached to the surface of magnesium chloride is through donation of electrons from the oxygen atoms to the magnesium ions.

Table I. Effect of rotation of methyl groups about the SiO bond

C(OMe)-Si-C (OMe) angle	Mg-C(OMe) Å	σ(iso)^{29}Si ppm	σ(iso)^{13}C(OMe) ppm
95°[a]		414	160
114°[a]		414	159
156°[a]		416	160
95°[b]	2.3	390	159
119°[b]	2.7	403	144
133°[b]	3.0	406	148
152°[b]	3.4	405	154

[a]Free dimethoxydimethylsilane. Basis set 6-311G(3d) on Si, 6-311G(d) on C, O, 6-311G on H, Gaussian94 [b]Dimethoxydimethylsilane with $(Mg_2Cl_2)^{2+}$ Basis set as above with 3-21G on Mg and Cl, Gaussian94

Interaction with magnesium ions The effect on the NMR chemical shielding of the presence of the magnesium chloride was studied using a series of small models to represent the surface. Following the results of the molecular mechanics given above we considered a group of four Mg^{2+} ions with the oxygen atoms between the ions. This produced a very large decrease in the shielding of silicon and a large decrease in the shielding of the carbon nuclei. With a large basis set on silicon, it was not feasible to include chloride ions in the calculation. The chosen model therefore had a large electrical charge. Since this in itself will affect the shielding, runs were also performed with point charges at the positions of the magnesium ions. Point charges of +2, the nominal charge on the magnesium and +0.9 , which is closer to the charge calculated (*11*) for the magnesium ions in β-$MgCl_2$ were chosen. The results of these calculations are given in Table II where it can be seen that the point charge +2 model leads to a greater increase in anisotropy than the magnesium ions and the point charge +0.9 model leads to an increase rather than the observed decrease in shielding of the silicon nucleus. The asymmetry is not very well reproduced, but shows an increase from the free silane value for point charges of 0.9 and a decrease for four magnesium ions and for point charges of +2. Inspection of the output from the ab initio calculation with four magnesium ions shows that there is some Mg-O bonding present and that the Mulliken charges on magnesium were 1.0 and 1.9. The very different results from the point charge model and from the inclusion of magnesium ions indicate that partial bonding to magnesium does play a role.

We also considered the two oxygens interacting with adjacent magnesium ions by running a model in which the two oxygens were symmetrically placed above a pair of magnesium ions. The results for these are also given in Table II and show much less dramatic changes from the shielding of the free silane. For two magnesium ions we were also able to consider a model consisting of a block of two magnesium ions and two chloride ions. The results in Table I show that this model produces changes in shielding close to that observed. With the magnesium ions present, rotation of the methoxy methyl groups about the Si-O bond has a greater effect on the ^{13}C shielding as it alters the distance of the C nuclei from the magnesium ions. The results in Tables I and II indicate that an Mg-O distance of around 3Å produces calculated results comparable to those found experimentally. At closer distances, the interaction of the methyl groups with the surface affects the results. In our previous paper, we suggested an Mg-O distance of around 2Å, but the model used there had only hydrogen attached to the oxygen atoms and these would be less effected by the presence of the surface than the bulkier methyl groups.

Table II. Effect of magnesium ions on chemical shielding

model	Mg-O distance Å	σ(iso)^{29}Si ppm	δ^{29}Si ppm	σ(iso)^{13}C ppm
$Si(MeO)_2Me_2$[a]		402	-18	150
$(Mg_4)^{8+}$[a]	3.0	241	-202	109
4 point charges +2[a]	3.0	278	-343	-10
4 point charges +0.9[a]	3.0	502	-69	-84
$(Mg_2)^{4+}$[a]	3.0	360	-43	84
$Si(OMe)_2Me_2$[b]		414	-18	160
$(Mg_2Cl_2)^{2+}$[b]	3.0	390	-40	159
$Si(MeO)_2Me_2$[c]		425	-41	186
$(Mg_2Cl_2)^{2+}$[c]	2.55	375	-74	204
$(Mg_2Cl_2)^{2+}$[c]	3.0	386	-66	204
$(Mg_2Cl_2)^{2+}$[c] (C-O distance = 1.565Å)	2.55	345	-95	178

[a] Basis set as in Table I, Gaussian94 [b] 6-311G* on Si, C, O, H, 3-21G on Mg, Cl, Gaussian94, [c] 6-311G(d) on Si, 6-31G(d) on C, O MINI (*23*) on H, Mg, Cl, GAMESS/RPAC11

Trimethoxymethylsilane

^{29}Si NMR results (*1*) on trimethoxyphenylsilane indicate a major product with a shift 4ppm downfield from that of the free silane and a minor product with a change in shift comparable to that obtained for dialkoxydialkylsilanes. It was proposed that the minor product was due to complexing in the same manner as the dialkoxydialkylsilanes and that the weaker σ-donor properties caused the major product to have a different form of attachment. ^{13}C NMR results (*2*) on trimethoxyalkylsilanes ($Si(OMe)_3R$ where R =

Et, Ph) show little change (<1ppm) in the NMR chemical shift on complexing with magnesium chloride. Tetramethoxysilane shows two signals, one shifted by 2ppm and one by 6ppm(2). Because of the less symmetrical nature of the trialkoxyalkylsilanes, we considered attachment to the magnesium chloride surface through one oxygen as well through two. Molecular modeling of methyltrimethoxysilane in the presence of a magnesium chloride surface in the manner described for dimethoxydimethylsilane led to a structure with one oxygen close to two magnesium ions at MgO distances of 3.3Å and 3.65Å and one oxygen at 3.76Å from one of these magnesium ions but over 5Å from any other magnesium ions. The results of our calculations of NMR chemical shielding tensors for trimethoxymethylsilane attached to one or two magnesiums are given in Table III.

Table III. NMR chemical shifts for trimethoxymethylsilane with magnesium chloride

model	Mg-O distance Å	$\sigma(iso)^{29}Si$ ppm	$\delta^{29}Si$	$\sigma(iso)^{13}C(OMe)$ ppm
$Si(OMe)_3Me^a$		448	-24	170/172/170
Mg^{2+}-O^a	2.46	439	-49	169/168/154
Mg^{2+}-O^a	3.0	441	-44	169/168/143
Mg^{2+}- $((O)_2{}^a$	2.45/2.49	434	-41	169/193/157
$Si(OMe)_3Me^b$		453	-22	173/175
$(Mg_2)^{4+b}$	3.0	421	-104	167/-57
Mg-Mg perpendicular to O-O				
$(Mg_2)^{4+b}$	3.0	434	-72	157/162
Mg-Mg directly below O-O				

[a]Basis set 6-311G(d) on Si, 6-31G(d) on C, O, MINI on H, 3-21G on Mg, Cl, optimized geometry for free trimethoxymethylsilane, GAMESS/RPAC [b]Basis set as in a but geometry simplified to add two magnesiums symmetrically, GAMESS/RPAC.

Even at an Mg-O distance of 3Å, the carbon of the attached methoxy group showed a considerable shift from that of the free silyl ether, but the other methoxy carbon shifts changed by only 1 ppm. With attachment through two oxygens, a noticable change in the methoxy shift was calculated. A shift in the silicon shielding of about 20ppm for attachment through two oxygens and about 10ppm for attachment through one oxygen were calculated corresponding to the minor and major peaks observed experimentally for methyltriphenylsilane.

Conclusions

Comparison of the calculated chemical shielding tensors with those observed for alkoxyalkylsilanes milled with magnesium chloride suggest that the most probable interaction of the dialkoxysilanes is through two oxygens forming weak dative bonds to two adjacent magnesium ions. The observation of anisotropy in the silicon NMR spectrum indicates that the silyl ether is not freely rotating. The calculated results show that a close approach of the oxygen to the magnesium would produce a greater change in the NMR chemical shielding tensor than that seen experimentally.

The calculated change in silicon NMR chemical shielding tensor when trimethoxymethylsilane attaches to one magnesium ion is closest to that observed for the major peak for trimethoxyphenylsilane. Attachment by two oxygens directly above two magnesium ions results in calculated changes close to those seen for dialkoxydialkylsilanes. These results suggest that the major complex of trialkoxyalkylsilanes with magnesium chloride involves attachment by one oxygen, but attachment in a manner similar to that of the dialkoxydialkylsilanes is also observed.

A possible role of the silyl ethers is to occupy sites on the 110 surface, encouraging coordination of titanium to 100 surfaces where precoordination of olefin is more favorable (*12*).

Acknowledgments One of us (EAM) would like to thank the Engineering and Physical Science Council (EPSRC) for a grant of computer time on the Convex C3800 (grant number GR/K/66307) and the UK Computational Chemistry Working Party for time on the Digital 8400 facility Columbus at RAL.

Literature Cited

1. Clayden, N. J.; Jones, P. J. V.; *J. Polym. Sci. A: Polym. Chem.*, **1993**, *31*, 2835
2. Pakkanen, T. T.; Vähäsarja, E.; Pakkanen, T. A.; Iiskola, E.; Sormunen, P.; *J. Catalysis* **1990**, *121*, 248
3. Kashiva, N.; Yoshitake, J.; *Polym. J.* **1980**, *12*, 603
4. Chien, J. C. W.; Wu, J.; *J. Polym. Sci.. Polym. Chem. Ed.* **1982**, *20*, 2445
5. Chien, J. C. W.; Wu, J.; Kuo, C. I.; *J. Polym. Sci.. Polym. Chem. Ed.* **1982**, *20*, 2019
6. Sergeev, S. A.; Bukatov, G. D.; Zakhorov, V. A.; Moroz, E. M.; *Makromol. Chem.***1983**, *184*, 2421
7. Busico, W.; Corradini, P.; DeMartino, L.; Proto, A.; Savino, V.; Albizzati, E.; *Makromol. Chem.* **1985**, *186*, 1279
8. Ciardelli, F.; Carlini, C.; Altomare, A.; Mendoni, F.; *J. Chem. Soc. Chem. Comm.* **1987**, 94
9. Tritto, I.; Sacchi, M. C.; Locatelli, P.; Zannoni, G.; *Macromolecules* **1988**, *21*, 384
10. Colbourn, E. A.; Cox, P. A.; Carruthers, B.; Jones, P. J. V.; *J. Mater. Chem.* **1994**, *4*, 805
11. Lin, J. S.; Catlow, C. R. A.; *J. Mater. Chem.* **1993**, *3*, 1217
12. Lin, J. S.; Catlow, C. R. A.; *J. Catalysis* **1995**, *157*, 145
13. Parodi, S.; Nocci, R. ; Gianninni, U.; Barbe, P. C.; Scata, U.; *Eur. Patent 45 977* Montedison, **1980**
14. Pakkanen, T. T.; Vähäsarja, E.; Pakkanen, T.A.; Iiskola, E.; Sormunen, P.; *J. Catal..*, **1990**, *121*, 248
15. Sormunen, P.; Pakkanen, T. T..; Vähäsarja, E.; Pakkanen, T.A.; Iiskola, E.; *Stud. Surf. Sci. Catal.*, **1990**, *56*, 139
16. Vähäsarja, E.; Pakkanen, T. T..; Pakkanen, T.A.; Iiskola, E.; Sormunen, P.; *J. Polym Sci A: Polym. Chem. Ed.*, **1987**, *25*, 3241
17. GAMESS version 26 July 1995. Schmidt, M. W.; Baldridge, K. K.; Boatz, J. A.; Elbert, S. T.; Gordon. M. S.; Jensen, J. H.; Koseki, S.; Matsunaga, N.; Nguyen, K. A.; Su, S. J.; Windus, T. A.; together with Dupuis, M.; Montgomery, J. A.; *J. Comput. Chem.*, **1993**, *14*, 1347

18. Bouman, T. D.; Hansen, Aa. E. ; Bak, K. L.; Pedersen, T. B.; Kirby, R. A.; *Program RPAC, Molecular Properties Package*, Version 11.0, University of Copenhagen, Denmark, **1996**
19. Krishnan, R.; Binkley, J. S.; Seeger, R.; Pople, J. A.; *J. Chem. Phys.*, **1980,** *72,* 650
20. McLean, A. D.; Chandler, G. S.; *J. Chem. Phys.*, **1980,** *72,* 5639
21. Gordon, M. S.; Binkley, J. S.; Pople, J. A.; Pietro, W. J.; Hehre, W. J.; *J. Am. Chem. Soc.*, **1982,** *104,* 2797
22. Clayden, N. J.; Moore, E. A.; Selina, W. L.; Roberts, D. R.; Chem. Phys. Lett. **1996**, *248*, 173
23. Huzinaga, S.; Andzelm, J.; Klobukowski, M.; Radzio-Andzelm, E.; Sakai, Y.; Tatewaki, H.; *Gaussian Basis sets for molecular calculations;* Elsevier, Amsterdam, Netherlands, **1984**

Chapter 19

Aluminum Magnetic Shielding Tensors and Electric Field Gradients for Aluminum(I) Hydride, Aluminum(I) Isocyanide, and the Aluminum(I) Halides: Ab Initio Calculations

Myrlene Gee and Roderick E. Wasylishen

Department of Chemistry, Dalhousie University, Halifax, Nova Scotia B3H 4J3, Canada

Ab initio calculations of aluminum nuclear magnetic shielding tensors for aluminum(I) hydride, aluminum(I) isocyanide and the aluminum(I) halides are presented. Accurate experimental ^{27}Al shielding tensors available for AlCl and AlNC from recent spin-rotation data provide a stringent test of the ab initio calculations. For these two molecules, RHF calculations employing large basis sets provide shielding tensors which are in good agreement with experiment. The calculations indicate that the ^{27}Al spin-rotation constant for AlH is relatively large, $C_{\perp}$ = 295 kHz. Calculated boron and gallium shielding tensors of the group 13 hydrides and halides exhibit trends analogous to the corresponding aluminum tensors. For example, the group 13 nucleus of the hydrides is least shielded, while that of the fluoride is most shielded. In all cases, the calculated B, Al and Ga shieldings perpendicular to the bond axis are found to correlate with the HOMO-LUMO gap. The component of the shielding tensor along the bond axis is insensitive to substituent and approximately equal to the free-atom value: 202.0 ppm for B, 789.9 ppm for Al and 2638.6 ppm for Ga. Calculated ^{27}Al quadrupolar coupling constants (C_Q) for AlF, AlCl and AlNC agree very well with experimental values. For AlH, the calculated values of C_Q are approximately -46 MHz while the reported experimental value is -36.72 MHz. This discrepancy may be due, in part, to the neglect of the ^{27}Al spin-rotation interaction in the analysis of the hyperfine structure in the high-resolution microwave spectrum of AlH.

Aluminum NMR spectroscopy is a well-established technique for the characterization of solid materials such as ceramics, cements, glasses and zeolites (*1-4*). It has also been used extensively to investigate aluminum complexes in aqueous and non-aqueous solutions (*5*, *6*). The widespread use of ^{27}Al (I = 5/2) NMR can be attributed to its favorable NMR properties: a natural abundance of 100%, a fairly large nuclear magnetic moment and a relatively small nuclear quadrupole moment. Isotropic aluminum chemical shifts are available for hundreds of compounds and from these data it is clear that the chemical shift range is approximately 300 ppm (*5*, *6*). In spite of this significant range, orientation-dependent aluminum chemical shifts have only recently been unambiguously characterized. Anisotropic aluminum chemical shifts have been reported for sapphire (α-Al_2O_3) (*7*), aluminum trichloride phosphoryl trichloride ($AlCl_3 \cdot OPCl_3$) (*8*) and a series of pseudo-octahedral aluminum complexes (*9*). While ^{27}Al NMR studies of stationary powder samples provide information about the relative orientation of the aluminum shielding and electric field gradient (EFG) tensors, information concerning the orientation of these tensors in the molecular frame is generally not available. However, reliable first-principles calculations of shielding and EFG tensors may be used to suggest likely orientations of these tensors in the molecular framework. Before tackling complex systems, it is important to establish the accuracy of ab initio methods for small molecules. With this goal in mind, the ab initio results for aluminum(I) hydride, aluminum(I) isocyanide and the aluminum(I) halides are presented. For AlCl and AlNC, the calculated magnetic shielding tensors can be compared to experimental values derived from ^{27}Al spin-rotation data (*vide infra*). For comparative purposes, calculated boron and gallium magnetic shielding tensors for the group 13 hydrides and halides are also presented. Finally, ^{27}Al nuclear quadrupolar coupling constants calculated for AlH, AlF, AlCl and AlNC are compared with accurate experimental values obtained from high-resolution microwave spectroscopy. One of the main benefits of comparing calculated and experimental ^{27}Al shielding and quadrupolar coupling data for these simple linear molecules is that the experimental data are obtained for "isolated" molecules in the gas phase. Furthermore, it is straightforward to correct calculated shielding and quadrupolar coupling data for rovibrational averaging.

Chemical Shielding

NMR spectroscopists generally measure the chemical *shift*, δ, which is defined by equation 1:

$$\delta = \frac{\nu_{sample} - \nu_{ref}}{\nu_{ref}} \times 10^6 \approx \sigma_{ref} - \sigma_{sample} \qquad (1)$$

where ν_{sample} and ν_{ref} are the resonance frequencies of the sample and the reference respectively; σ_{sample} and σ_{ref} are their respective nuclear magnetic shielding constants. The $Al(H_2O)_6^{3+}$ cation of 0.1 M $Al(NO_3)_3$(aq) serves as a chemical shift reference for ^{27}Al NMR studies. The aluminum absolute shielding constant for $Al(H_2O)_6^{3+}$ is not known but ab initio calculations indicate a value of approximately 612 ppm (*10*). Experimental techniques for measuring absolute shielding constants (i.e., relative to the bare nucleus) have been summarized (*11*, *12*). If accurate nuclear spin-rotation data are available for one or more molecules containing the nucleus of interest, reliable "experimental" absolute shielding constants can be determined. For ^{27}Al, accurate spin-rotation constants are only available for AlCl and AlNC.

In general, nuclear magnetic shielding is described by a second-rank tensor. For linear molecules, only two components are unique: $\sigma_\parallel$, the shielding when the C_∞-axis of the molecule is parallel to the applied magnetic field, and $\sigma_\perp$ for the perpendicular orientation. The isotropic shielding constant is given by $\sigma_{iso} = (\sigma_\parallel + 2\sigma_\perp)/3$ and the span, Ω, of the shielding tensor is $|\sigma_\parallel - \sigma_\perp|$.

It is often convenient to describe the shielding at a particular nucleus as arising from two mechanisms, as originally proposed by Ramsey (*13*):

$$\sigma = \sigma^{d} + \sigma^{p} \tag{2}$$

The diamagnetic portion of the shielding tensor, σ^{d}, depends only on the ground electronic state of the molecule, whereas the paramagnetic part, σ^{p}, depends on both the ground and excited electronic states of the molecule. Because virtual molecular electronic states must be accurately represented, the calculation of reliable shielding tensors remains a challenging problem. Generally, the inclusion of electron correlation together with highly polarized basis sets is essential for reliable ab initio calculations of shielding tensors. With the currently available technology, such demanding calculations are generally feasible only for moderately small molecules. The current status of ab initio magnetic shielding calculations has been reviewed recently (*14-17*).

It has been known for many years that information about the paramagnetic term can be derived from experimental spin-rotation tensors (*18*). Nuclear spin-rotation tensors are measured using molecular beam magnetic resonance (MBMR) and molecular beam electric resonance (MBER) methods as well as high-resolution microwave spectroscopy (*19, 20*). Ramsey (*13, 18*) and Flygare (*21, 22*) have shown how the diagonal components of the spin-rotation tensor are related to the paramagnetic shielding tensor. For a linear molecule, the shielding component perpendicular to the molecular axis is given by:

$$\sigma_{\perp} \simeq \frac{-m_{p}C_{\perp}}{2mg_{N}B} + \sigma^{d}_{atom} \tag{3}$$

where m and m_p are the masses of the electron and proton, respectively, g_N is the nuclear g-value, $C_{\perp}$ is the perpendicular component of the spin-rotation tensor, B is the molecular rotation constant and σ^{d}_{atom} is the shielding constant of the free atom. For boron, aluminum and gallium, the following values have been calculated for σ^{d}_{atom}: 202.0, 789.9 and 2638.6 ppm, respectively (*23, 24*). The sign convention used for equation 3 follows that of Ramsey (*18*) where $C_{\perp}$ is positive if $g_N > 0$ and negative if g_N is < 0. It is important to recognize that, with few exceptions (*25*), the sign convention must result in a negative sign for the first term in equation 3. Recently, there has been a renewed interest in the accurate calculation of spin-rotation tensors for diatomics using modern quantum chemistry techniques (*26, 27*).

The paramagnetic component of the shielding tensor along the molecular axis is zero; thus $\sigma_{\parallel}$ can be accurately calculated by ab initio methods. Flygare (*22*) has shown that, to a good approximation:

$$\sigma_{\parallel} = \sigma^{d}_{\parallel} \simeq \sigma^{d}_{atom} \tag{4}$$

In order to make a rigorous comparison between calculated and experimental shielding tensors, it is necessary to correct the calculated shielding for rovibrational averaging (*28-33*). Typically, ab initio calculations are carried out on an isolated rigid molecule at the equilibrium geometry, while the spin-rotation tensor, which is used to obtain $\sigma_\perp$, is a weighted average over the rotational and vibrational coordinates of the molecule. Generally, spin-rotation constants are measured for molecules in their ground vibrational state. For diatomic molecules, one can estimate vibrational corrections using the following expressions (*28-31*):

$$\langle\sigma\rangle^T = \sigma_e + f(T)\left(\frac{d\sigma}{dr}\right)_e + g(T)\left(\frac{d^2\sigma}{dr^2}\right)_e \tag{5}$$

where σ_e is the calculated shielding at r_e, and $f(T)$ and $g(T)$ are defined as:

$$f(T) = \frac{3h}{16\pi^2 r_e \mu c\omega_e}\left(1 + \frac{\alpha_e\omega_e}{6B_e^2}\right)\coth\left(\frac{hc\omega_e}{2kT}\right) + \frac{kT}{2\pi^2 r_e \mu c^2 \omega_e^2} \tag{6}$$

$$g(T) = \frac{h}{16\pi^2 \mu c\omega_e}\coth\left(\frac{hc\omega_e}{2kT}\right) \tag{7}$$

where B_e, ω_e and α_e are the usual molecular spectroscopic constants and μ is the reduced mass (*34*).

Recent reports of spin-rotation constants for aluminum(I) chloride (*35*) and aluminum(I) isocyanide (*36*) have made possible the comparison of "experimental" and ab initio calculated shielding results. If one were able to measure the ^{27}Al chemical shift of one or both these compounds, it would be possible, in principle, to establish an absolute shielding scale for aluminum; however, the high reactivity of these compounds has so far precluded such measurements. High-resolution microwave measurements have also been recently carried out on AlH (*37*); however, analysis of the data did not consider the ^{27}Al spin-rotation interaction (*vide infra*).

There are numerous reports in the literature involving ab initio calculations of ^{27}Al nuclear shielding (*10, 38-44*). Of particular relevance to this work are the results of Gauss et al. (*42*) where ab initio calculations of isotropic aluminum shielding constants for AlH, AlF and AlCl using self-consistent field (SCF) and Møller-Plesset (MP2) methods are presented. As well, the relationship between the HOMO-LUMO gap and isotropic values of the nuclear shielding constants was demonstrated. Calculations of aluminum shielding constants for AlX_4^- (X = H, F, Cl, Br, or I) have shown that inclusion of the spin-orbit interaction improves considerably the agreement between experimental and calculated results (*44*), particularly for the heavier halides.

Computational Details. Restricted Hartree-Fock (RHF) calculations were carried out using Gaussian 94 (*45*) and ACES II (*46*) on an IBM RISC/6000 computer. The gauge independent atomic orbitals (GIAO) method was used for the shielding calculations (*47*). All second-order many-body perturbation theory (MBPT2, also referred to as MP2) calculations were performed with ACES II (*46*).

The bond lengths employed for the calculations are summarized in Table I. For AlH, AlF, AlCl and AlNC, a detailed study of the basis set dependence of the nuclear shielding was carried out, the smallest basis set being 6-31G and the largest being 6-311++G(3df,3pd). For the largest basis set, the notation indicates that the 6-311G basis set is augmented by three sets of d and f polarization functions in addition to diffuse functions for atoms other than hydrogen for which three sets of p and d polarization functions and diffuse functions are employed. Calculations were performed at both RHF and MP2 levels of theory for AlH, AlF, AlCl and AlNC. First and second derivatives of the shielding with respect to internuclear separation were estimated by calculating the shielding at $r_e \pm 0.01$ Å.

For comparative purposes, calculated boron and gallium shielding tensors of their respective hydrides and halides were also investigated at the RHF level of theory with the 6-311G(d,p) basis set, which was the largest available for all the elements involved, with the exception of iodine for which a [10s,8p,4d] basis set (*48*) with two sets of f functions (orbital exponents of 4.0 and 1.0) (*49*) was used. Calculations were also performed at the MP2 level, except for molecules containing Ga, Br and I.

Results and Discussion. The calculated aluminum shielding tensors for AlH, AlF, AlCl and AlNC for a variety of basis sets are presented in Table II. This particular family of basis sets was chosen as it is the one most commonly used in shielding calculations, especially for large systems. The "experimental" magnetic shielding tensors are obtained from

Table I : Geometries Employed in Aluminum Shielding and EFG Tensor Calculations

Molecule	Bond Length(s) (Å)		Reference
AlH	1.645 362 2		(*50*)
	1.645 366 92		(*51*)
AlF	1.654 360		(*52*)
AlCl	2.130 11		(*53*)
AlNC	1.849 (Al, N)	1.171 (N, C)	(*54*)
	1.8785 (Al, N)	1.1938 (N, C)	optimized
AlBr	2.294 80		(*53*)
AlI	2.537 09		(*53*)
BH	1.232 4		(*55*)
BF	1.262		(*55*)
BCl	1.716		(*56*)
BBr	1.888		(*55*)
BI	2.10		(*57*)
GaH	1.662 120 7		(*58*)
GaF	1.774 361 9		(*52*)
GaCl	2.201 690		(*55*)
GaBr	2.352 48		(*55*)
GaI	2.574 673		(*59*)

Table II: Aluminum Magnetic Shielding Tensors for AlH, AlF, AlCl and AlNC[a]

molecule and basis set	RHF σ_{iso}	RHF $\sigma_{\perp}$	RHF $\sigma_{\parallel}$	MP2 σ_{iso}	MP2 $\sigma_{\perp}$	MP2 $\sigma_{\parallel}$
AlH : 6-31G	182.5	-120.0	787.4	223.2	-58.9	787.5
6-31G(d,p)	205.9	-84.8	787.3	212.7	-74.6	787.4
6-31++G(d,p)	203.7	-88.1	787.3	208.5	-81.0	787.4
6-311G(d,p)	129.4	-199.5	787.3	138.1	-186.7	787.6
6-311G(3df,3pd)	137.0	-188.2	787.4	-[b]	-	-
6-311++G(3df,3pd)	136.9	-188.4	787.4	-[b]	-	-
Gauss et al.:[c] DZP	126.8	-	-	137.0	-	-
DZPCS	131.5	-	-	136.6	-	-
AlF : 6-31G	597.6	500.6	791.5	599.6	503.3	792.2
6-31G(d,p)	598.3	501.6	791.8	592.7	492.9	792.4
6-31++G(d,p)	608.3	516.5	791.8	604.8	511.1	792.3
6-311G(d,p)	572.2	462.3	791.9	564.7	450.8	792.6
6-311G(3df,3pd)	576.8	469.3	791.9	-[b]	-	-
6-311++G(3df,3pd)	577.6	470.5	791.9	-[b]	-	-
Gauss et al.:[c] DZP	571.9	-	-	563.3	-	-
DZPCS	574.2	-	-	565.5	-	-
AlCl : 6-31G	512.9	372.4	793.9	520.0	382.8	794.3
6-31G(d,p)	512.5	371.8	793.8	511.0	369.5	794.1
6-31++G(d,p)	509.8	367.8	793.7	-[b]	-	-
6-311G(d,p)	468.7	306.2	793.8	465.0	300.3	794.3
6-311G(3df,3pd)	473.7	313.7	793.8	-[b]	-	-
6-311++G(3df,3pd)	473.1	312.8	793.8	-[b]	-	-
Gauss et al.:[c] DZP	466.0	-	-	463.5	-	-
DZPCS	469.1	-	-	464.7	-	-
Experimental	473.4	313(17)	794.3	473.4	313(17)	794.3
AlNC:[d] 6-31G	579.5	473.5	791.5	577.5	470.4	791.6
6-31G(d,p)	570.5	459.9	791.6	563.3	449.1	791.7
6-31++G(d,p)	573.8	464.8	791.7	566.6	454.1	791.7
6-311G(d,p)	540.1	414.3	791.7	532.6	403.0	791.9
6-311G(3df,3pd)	542.4	417.8	791.7	-[b]	-	-
6-311++G(3df,3pd)	543.3	419.2	791.7	-[b]	-	-
Experimental	520.0	384.3(8.9)	791.9	520.0	384.3(8.9)	791.9

a. Vibrational corrections have not been applied to the calculated shielding, as these corrections are small - see text.
b. MP2 calculations for these basis sets were not possible with our computer facilities.
c. Gauss et al. (*42*) employed double-zeta plus polarization functions (DZP) and DZP where the inner p functions of the valence shell are uncontracted (DZPCS). Values for $\sigma_{\perp}$ and $\sigma_{\parallel}$ are not reported.
d. Calculated using the optimized geometry.

the spin-rotation constants using equations 3 and 4. In all cases, $\sigma_{\parallel} \approx \sigma^{d}_{atom} = 789.9$ ppm (*23*), as anticipated from equation 4. While Flygare et al. (*60*) have proposed semi-empirical methods of calculating $\sigma_{\parallel}$, we prefer to use the ab initio $\sigma_{\parallel}$ as the "experimental" value. The size of the basis set and the inclusion of electron correlation results in minor variations in $\sigma_{\parallel}$. For AlCl and AlNC, the $C_{\perp}(^{27}Al)$ are 5.54(16) kHz (*35*) and 3.850(84) kHz (*36*), respectively. The experimental values of $\sigma_{\perp}$ for AlCl and AlNC in Table II were obtained using these spin-rotation constants and equation 3. The experimental uncertainties in $C_{\perp}(^{27}Al)$ result in $\sigma_{\perp}$ values with errors of 17 ppm for AlCl and 8.9 ppm for AlNC. Given these uncertainties, the agreement between experimental and calculated results is excellent. The aluminum shielding tensor of AlNC was calculated using an optimized geometry, as the experimental equilibrium geometry is not well-determined at this time (*54*). The rovibrational correction, $\langle\sigma_{iso}\rangle_{v,J} - \sigma_{iso}(r_e)$, for AlCl is 2.12 ppm, which is relatively small.

The ^{27}Al spin rotation constant for AlF has also been reported, 8(6) kHz (*52*); however, given the experimental error, $\sigma_{\perp} = 484 \pm 668$ ppm! Assuming that the ab initio calculations for AlF are of the same caliber as those for AlCl and AlNC, one can predict $C_{\perp}$. Using equations 5 to 7 and published molecular parameters (*61*), the vibrational correction amounts to +0.7 ppm, resulting in $\sigma_{\perp}$ = 470 ppm (for RHF with the 6-311++G(3df,3pd) basis set). Using this value and equation 3, the spin-rotation constant is calculated to be 8.37 kHz, in excellent agreement with the reported value.

Unfortunately, there are no experimental spin-rotation data for AlH; however, $C_{\perp}$ can also be predicted from the ab initio data. Using $\sigma_{\perp}$ = 138.0 ppm (from MP2/6-311G(d,p)) and the calculated vibrational correction, 6.76 ppm, one obtains $C_{\perp} = 295$ kHz, which agrees well with a much earlier ab initio calculation, $C_{\perp}$ = 260 kHz (*39*). Interestingly, $\sigma_{\perp}$ is predicted to be negative for AlH, consistent with the prediction of a positive (i.e., paramagnetic) perpendicular component of the magnetic susceptibility tensor ($\chi_{\perp}$) of the Van Vleck type - independent of temperature (*39*). Recently, Sauer and Ogilvie reported experimental evidence of a marginally positive value for $\chi_{\perp}$ in AlH (*51*). Similar predictions are made for BH (*62*), for which $\sigma_{\perp}$(B) is also calculated to be negative (Table III).

In order to improve on the RHF values, electron correlation is included in the ab initio method by using MP2. Comparing the results for the 6-311G(d,p) basis set for all the diatomics, electron correlation results in a sizable correction to the RHF results; differences of 8.7 ppm for AlH, 8.0 ppm for AlF, 3.7 ppm for AlCl and 7.4 ppm for AlNC are noted. In the case of AlH, the isotropic value of the aluminum shielding calculated at the MP2 level is greater than the result from RHF theory; however, for the fluoride, chloride and isocyanide, the opposite trend holds. Gauss et al. (*42*) also observed this trend for the hydride, fluoride and chloride.

Finally, the basis set dependence of the isotropic shielding is illustrated in Figure 1, where both the RHF and MP2 data are plotted for AlH. It is evident that at least the 6-311G(d,p) basis set is essential for aluminum in order to obtain meaningful shielding constants; smaller basis sets give erroneous results, and any apparent agreement with experimental values is fortuitous. As well, one can be led to a false conclusion that the nuclear shielding calculated using the smaller basis sets is converging. For example, the RHF results for AlCl using the three smallest basis sets average to about 510 ppm; however, the larger basis sets average to about 470 ppm, a difference of 40 ppm. The results from the larger basis sets agree well with the available experimental data, and with the calculations reported by Gauss et al. (*42*). This feature is more dramatic in the case of AlH, where the difference between 6-31++G(d,p) and 6-311G(d,p) is about 70 ppm.

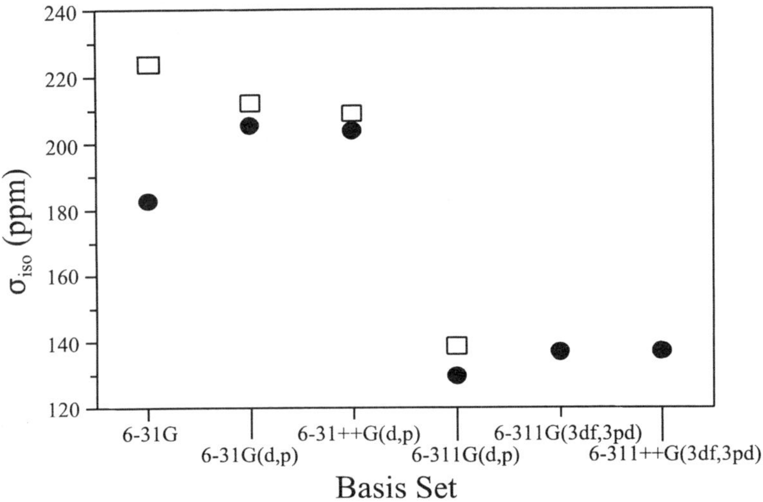

Figure 1. Basis set and method dependence of the calculated isotropic aluminum magnetic shielding for A1H. The RHF data are denoted by the filled circles and the MP2 data are denoted by the open squares. The horizontal scale is a qualitative representation of the basis set size.

This disparity is evident for all the molecules considered, at both the RHF and MP2 levels of theory. For larger systems, where one might be tempted to choose the smaller basis sets to reduce computation time, this investigation of model systems indicates it is necessary to use at least the 6-311G(d,p) basis set in order to obtain reliable results.

Comparison of the isotropic shielding for the aluminum diatomics (Tables II and III) reveals that the aluminum nucleus of AlF is the most shielded while that of AlH is least shielded, contrary to what one might naively expect on the basis of electron density arguments. As well, the calculated trend follows the "inverse halogen dependence" (IHD) where the fluorine substituent leads to greatest aluminum shielding while the iodine substituent leads to the least shielding (*63*). This observation can be rationalized by Ramsey's theory (*13*), where the paramagnetic part of the nuclear shielding is described in terms of the energies and electronic wavefunctions of excited molecular states. In many cases, the HOMO-LUMO gap dominates the shielding trend. Gauss et al. (*42*) have demonstrated a linear dependence of the isotropic aluminum shielding on the HOMO-LUMO gap for a series of simple aluminum compounds that included AlH, AlF and AlCl. As discussed by these authors, such correlations must be treated with care, as the calculated orbital energies have no physical meaning. Nevertheless, qualitative discussions of this nature are extremely valuable. In Figure 2, $\sigma_{\perp}$(Al) is plotted as a function of the HOMO-LUMO gap. In each case, the mixing involves a σ HOMO and a π^* LUMO. Perpendicular components of the boron and gallium shielding tensors of the group 13 hydrides and halides also show the same shielding trend (see Table III and Figure 2). The slope of the line increases proceeding down the group from B to Ga, consistent with the increasing chemical shift range on going from boron to gallium (e.g., see Figure 1 in

Table III: Nuclear Magnetic Shielding Tensors (in ppm) for Group 13 Nuclei of Several Diatomic Molecules[a]

molecule	RHF			MP2		
	σ_{iso}	$\sigma_{\perp}$	$\sigma_{\parallel}$	σ_{iso}	$\sigma_{\perp}$	$\sigma_{\parallel}$
BH	-259.6	-488.8	198.8	-212.7	-418.7	199.2
BF	78.2	12.6	209.4	86.7	24.8	210.6
BCl	-22.6	-139.3	210.6	1.6	-103.5	211.7
BBr	-58.2	-193.5	212.3	-	-	-
BI	-138.5	-315.1	214.7	-	-	-
AlH	129.4	-199.5	787.3	138.1	-186.7	787.6
AlF	572.2	462.3	791.9	564.7	450.8	792.6
AlCl	468.7	306.2	793.8	465.0	300.3	794.3
AlBr	432.8	251.6	795.2	-	-	-
AlI	352.7	130.6	796.9	-	-	-
GaH	1034.6	234.0	2636.0	-	-	-
GaF	2088.4	1812.9	2639.4	-	-	-
GaCl	1861.6	1471.5	2641.9	-	-	-
GaBr	1774.1	1339.5	2643.3	-	-	-
GaI	1594.9	1069.8	2645.2	-	-	-

a. The 6-311G(d,p) basis set is used for all calculations. Nuclear shieldings are not vibrationally corrected.

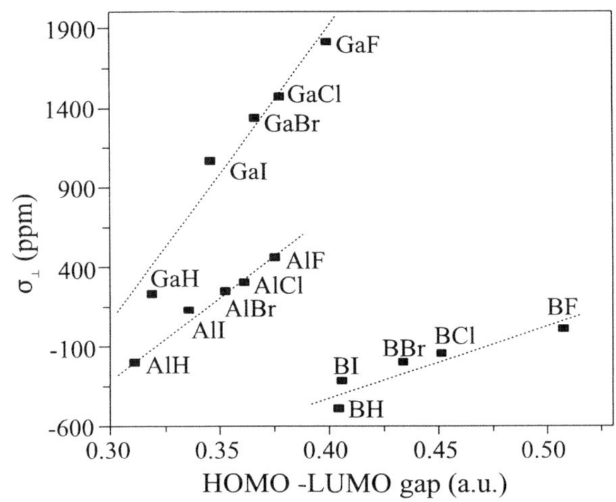

Figure 2. The nuclear magnetic shielding component perpendicular to the molecular bond for group 13 nuclei of several diatomic molecules as a function of the HOMO-LUMO gap.

reference *11*). It should be noted that the shielding calculations do not include relativistic effects, which may be quite significant for the molecules composed of heavier elements, particularly the iodides (*63*). Furthermore, the calculations presented in Table III are for comparative purposes. Boron magnetic shielding for BH approaching full configuration interaction (CI) has been reported (*64*, *65*).

Aluminum Electric Field Gradients

Classically, the electric field gradient at a nucleus is produced by the arrangement of charges (i.e., other nuclei and electrons) about that nucleus (*66*). If the nucleus is quadrupolar, as in the case of ^{27}Al, then the interaction of its nuclear quadrupole moment, eQ, with the largest component of the EFG tensor, V_{33}, is defined as the quadrupolar coupling constant, C_Q:

$$C_Q = \frac{eQV_{33}}{h} \tag{8}$$

For ^{27}Al the accepted value of the nuclear quadrupole moment is $0.1403(10) \times 10^{-28}$ m^2 (*67*, *68*). Recently, a new value of $0.1466(10) \times 10^{-28}$ m^2 has been suggested, based on values of C_Q from microwave spectroscopy (*69*) and accurate ab initio calculations of the EFG at aluminum in AlF and AlCl (Kellö, V.; Sadlej, A. J.; Pyykkö, P.; Sundholm, D.; Tokman, M. *Chem. Phys. Lett.*, submitted).

Most ^{27}Al quadrupolar coupling constants have been obtained from ^{27}Al NMR studies of solids (*70*). Nuclear quadrupolar coupling constants for quadrupolar nuclei in small molecules are sometimes available from high-resolution microwave spectroscopy (*34*, *71-73*). Consider an aluminum-containing diatomic where the neighboring nucleus does not possess the property of spin. The quadrupolar interaction provides a mechanism by which the quadrupolar nucleus, with spin I, couples with the rotational motion of the molecule, described by the quantum number J, resulting in the total quantum number, F. While the $J = 0$ level is not perturbed, the $J = 1$ rotation level is split by this interaction. The energies of these states are further modified by the spin-rotation interaction, so that for ^{27}Al ($I = 5/2$) the three frequencies and relative intensities associated with the $J = 1 \leftarrow 0$ transition are:

$$\nu(F{=}5/2 \leftarrow 5/2) = \nu_0 + 0.16\, C_Q - C_\perp \qquad \text{Rel. Int.} = 16 \tag{9}$$

$$\nu(F{=}7/2 \leftarrow 5/2) = \nu_0 - 0.05\, C_Q + 2.5 C_\perp \qquad \text{Rel. Int.} = 33 \tag{10}$$

$$\nu(F{=}3/2 \leftarrow 5/2) = \nu_0 - 0.14\, C_Q - 3.5 C_\perp \qquad \text{Rel. Int.} = 6 \tag{11}$$

The frequency, ν_0, is approximately $2B_0 - 4D_0$ where B_0 and D_0 are the usual spectroscopic constants (*34*). If all three transitions are observed, one can solve these equations for ν_0,

C_Q and $C_\perp$. Aluminum-27 quadrupolar coupling constants have been measured by microwave spectroscopy for AlH (*37*), AlF (*52*), AlCl (*35*, *74*) and AlNC (*36*).

The calculation of EFGs at a nucleus by ab initio methods has recently been discussed (*75-78*). Since the EFG depends only on the ground electronic state (i.e., a first-order property), it is often considered to be an easier parameter to calculate than the nuclear magnetic shielding. Nevertheless, accurate calculations of EFGs are not straight-forward because basis sets must be of high quality near the nucleus of interest. It is generally recognized that ab initio methods which include electron correlation, in conjunction with large basis sets, must be employed. As in the case of magnetic shielding, vibrational corrections to the calculated EFGs are required. The corrected V_{33} in the vth vibrational state is (*79*, *80*):

$$V_{33}^{v} \simeq V_{33}^{eq} + V_{33}^{vib}\left(v + \frac{1}{2}\right) \tag{12}$$

where V_{33}^{eq} is calculated by ab initio methods for r_e and the vibrational correction, V_{33}^{vib}, is determined using equations analogous to 6 and 7 (*79*, *80*).

Computational Details. All calculations of EFGs were performed using Gaussian 94 (*45*). The dependence of the EFG on the level of theory was investigated using the 6-311++G(3df,3pd) basis set for AlH and AlF. For AlCl and AlNC, the largest basis sets we could employ were 6-311G(3df,3pd) and 6-311G(d,p) respectively. The methods employed include RHF, density functional theory (DFT) using Becke's three parameter hybrid (*81*) and the correlation functional of Lee, Yang and Parr (B3LYP) (*82*), and the MP2, MP3, and MP4-SDQ (i.e., with single, double and quadruple excitations) series of perturbation methods. As well, some configuration interaction methods with double (CID), single and double (CISD), and single, double, and quadruple (QCISD) excitations were used. Finally, a coupled cluster calculation which includes double excitation (CCD) was performed. The basis set dependence of the ^{27}Al EFG for AlH was examined at the MP2 level of theory using basis sets ranging from 6-31G to 6-311++G(3df,3pd). Calculations were carried out using equilibrium geometries for AlH, AlF and AlCl, while the optimized bond lengths were used for AlNC (see Table I).

Calculated EFGs were converted from atomic units to Vm^{-2} using the factor 9.7177×10^{21} Vm^{-2} per atomic unit (*83*). The accepted value of eQ for ^{27}Al was used for all calculations (*68*). Use of the recently suggested value, $0.1466(10) \times 10^{-28}$ m^2, will result in C_Q values that are 4.5% larger than the calculated values presented in Tables IV and V. Vibrational corrections were estimated by determining the EFG at $r = r_e \pm 0.01$ Å.

Results and Discussion. The experimental and calculated ^{27}Al C_Qs for AlH, AlF, AlCl and AlNC are presented in Table IV. The dependence of C_Q on the level of theory is illustrated in Figure 3 for AlH. It is evident that the theoretical results converge to about -45.8 MHz, which is quite different than the experimental C_Q of -36.72 MHz (*37*). The vibrational correction to the calculated C_Q value, +0.71 MHz, does not account for the 9 MHz difference. To check the accuracy of the calculations, the same ab initio methods were used to determine C_Q for ^{27}Al in AlF, AlCl and AlNC (Table IV) where experimental data are available. Vibrational corrections of +0.39 MHz for AlF and +0.28 MHz for AlCl have been applied. For AlNC, no vibrational correction has been calculated, as

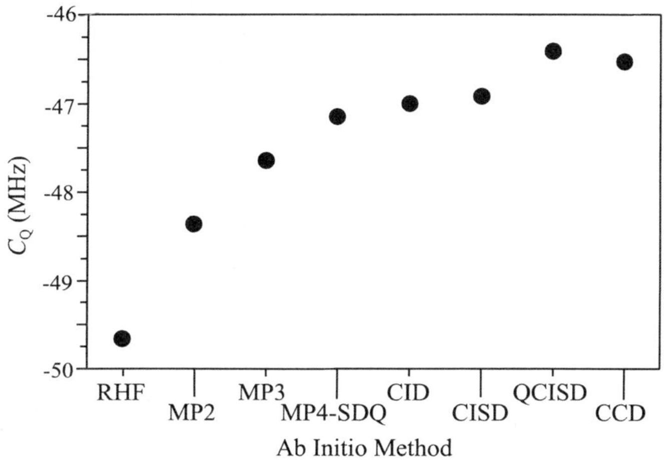

Figure 3. The dependence of the aluminum C_Q in A1H on the level of theory. The horizontal scale is strictly qualitative.

Table IV: Aluminum Quadrupolar Coupling Constants for AlH, AlF, AlCl and AlNC[a]

Level	C_Q (MHz)			
	AlH	AlF	AlCl	AlNC[b]
DFT (B3LYP)	-50.39	-37.56	-31.48	-35.6
RHF	-48.94	-39.13	-32.77	-36.0
MP2	-47.65	-36.31	-30.08	-34.1
MP3	-46.93	-36.57	-30.03	-33.9
MP4-SDQ	-46.44	-35.75	-29.72	-33.6
CID	-46.30	-37.39	-31.85	-34.9
CISD	-46.20	-37.24	-30.79	-34.9
QCISD	-45.70	-35.48	-29.34	-33.1
CCD	-45.82	-35.83	-29.35	-33.1
Experimental[c]	-36.72	-37.75	-30.4081	-35.6268

a. For AlH and AlF, the 6-311++G(3df,3pd) basis set was used, while the 6-311G(3df,3pd) and 6-311G(d,p) basis sets were used for AlCl and AlNC respectively.
b. No vibrational correction applied.
c. See the discussion in the text for references to experimental C_Qs.

equation 12 is not strictly valid for a triatomic molecule; however, it is estimated to be about 0.4 MHz. For these aluminum-containing molecules, the results are in good agreement with experimental values, suggesting that the currently recommended value of eQ for ^{27}Al (*68*) is reasonably accurate.

Since the ab initio calculations perform well for AlF, AlCl and AlNC, the large discrepancy between the experimental and calculated ^{27}Al C_Q in AlH may indicate an error in the analysis of the experimental data. In particular, consideration of the ^{27}Al spin-rotation interaction is clearly important. While the value of $C_\perp(^{27}\text{Al})$ obtained for AlH from our magnetic shielding calculations is relatively small (295 kHz) compared to C_Q, this can affect the analysis significantly, since $C_\perp$ is multiplied by factors of 2.5 and 3.5 in equations 10 and 11. On the other hand, C_Q is scaled by a factor of 0.16 or less in the same equations. In principle, $C_\perp$ for hydrogen is also involved; however, our ab initio results indicate that $C_\perp$ is only 0.93 kHz for ^{1}H, two orders of magnitude smaller than $C_\perp$ for ^{27}Al. Table V shows the basis set dependence of C_Q for AlH. As in the aluminum shielding calculations, there is a significant difference (about 9 MHz) between the three small basis sets and the larger ones.

Conclusions

The quality of aluminum nuclear magnetic shielding tensors calculated from ab initio methods can be assessed by comparison with experimental values obtained from spin-rotation constants. An interesting "inverse halogen dependence" of the shielding for the group 13 nuclei in the diatomic halides is presented and rationalized in terms of the HOMO-LUMO gap. The availability of EFGs from quadrupolar coupling constants obtained from high-resolution microwave spectroscopy allows for a good evaluation of theoretical methods. For AlF, AlCl and AlNC, the ab initio methods perform very well. The calculated quadrupolar coupling constant for AlH differs significantly from experi-

Table V : Basis Set Dependence of C_Q for AlH[a]

Basis Set	C_Q (MHz), vibrationally corrected	C_Q (MHz), not vibrationally corrected
6-31G	-36.97	-37.70
6-31G(d,p)	-38.83	-39.54
6-31++G(d,p)	-39.60	-40.32
6-311G	-47.75	-48.46
6-311G(d,p)	-48.75	-49.47
6-311++G(d,p)	-48.65	-49.37
6-311G(3df,3pd)	-47.57	-48.28
6-311++G(3df,3pd)	-47.65	-48.36

a. All calculations were carried out at the MP2 level.

mental results, a possible consequence of neglecting the ^{27}Al spin-rotation interaction in the analysis of the high-resolution microwave spectrum. It is interesting to point out that AlH, AlF and AlCl have been detected in interstellar space (*84-89*). The results of this study demonstrate that ab initio calculations can serve as a useful guide in the analysis of microwave spectroscopic data. More important to the NMR community, it is clear that ab initio methods have a bright future in complementing experimental studies.

Acknowledgments

We are grateful to all the members of the solid-state NMR group at Dalhousie University for helpful discussions and suggestions. We thank R. J. Bartlett and A. Perera for providing a copy of the ACES II software. As well, we are grateful to P. Pyykkö for providing the results of high-level calculations on AlF and AlCl. R. E. W. acknowledges financial support from the Natural Sciences and Engineering Research Council (NSERC) of Canada. M. G. wishes to thank NSERC and the Izaak Walton Killam Trust for post-graduate scholarships. Finally, special thanks to R. W. Schurko for contributing to the early stages of this project.

References

1. Smith, M. E. *Appl. Magn. Reson.* **1993**, *4*, 1.
2. Bastow, T. J. *Z. Naturforsch.* **1994**, *49a*, 320.
3. Klinowski, J. *Anal. Chim. Acta* **1993**, *283*, 929.
4. Alemany, L. B. *Appl. Magn. Reson.* **1993**, *4*, 179.
5. Akitt, J. W. *Prog. NMR Spectrosc.* **1989**, *21*, 1.
6. Delpuech, J. J. In *NMR of Newly Accessible Nuclei*; Laszlo, P., Ed.; Academic: New York, 1983; Vol. 2, pp 153-195.
7. Vosegaard, T.; Jakobsen, H. J. *J. Magn. Reson.* **1997**, *128*, 135.
8. Schurko, R. W.; Wasylishen, R. E.; Phillips, A. D. *J. Magn. Reson.* **1998**, *133*, 388.
9. Schurko, R. W.; Wasylishen, R. E.; Foerster, H. *J. Phys. Chem. A* **1998**, *102*, 9750.
10. Sykes, D.; Kubicki, J. D.; Farrar, T. C. *J. Phys. Chem. A* **1997**, *101*, 2715.
11. Jameson, C. J.; Mason, J. In *Multinuclear NMR*; Mason, J., Ed.; Plenum: New York, 1987; pp 51 - 88.

12. Jameson, C. J. In *The Encyclopedia of NMR*; Harris, R. K.; Grant, D. M., Eds.; John Wiley and Sons: New York, 1996; pp 1273 - 1281.
13. Ramsey, N. F. *Phys. Rev.* **1950**, *78*, 699.
14. De Dios, A. C. *Progr. NMR Spectrosc.* **1996**, *29*, 229.
15. Fukui, H. *Progr. NMR Spectrosc.* **1997**, *31*, 317.
16. Jameson, C. J. In *Nuclear Magnetic Resonance - A Specialist Periodical Report*; Webb, G. A., Ed.; Royal Soc. Chem.: Cambridge, 1998; Vol. 27, pp 44 - 82.
17. Schreckenbach, G.; Ziegler, T. *Theor. Chem. Acc.* **1998**, *99*, 71.
18. Ramsey, N. F. *Molecular Beams*; Oxford University: London, 1956.
19. English, T. C.; Zorn, J. C. In *Methods of Experimental Physics*; Williams, D., Ed.; Academic: New York, 1974; Vol. 3B, pp 669 - 846.
20. White, R. L. *Rev. Mod. Phys.* **1955**, *27*, 276.
21. Flygare, W. H. *J. Chem. Phys.* **1964**, *41*, 793.
22. Flygare, W. H. *Chem. Rev.* **1974**, *74*, 653.
23. Malli, G.; Froese, C. *Int. J. Quantum Chem.* **1967**, *IS*, 99.
24. Fraga, S.; Malli, G. *Many - Electron Systems: Properties and Interactions*; Studies in Physics and Chemistry 7; W. B. Saunders: Philadelphia, 1968; p 110.
25. Gatehouse, B.; Müller, H. S. P.; Gerry, M. C. L. *J. Chem. Phys.* **1997**, *106*, 6916 and references therein.
26. Gauss, J.; Ruud, K.; Helgaker, T. *J. Chem. Phys.* **1996**, *105*, 2804.
27. Gauss, J.; Sundholm, D. *Mol. Phys.* **1997**, *91*, 449.
28. Buckingham, A. D. *J. Chem. Phys.* **1962**, *63*, 3096.
29. Jameson, C. J. *J. Chem. Phys.* **1977**, *66*, 4977.
30. Jameson, C. J. *Bull. Magn. Reson.* **1980**, *3*, 3.
31. Jameson, C. J.; de Dios, A. C. *J. Chem. Phys.* **1993**, *98*, 2208.
32. Raynes, W. T.; Panteli, N. *Mol. Phys.* **1983**, *48*, 439.
33. Buckingham, A. D.; Olegário, R. M. *Mol. Phys.* **1997**, *92*, 773.
34. Gordy, W.; Cook, R. L. *Microwave Molecular Spectroscopy* (3rd edition); Wiley: New York, 1984.
35. Hensel, K. D.; Styger, C.; Jäger, W.; Merer, A. J.; Gerry, M. C. L. *J. Chem. Phys.* **1993**, *99*, 3320.
36. Walker, K. A.; Gerry, M. C. L. *Chem. Phys. Lett.* **1997**, *278*, 9.
37. Goto, M.; Saito, S. *Astrophys. J.* **1995**, *452*, L147.
38. O' Reilly, D. E. *J. Chem. Phys.* **1960**, *32*, 1007.
39. Laws, E. A.; Stevens, R. M.; Lipscomb, W. N. *J. Chem. Phys.* **1971**, *54*, 4269.
40. Weller, Th.; Meiler, W.; Köhler, H. -J.; Lischka, H.; Höller, R. *Chem. Phys. Lett.* **1983**, *98*, 541.
41. Chesnut, D. B. *Chem. Phys.* **1986**, *110*, 415.
42. Gauss, J.; Schneider, U.; Ahlrichs, R.; Dohmeier, C.; Schnöckel, H. *J. Am. Chem. Soc.* **1993**, *115*, 2402.
43. Schneider, U.; Ahlrichs, R. *Chem. Phys. Lett.* **1994**, *226*, 491.
44. Nakatsuji, H.; Hada, M.; Tejima, T.; Nakajima, T.; Sugimoto, M. *Chem. Phys. Lett.* **1996**, *249*, 284.
45. Gaussian 94, Revision B.2, Frisch, M. J.; Trucks, G. W.; Schlegel, H. B.; Gill, P. M. W.; Johnson, B. G.; Robb, M. A.; Cheeseman, J. R.; Keith, T. A.; Petersson, G. A.; Montgomery, J. A.; Raghavachari, K.; Al-Laham, M. A.; Zakrzewski, V. G.; Ortiz, J. V.; Foresman, J. B.; Cioslowski, J.; Stefanov, B. B.; Nanayakkara, A.; Challacombe, M.; Peng, C. Y.; Ayala, P. Y.; Chen, W.; Wong, M. W.; Andres, J. L.; Replogle, E. S.;

Gomperts, R.; Martin, R. L.; Fox, D. J.; Binkley, J. S.; Defrees, D. J.; Baker, J.; Stewart, J. P.; Head-Gordon, M.; Gonzalez, C.; Pople, J. A. Gaussian, Inc., Pittsburgh PA, **1995**.

46. ACES II, an ab initio program system, authored by J. F. Stanton, J. Gauss, J. D. Watts, W. J. Lauderdale, and R. J. Bartlett, also containing modified versions of the MOLECULE Gaussian integral program of J. Almlöf and P. R. Taylor, the ABACUS integral derivative program written by T. Helgaker, H. J. Aa. Jensen, P. Jørgensen, and P. R. Taylor, and the PROPs property integral code of P. R. Taylor. See also: Stanton, J. F.; Gauss, J.; Watts, J. D.; Lauderdale, W. J.; Bartlett, R. J. *Int. J. Quantum Chem. Symp.* **1992**, *26*, 879.
47. Ditchfield, R. *Mol. Phys.* **1974**, *27*, 789.
48. Poirier, R.; Kari, R.; Csizmadia, I. G. *Handbook of Gaussian Basis Sets*; Physical Sciences Data 24; Elsevier: Amsterdam, 1985; p G498.
49. Fukui, H.; Baba, T.; Inomata, H. *J. Chem. Phys.* **1996**, *105*, 3175.
50. White, J. B.; Dulick, M.; Bernath, P. F. *J. Chem. Phys.* **1993**, *99*, 8371.
51. Sauer, S. P. A.; Ogilvie, J. F. *J. Phys. Chem.* **1994**, *98*, 8617.
52. Hoeft, J.; Lovas, F. J.; Tiemann, E.; Törring, T. *Z. Naturforsch.* **1970**, *25a*, 1029.
53. Wyse, F. C.; Gordy, W. *J. Chem. Phys.* **1972**, *56*, 2130.
54. Robinson, J. S.; Apponi, A. J.; Ziurys, L. M. *Chem. Phys. Lett.* **1997**, *278*, 1.
55. Huber, K. P.; Herzberg, G. *Constants for Diatomic Molecules*; Van Nostrand Reinhold: New York, 1979.
56. Verma, R. D. *J. Mol. Spectrosc.* **1995**, *169*, 295.
57. Coxon, J. A.; Naxakis, S. *J. Mol. Spectrosc.* **1987**, *121*, 453.
58. Urban, R. -D.; Jones, H. *Chem. Phys. Lett.* **1992**, *190*, 609.
59. Barrett, A. H.; Mandel, M. *Phys. Rev.* **1958**, *109*, 1572.
60. Gierke, T. D.; Flygare, W. H. *J. Am. Chem. Soc.* **1972**, *94*, 7277.
61. Wyse, F. C.; Gordy, W.; Pearson, E. F. *J. Chem. Phys.* **1970**, *52*, 3887.
62. Stevens, R. M.; Lipscomb, W. N. *J. Chem. Phys.* **1965**, *42*, 3666.
63. Kaupp, M.; Malkina, O. L.; Malkin, V. G.; Pyykkö, P. *Chem. Eur. J.* **1998**, *4*, 118.
64. Gauss, J.; Ruud, K. *Int. J. Quantum Chem. Symp.* **1995**, *29*, 437.
65. Gauss, J.; Stanton, J. F. *J. Chem. Phys.* **1996**, *104*, 2574.
66. Han, O. H.; Oldfield, E. *Inorg. Chem.* **1990**, *29*, 3666.
67. Sundholm, D.; Olsen, J. *Phys. Rev.* **1993**, *47*, 2672.
68. Pyykkö, P. *Z. Naturforsch.* **1992**, *47a*, 189.
69. Lovas, F. J.; Tiemann, E. *J. Phys. Chem. Ref. Data* **1974**, *3*, 609.
70. Freude, D.; Haase, J. *NMR Basic Princ. and Prog.* **1993**, *29*, 1.
71. Townes, C. H.; Schawlow, A. L. *Microwave Spectroscopy*; McGraw-Hill: New York, 1955; pp 149 - 173.
72. Tatum, J. B. *Astrophys. J. Suppl. Ser.* **1986**, *60*, 433.
73. Ogilvie, J. F. *The Vibrational and Rotational Spectrometry of Diatomic Molecules*; Academic: San Diego, 1998.
74. Lide, D. R. *J. Chem. Phys.* **1965**, *42*, 1013.
75. Bailey, W. C. *J. Mol. Spectrosc.* **1997**, *185*, 403.
76. Pyykkö, P.; Seth, M. *Theor. Chem. Acc.* **1997**, *96*, 92.
77. Rizzo, A.; Ruud, K.; Helgaker, T.; Jaszuński, M. *J. Chem. Phys.* **1998**, *109*, 2264.
78. Palmer, M. H.; Blair-Fish J. A.; Sherwood, P.; Guest, M. F. *Z. Naturforsch.* **1998**, *53a*, 383.
79. Lucken, E. A. C. *Z. Naturforsch.* **1993**, *47a*, 133.

80. Kellö, V.; Sadlej, A. J. *Mol. Phys.* **1996**, *89*, 127.
81. Becke, A. D. *J. Chem. Phys.* **1993**, *98*, 5648.
82. Lee, C.; Yang, W.; Parr, R. G. *Phys. Rev. B* **1988**, *37*, 785.
83. Brown, R. D.; Head-Gordon, M. P. *Mol. Phys.* **1987**, *61*, 1183.
84. Merrill, P. W. *Astrophys. J.* **1953**, *118*, 453.
85. Herbig, G. *Pub. Astron. Soc. Pacific*, **1956**, *68*, 204.
86. Johnson, H. R.; Sauval, A. J. *Astron. Astrophys. Suppl. Ser.* **1982**, *49*, 77.
87. Sotirovski, P. *Astron. Astrophys. Suppl.* **1972**, *6*, 85.
88. Ziurys, L. M.; Apponi, A. J.; Phillips, T. G. *Astrophys. J.* **1994**, *433*, 729.
89. Cernicharo, J.; Guélin, M. *Astron. Astrophys.* **1987**, *183*, L10.

Chapter 20

Modeling ^{17}O NMR Tensors—efg and Chemical Shifts—in Oxides and Polyoxometallates

Marc Henry

Laboratoire de Chimie Moléculaire de l'État Solide (UMR CNRS 7513), Université Louis Pasteur, 4, Rue Blaise Pascal, 67070 Strasbourg, France

An empirical model is described allowing the easy computation of efg tensors and ^{17}O NMR shielding tensors in small molecules, molecular crystals and inorganic crystalline frameworks. It is based on the identification of the electronegativity with the electronic chemical potential, and of the chemical hardness with the HOMO-LUMO or band gap, allowing an easy computation of partial atomic charges and frontier indexes. Reliable values for the asymmetry parameters η_Q of the efg tensors can be obtained from these charge distributions. Calculations of C_Q values require the use of a Sternheimer antishielding factor and are in much poorer agreement with experiments. Reliable evaluation of isotropic chemical shifts values directly from the molecular or crystalline structure is demonstrated in absence of any π-bonding effects. Reliable evaluation of NMR shielding tensors is not yet possible, but very encouraging preliminary results have been obtained concerning the qualitative estimation of shielding anisotropies and asymmetries.

Multinuclear NMR techniques both in solution and in the solid-state are nowadays established on a routine basis for most nuclei. This leads to an accumulation of experimental data of very good quality on nuclear magnetic shielding and electric field gradient (efg) tensors. Faced with such a large amount of data, experimentalists heavily rely on empirical models to link the measured NMR properties and the molecular and/or crystalline structure. These empirical models though very useful, are however full of adjustable parameters whose physical meaning is very doubtful. On the other hand, *ab initio* calculations of NMR parameters have met considerable success, but are too often limited to very small molecules. Moreover, the mathematical requirements of these theories are so high that only a very small number of experimentalists are able to handle the softwares and make their own calculations. It is the aim of this paper to propose a robust and simple interface based on the widespread Windows environment allowing to switch very quickly between the molecular or crystalline structure and the NMR tensors. To be of the most widespread significance for both experimentalists and theoricists, the software is able to:

i) Treat small molecules and reticulated networks on the same theoretical grounds. The time-scale for the calculations on a Pentium-II is typically of a few minutes for very large zeolite networks, a few seconds for most crystalline structures and virtually no delay for isolated molecules. This allows the user to handle the full chemical diversity (protecting groups, disorder, substitutions...) and not a frustrating idealized model.

ii) Keep the amount of adjustable parameters to a strict minimum. Each parameter having a clear and well-defined physical meaning, the observed trends are easily analyzed in terms of basic chemical knowledge.

iii) Treat all nuclei, quadrupolar or not, light or heavy without any drastic changes to the formalism. This is done by a judicious choice of the variables needed to compute the NMR parameters. Consequently, the necessary *ab-initio* atomic informations are accessible to the NMR tensors only through programmed functions, and any changes inside the function (parameter refinements, relativistic corrections...) will alter the computed values but not the general philosophy of the model.

We have chosen to demonstrate the validity of our approach with the only magnetic isotope of oxygen ^{17}O ($I = 5/2$), whose magnetic resonance has been observed since 1951 despite a natural abundance of only 0.037% and a moderate quadrupole moment ($Q = -0.0256\times10^{-28}$ m^2) *(1)*. This interest in ^{17}O is beyond any doubts due to the wide range of chemical compounds formed by oxygen. Combined with metallic elements, it gives rise to an unlimited number of inorganic materials displaying very interesting electrical, optical or magnetical properties. Combined with non-metallic elements, it forms through strong covalent bonding, another unlimited number of organic molecules at the very basis of life chemistry. It is thus not exaggerated to say that oxygen is a corner-stone element of the periodic table and that the structure of oxygen-containing compounds is a subject of widespread significance bringing together scientists coming from very different research fields (chemistry, biology, materials science, mineralogy, earth science, astronomy...). Moreover, with a chemical shift range of -300...+1600 ppm and a Nuclear Quadrupole Coupling Constants (NQCC) ranging from -8 to +21 MHz, detailed interpretation of ^{17}O NMR parameters still remains a real challenge for both theoricists and experimentalists. Any advance in this field should then be welcomed by the whole NMR community.

The Partial Charge Model (PCM)

As explained above, an encapsulation of all the atomic details is needed in order to express the NMR tensors in terms of functional quantities embedded into an inalterable theoretical frame. It is our feeling that this encapsulation is best done using the two dyads which are electronegativity/hardness on one hand, and partial charge/frontier indexes on the other hand.

Electronegativity. First, for any system with N electrons having an energy E in an external potential v one may define its electronic chemical potential as $\mu_e = (\partial E/\partial N)_v$. For an isolated molecule, this allows to identify the HOMO and LUMO levels, whereas for a solid it is just the definition of the Fermi level of the electrons. Consequently, the electronegativity χ of this system which is just the opposite of its electronic chemical potential ($\chi = -\mu_e$) according to density-functional theory *(2)* should be an obvious partner for the description of any chemical bond. With this identification, the electronegativity of an atom $\chi°$ should be the mean value between its first ionization potential I and its electronic affinity A: $\chi° = ½(I+A)$, in agreement with the Mulliken definition *(3)*. For a molecule or a solid, it becomes a mean value $<\chi>$ common to all constituent atoms, in agreement with the electronegativity equalization principle of R.T. Sanderson *(4)*. Work functions of metals and non-metals, Fermi energies and photoelectric thresholds are accordingly highly related to the electronegativity concept *(5-6)*.

Hardness. The second derivative of energy against the number of electrons at constant potential $\eta = (\partial^2E/\partial N^2)_v$ is a second obvious chemical bonding parameter. This is because electrons have to flow between bonded atoms in order to equalize their electronegativities. Consequently, it is important to know for each atom the resistance of the electronic chemical potential to any change in the total number of electrons: $\eta = (\partial\mu_e/\partial N)_v = (\partial^2E/\partial N^2)_v$. For an isolated atom η = (I - A) and is just what the chemist calls its hardness (or its softness $\sigma = 1/\eta$) in agreement with the HSAB principle of R.G. Pearson *(7)*. Moreover, if an atom is visualized as a collection of electrons filling a sphere of radius r_{at}: $\eta = 1/(4\pi\varepsilon_0 r_{at})$ *(8)*, that is to say that hard atoms are small in opposition to soft atoms which are much bigger. For a molecule or a solid, the global hardness is most easily visualized as the HOMO-LUMO or band gap ΔE *(7)*. The higher the gap, the harder the compound and the higher the stability *(9)*.

Mean electronegativity. The partial charge q of an atom is merely the difference between the atomic number Z and the total number of electrons N assigned to that atom: q = Z - N. As explained above chemically bonded atoms have to give (q > 0) or gain electronic density (q < 0) in order to maintain a constant electronic chemical potential over the whole compound. For a compound having n constituent atoms, it is easy to show using classical electrostatics *(10)*, MO theory *(11)* or point-charge approximation of density-functional equations *(12)*, that each atom i in the compound should obey the following linear system:

$$\chi_i = <\chi> = \chi_i^0 + \eta_i \times (eq_i) + \frac{1}{4\pi\varepsilon_0}\sum_{j=1}^{n} M_{ij} \times (eq_j) \qquad \text{with} \qquad \sum_{i=1}^{n} q_i = z \qquad (1)$$

In this relationship e and ε_0 are respectively the elementary charge and vacuum permittivity, χ_i electronegativities, η_i chemical hardnesses and q_i partial charges, whose sum should give the global electric charge z carried by the compound. The remaining term M_{ij} is the Madelung contribution to the potential. For an isolated molecule it is just equal to $1/R_{ij}$, the reciprocal of the distance between atoms i and j. For a crystalline network, it is a sum over the whole lattice of such inverse distances *(13-14)*. Details about computation of this parameter have been given elsewhere *(15)*. Consequently, the knowledge of the free-atom electronegativities $\chi_i°$, their chemical hardnesses η_i and the molecular or crystalline structure M_{ij} uniquely determine through equations (1) the mean electronegativity $<\chi>$ and the associated partial charge distribution. In more physical words, electronic Fermi levels and gross atomic populations are directly accessible knowing relative orbital energies, atomic sizes and spatial atomic coordinates independently of any LCAO-MO formalism. From this charge distribution, it is very easy to characterize the chemical compound as a whole entity using a global ionicity index I, ranging from 0 ($q_i = 0$) up to 100% when the q_i reach their integral oxidation state z_i:

$$I(\%) = 100\sqrt{\frac{\sum_{i=1}^{n}\sum_{j=1}^{n} M_{ij} \times q_i \times q_j}{\sum_{i=1}^{n}\sum_{j=1}^{n} M_{ij} \times z_i \times z_j}} \qquad (2)$$

Global hardness. The partial charge concept helps the chemist to visualize within a molecule or a network, how the electronic density changes as a function of the spatial location of the various atomic constituents. Another useful information for chemical reactivity would be to visualize where the electronic chemical potential variation should be the largest or the lowest. Such a parameter is called a frontier index and may be defined in density-functional theory as $f = \partial\varrho_{HOMO/LUMO}/\partial N$ *(16)*. Point-charge approximation of this relation shows that each atom of a chemical compound should have a frontier index f_i such that *(17)*:

$$\eta = \eta_i \times f_i + \frac{1}{4\pi\varepsilon_0}\sum_{j=1}^{n} M_{ij} \times f_j \quad \text{with} \quad \sum_{i=1}^{n} f_i = 1 \qquad (3)$$

The global hardness or HOMO-LUMO gap η and the dimensionless frontier indexes f_i are again uniquely determined by the knowledge of the individual chemical hardnesses η_i and the structure (M_{ij} matrix). Being dimensionless and always positive quantities, the set of frontier indexes are also constrained by the normalization of the total wave function of the system $<\Psi|\Psi> = 1$. Negative f_i values point to an ill-balanced choice of chemical hardnesses η_i and should be, if possible, avoided. At this stage, it must be stressed that if the mean electronegativity and the charge distribution of a neutral compound ($z = 0$) remain both invariant under an origin change of the Madelung contribution ($M'_{ij} = M_{ij} + \Omega$), this will be never the case for the global hardness η. This merely comes from the fact that $\Sigma_i\, q_i = 0$ whereas $\Sigma_i\, f_i \neq 0$. Consequently, great care must be taken for evaluating η in crystalline solids as M_{ij} values are computed according to the Ewald procedure with a variable convergence parameter K allowing optimum convergence of the summation process *(18)*. If the M_{ij} matrix is not properly normalized before evaluating the global hardness η, this may well be found to be very high or worst negative. Yet another problem related to the use of equation (5) for solids lies is that the number of formula units per unit-cell Z is dictated by the space-group chosen for solving the structure. As $\Sigma_i f_i \neq 0$, it does matter for the global hardness η if Z is two or four. At the present time we have no really satisfactory answers to these problems. We have thus been obliged to proceed empirically and found that using the constraint $\Sigma_i f_i = Z$ instead of $\Sigma_i f_i = 1$ and removing from each M_{ij} value the constant $\Omega = \min(M_{ij})/Z$ leads to reasonable positive η values in almost every case.

From a practical viewpoint the free-atom electronegativities have been approximated by $\chi° = \chi_{p,d} + s\times[\chi_s - \chi_{p,d}]$, where $\chi_{p,d}$ and χ_s are the Mulliken electronegativities of p-, d- or s-orbitals. These values have been tabulated for non-transition elements *(19)* and for the first series of transition elements *(20)*. The s-orbital mixing coefficients s have been evaluated from the orbital radii of valence atomic orbitals. These values have also been tabulated for all elements using relativistic wave functions *(21)*. For oxygen $e\chi_s = 27.19$ eV, $e\chi_p = 9.63$ eV, $r_s = 45$ pm and $r_p = 41.35$ pm leading to $s = 0.521$ and $\chi°(O) = 18.78$ eV. The chemical hardnesses have been approximated as $\eta = 1/(4\pi\varepsilon_0 \times r_{at})$, where r_{at} are Bragg-Slater radii approximating atomic sizes in crystals. These values are also known for almost all elements *(22)*. For oxygen $r_{at} = 60$ pm leading to $e^2\eta(O) = 24.0$ eV. Although this necessary atomic parametrization is by no means perfect and may be criticized, it has a clear physical meaning and is readily applicable to all chemical elements. If some highly unrealistic charge distributions are encountered, simple adjustments of the s-orbital mixing coefficients or of the atomic radii remain possible. This confers a great flexibility to the model, but within a very well-defined and strict limits for the variations of the parameters s $(0 \leq s \leq 1)$ and r_{at} (smallest ionic radius as a lower limit and van der Waals radius as an upper limit).

The electric field gradient (efg) tensor

With a realistic charge distribution in one hand and a well-defined molecular or crystalline structure in the other hand, it is straightforward to choose one atom as the origin and compute the following tensor:

$$V_{\alpha\beta} = \frac{e}{4\pi\varepsilon_0} \sum_k \frac{3r_k^\alpha r_k^\beta - r_k^2 \delta_{\alpha\beta}}{r_k^5} \quad \alpha, \beta = x, y, z \tag{4}$$

where the summation over the k neighbouring atoms is stopped as soon as convergence has been reached. For a crystal, summation over about 1000 unit-cells usually leads to stable values. Once computed, the tensor is readily diagonalized to find its principal axis system (PAS), giving a set of five parameters: the V_{zz} value witch is such that $|V_{zz}| > |V_{yy}| > |V_{xx}|$, the asymmetry parameter $\eta_Q = |V_{xx} - V_{yy}|/|V_{zz}|$ and the three Euler angles (α, β, γ) giving the relative orientation of the PAS and of the current coordinate system. The V_{zz} value is related to the Nuclear

Quadrupole Coupling Constant (NQCC) C_Q measured by NMR or NQR techniques through the following relation *(23)*:

$$C_Q = \left|(1-\gamma_\infty)\frac{e^2QV_{zz}}{h}\right| \qquad (5)$$

where h is the Planck constant, Q the quadrupole moment of the nucleus and γ_∞ the antishielding factor taking into account the amplification of the gradient induced by the polarized core *(24)*. For the ^{17}O nucleus $Q = -0.0256\times10^{-28}\ m^2$, leading to:

$$C_Q\ (MHz) = 0.0619\times|(1 - \gamma_\infty)V_{zz}(V.Å^{-2})| \qquad (6)$$

Corundum α-Al_2O_3 structure. Let us take as a first example the case of ruby where the NMR signals due to ^{17}O nuclei have been observed leading to: $C_Q = (2.167 \pm 0.005)$ MHz, $\eta_Q = 0.517 \pm 0.005$ and $\beta = (45.85 \pm 0.2)°$ *(25)*. Table I shows the charge distribution and the efg tensor computed using four different set of atomic radii. Set #3 which uses Pauling covalent radii for approximating the atomic hardnesses predicts a fully ionic structure (I = 100%), whereas set #4 which uses s-orbital atomic radii describes on the contrary a rather covalent network (I = 30%). Set #1 which uses Bragg-Slater radii leads to I = 65%, a network where covalency and ionicity are rather well-balanced. Set #2 is a mixed parametrization (Bragg-Slater radius for oxygen and s-orbital radius for aluminum) which best fits the experimental values q(Al) = +1.32 (5) and q(O) = -0.88 (8) found from a $\varkappa$-refinement of the corundum structure *(26)*. It is very interesting to see that all sets lead to a C_2 site symmetry (one non-zero off-diagonal component) with the same values for the angle β between the z-direction and the c-axis, and the asymmetry parameter η_Q. If the predicted values are found to be outside the experimental error range, they are however of the right order of magnitude. Taking into account the fact that the NMR measurements have been made on a single-crystal containing 0.13 wt% Cr^{3+} ions and that the crystal structure used for the computation was solved for a single-crystal having a Cr content of less than 0.05%, the agreement between theory and experiment is quite satisfactory. For this kind of crystal structure, it is then found that the charge distribution has no effect on the η_Q, α, β and γ values affecting only V_{zz} and γ_∞. This is obviously an important result for the simulation of NMR powder spectra of quadrupolar nuclei as it leaves just the NQCC as a variable parameter for fitting the NMR line shape of known crystalline structures. Table I also shows the difficulty met with the normalization of the Madelung matrix M_{ij}. If we consider the triple hexagonal cell (Z = 6) we get $\langle e^2\eta\rangle = 25.5$ eV against only 5.1 eV for the primitive rhombohedral cell (Z = 2). Other parameters are not affected by this unit-cell change.

Water and ice polymorphs. Water is a very interesting compound as the oxygen quadrupole coupling tensor of the free molecule has been accurately measured by beam-maser spectroscopy *(27)* and various ice polymorphs have been studied by NMR or NQR *(28)*. This gives us an opportunity to show that the PCM formalism is able to switch easily between isolated molecules and fully reticulated networks. Table II gives our results obtained using the following parametrization for

Table I. Effect of the PCM parametrization[a] on the computed efg tensor of the ^{17}O nucleus in the corundum (α-Al_2O_3) structure.

Parameter	*First set*	*Second set*	*Third set*	*Fourth set*
r_{at}(Al)/pm	125	104	120	104
r_{at}(O)/pm	60	60	73	45
e<χ> /eV	4.85	8.51	5.56	7.62
q(Al)	+1.933	1.426	+3.0	+0.890
q(O)	-1.289	-0.95	-2.0	-0.593
I/%	64.5	47.5	100.0	29.7
$e^2\eta(Z = 6)^b$	24.9	26.0	25.2	26.0
$e^2\eta(Z = 2)^c$	4.4	5.6	4.7	5.6
f(Al)	+0.552	+0.407	+0.523	+0.442
f(O)	-0.034	+0.062	-0.016	+0.039
V_{ZZ}/V.Å^{-2}	-7.59	-5.60	-11.90	-3.49
V_{yy}/V.Å^{-2}	6.08	4.48	9.55	2.80
V_{xx}/V.Å^{-2}	1.51	1.11	2.35	0.69
η_Q	0.6	0.6	0.6	0.6
α/°	90.01	90.4	90.1	90.1
β/°	44.4	44.4	44.4	44.4
γ/°	90.1	90.2	90.1	90.1
$\|1 - \gamma_\infty\|$	4.6	6.3	2.9	10.0

[a] Atomic electronegativity for aluminum: eχ°(Al) = 6.152 eV (s = 0.443)

[b] Hexagonal cell [c] Rhombohedral cell

hydrogen atoms: s(H) = 1.0 i.e. $e\chi(H)$ = 7.18 eV and $r_{at}(H)$ = 25 pm (Bragg-Slater radius).

The free water molecule was assumed to lie in the yz plane with its C_2 axis pointing along the z-direction and with d(O-H) = 95.7 pm and ∠HOH = 104.57°. A perfect agreement with experimental values is obtained for the asymmetry parameter η_Q and an antishielding factor $(1 - \gamma_\infty) = 22.7$ is required for matching the coupling constant C_Q. As for corundum, changing atomic radii alters only the γ_∞ value.

Above 72K, ice adopts a fully disordered crystal structure (space-group $P6_3/mmc$), whereas below that temperature the structure becomes ordered (space-group $Cmc2_1$) *(29)*. In order to treat the disordered polymorph, our software does not apply the full space-group symmetry to fractional atoms but uses instead suitable symmetry masks specially designed to minimize atomic overlaps. In the case of hexagonal ice, atoms H1 and H2 are placed in 4f and 12k positions, both sites having an occupancy factor of 0.5. By applying only the following symmetry operators: x,y,z (H1 and H2); -x, -y, ½+z (H1 and H2); y, y-x, -z (H2); x-y, x, -z (H2), ; -y, x-y, ½-z (H2); y-x, -x, ½-z (H2), it is possible to generate a unit-cell having the H_2O stoichiometry (Z = 4), where each oxygen atom is chemically bonded with only two hydrogen atoms instead of four if the whole space-group symmetry is applied. Table II clearly shows that even after extensive hydrogen bonding, the whole network effect is to increase the global ionicity by a little 1%. Atoms H1 and H2 occupying different crystalline sites are found to have the same charge, as expected for a fully disordered crystal structure. The efg tensor is found to have C_2 symmetry and assuming that $(1 - \gamma_\infty) = 22.7$ as in the free water molecule leads to good agreement between theory and experiments for η *(30)*. The neat increase of η_Q and decrease of C_Q on going from the free molecule to the solid-state is reproduced even if our C_Q value is 1 MHz too high.

The importance of having well-refined crystal structures for computing efg tensors is clearly demonstrated in the case of Ice-II, a fully deuteron-ordered structure with two independent types of water molecules. Table II shows the results obtained with two crystal structures using X-ray diffraction (R = 0.08) *(31)*, or single-crystal neutron diffraction (R = 0.098) *(32)*. Only one NQR oxygen signal was detected for this structure *(33)*. Despite rather similar charge distributions, only the neutron structure (site O2) is able to reproduce the experimental value. The sequence $C_Q(H_2O) > C_Q(\text{Ice-II}) > C_Q(\text{Ice } I_h)$ is confirmed by experiment, but not the absolute values.

Among the other ice polymorphs only ice-VI and ice-VIII have been studied by NMR or NQR *(34)*. Ice-VI is another example of a disordered structure *(35)* which can be easily handled using symmetry masks: x,y,z (H1, H3, H4), ½-x,y,z (H4), x,½-y,z (H1, H4), -x,-y,-z (H2), -x,½+y,-z (H2, H3), ½+x,½+y,-z (H4), y,x,½+z (H1), ½-y,x,½+z (H1, H3, H4), y, ½-x,½+z (H4), ½-y,½-x,½+z (H4), -y,-x,½-z (H2, H3, H4) and ½+y,-x,½-z (H2). While experiment shows two oxygen sites with η_Q = 0.9 and 0.2, our model gives η_Q = 0.86 and 0.33, a reasonable agreement. Ice-VIII is on the contrary a fully ordered structure (Z = 8) with only one oxygen site *(35-36)*. The values reported in table II are for *(36)*, structural data from *(35)* leading to η_Q = 0.67 and C_Q = 8.1 MHz. Consequently, none of these two structures can be considered to

Table II. PCM modelization of ^{17}O efg tensors of water molecule and of some selected ice polymorphs.

Value	H_2O (g)	*Ice-I*	*Ice-II* $O1_{ND}$	*Ice-II* $O2_{ND}$	*Ice-II* $O1_{RX}$	*Ice-II* $O2_{RX}$	*Ice-VI* O1	*Ice-VI* O2	*Ice-VIII*
$e\langle\chi\rangle^a$	15.0	15.25	15.35		15.28		15.45		15.4
q(O)	-0.42	0.44	-0.45	-0.45	-0.43	-0.43	-0.44	-0.46	-0.44
q(H1)	+0.21	+0.22	+0.23	+0.22	+0.21	+0.21	+0.23	+0.23	+0.22
q(H2)	+0.21	+0.22	+0.22	+0.22	+0.22	+0.22	+0.22	+0.23	+0.22
I /%	21.1	22.1	22.4		21.5		22.7		22.1
$e^2\eta^a$	21.1	26.3	33.3		33.3		41.2		23.4
f(O)	+0.67	+0.70	+0.70	+0.71	+0.70	+0.70	+0.72	+0.71	+0.71
f(H1)	+0.16	+0.15	+0.15	+0.14	+0.15	+0.16	+0.16	+0.15	+0.15
f(H2)	+0.16	+0.15	+0.16	+0.14	+0.14	+0.16	+0.15	+0.14	+0.15
V_{xx}^b	-6.94	-5.33	5.26	6.16	-3.49	-5.31	4.03	-6.89	-5.76
V_{yy}^b	6.08	-0.18	-6.53	-6.59	2.80	3.78	-6.07	7.42	4.80
V_{zz}^b	0.86	+5.51	1.27	0.43	0.69	1.52	2.04	-0.53	0.95
η_Q	0.75	0.94	0.61	0.87	0.43	0.31	0.33	0.86	0.78
η_Q^c	0.75	0.94	0.87		0.87		0.2	0.9	0.97
α^d	90	90	133	29	180	99.6	91.0	90.4	90
β^d	90	35	67	61	47	42.7	0.3	33.8	90
γ^d	90	90	118	0	180	93.5	91.0	90.7	90
C_Q^e	10.17	7.7	9.18	9.26	7.46	7.88	8.5	10.4	8.3
$C_Q^{c,e}$	10.17	6.66	6.89		6.89				7.14

[a] Values in eV [b] Values in V.Å^{-2} [c] Experimental values
[d] Angles in degrees [e] Values in MHz

give a good agreement. This may just indicate that powder neutron diffraction data is unable to reach the same accuracy as single-crystal neutron diffraction.

Silicate structures. We turn now to the study of more complex crystal structures such as those found in silicate chemistry. Using DAS experiments, K.T. Mueller et al. have been able to determine the efg tensors of the three different crystallographic sites of diopside *(37)*. Table III shows our results obtained for a well-refined crystal structure of this mineral *(38)*. Partial charges not higher than +2 for alkaline-earth cations and non negative oxygen frontier indexes are obtained by selecting the 3s-orbital radius for Mg *(21)* and the crystalline ionic radius of Ca^{2+} in twelve fold coordination *(39)*. The structure is found to be slightly less ionic than corundum, suggesting that C_Q values should be computed using the antishielding factor of α–Al_2O_3. Our η_Q values agree rather well with those found experimentally: 0.13 ± 0.10 (δ_{iso} = 86 ppm), 0.00 ± 0.10 (δ_{iso} = 64 ppm) and 0.36 ± 0.05 (δ_{iso} = 69 ppm) with the site of highest asymmetry having the highest C_Q value. These values belong to site O2 which is threefold coordinated (OSiMgCa) the two others sites O3 and O1 being in fourfold coordination (OSi_2Ca_2) and ($OSiMg_2Ca$) respectively, with a logical higher asymmetry for O1. Consequently, the systematic assignment of the highest C_Q values to "bridging" oxygen atoms in silicate structures is not supported by our model. This may be related to the very weak charge difference between oxygen atoms linked to one silicon atom (O1 and O2) or two silicon atoms (O3). Table III shows also that both the Euler angles and the asymmetry parameters η_Q are now very sensitive to the charge distribution. Increasing the atomic radius of the magnesium atom up to 138 pm leads to q(Mg) ~ q(Ca) ~ +2 with magnesium electrons flowing preferentially towards atoms O1 and O2. Accordingly, efg tensors of O1 and O2 are greatly affected both in orientation and magnitude, while that of O3 remains virtually unchanged.

Forsterite Mg_2SiO_4 has also three different oxygen sites, but here all oxygen atoms are of the $OSiMg_3$ type *(40)*. Table III shows that the lowering of C_Q on going from diopside to forsterite is well-reproduced and that one η_Q value (O2) is significantly higher than the two other ones (O1 and O3) as observed experimentally *(41)*. Any attempt to increase the atomic radius of magnesium atoms in order to reach q(Mg) ~ +2, changes completely the orientation and the magnitude of the three efg tensors. For example using r_{at}(Mg) = 143 pm instead of 128 pm leads to q(Mg2) = +2.00, q(O) ~ -1.22, but also to η_Q(O1) = 0.23, η_Q(O2) = 0.27 and η_Q(O3) = 0.17 in poor agreement with experimental observations.

Table IV shows that our model is able to discriminate between a cyclosilicate such as: pseudowollastonite (α-$CaSiO_3$) *(42)* and an inosilicate such as parawollastonite (β-$CaSiO_3$) *(43)*. For the cyclosilicate structure, a clear distinction is possible between "terminal" oxygen atoms, which are found to have C_Q = (3.0 ± 0.4) MHz and η_Q = 0.08 ± 0.03, and "bridging" oxygen atoms which show much less dispersion: C_Q = (2.9 ± 0.1) MHz and η_Q = 0.39 ± 0.02. If the experimental difference in C_Q values is not well reproduced *(44)*, the agreement is much better on η_Q values. The inosilicate structure is characterized by a very large dispersion for

Table III. PCM modelization of ^{17}O efg tensors in diopside $CaMgSi_2O_6$ and forsterite Mg_2SiO_4.Units for V_{aa} are V.Å^{-2} and $|1 - \gamma_\infty| = 6.3$ as for the corundum structure.

Value	Diopside[a]	Diopside[b]	Forsterite[a]
e<χ> /eV	6.24	5.85	6.37
q(O1)	-1.13	-1.20	-1.10
q(O2)	-1.06	-1.10	-1.10
q(O3)	-1.05	-1.05	-1.08
q(Si)	+1.39	+1.38	+1.21
q(Mg1)	+1.69	+1.98	+1.96
q(Ca, Mg2)	+2.03	+1.96	+1.98
I /%	51.3	53.1	52.0
$e^2\eta$ /eV	7.6	7.5	9.1
f(O1)	-0.01	-0.02	0.00
f(O2)	0.00	0.00	0.00
f(O3)	0.00	0.00	0.00
f(Si)	0.18	0.18	0.23
f(Mg1)	0.25	0.29	0.36
f(Ca, Mg2)	0.40	0.39	0.42
$[V_{xx}, V_{yy}, V_{zz}]$(O1)	6.47, -2.78, -3.69	6.40, -4.04, -2.35	5.63, -2.48, -3.14
$[\alpha, \beta, \gamma]$(O1)	180, 29, 180	180, 47, 140	90, 0, 90
$[C_Q$/MHz, $\eta_Q]$(O1)	2.5, 0.14	2.5, 0.26	2.2, 0.12
$[V_{xx}, V_{yy}, V_{zz}]$(O2)	-6.68, 9.05, -2.38	-7.03, 9.83, -2.83	-0.75, 3.93, -3.17
$[\alpha, \beta, \gamma]$(O2)	76, 18, 85	107, 15, 96	89, 0, 89
$[C_Q$/MHz, $\eta_Q]$(O2)	3.5, 0.48	3.8, 0.43	1.5, 0.62
$[V_{xx}, V_{yy}, V_{zz}]$(O3)	-3.66, -4.52, 8.18	-3.67, -4.41, 8.09	-1.58, -3.05, 4.63
$[\alpha, \beta, \gamma]$(O3)	72, 36, 106	75, 36, 106	96, 23, 97
$[C_Q$/MHz, $\eta_Q]$(O3)	3.2, 0.10	3.2, 0.09	1.8, 0.32

[a] (χ°/eV, r_{at}/pm) Si = (8.575, 110), Mg = (5.84, 128) and Ca = (4.48, 148)
[b] r_{at}(Mg) = 138 pm.

Table IV. PCM modelization of ^{17}O efg tensors in pseudowollastonite α-$CaSiO_3$ and parawollastonite β-$CaSiO_3$. Same parametrization as in table III. Units for V_{aa} are $V.Å^{-2}$ and $|1 - \gamma_\infty| = 6.3$ as for the corundum structure.

Site	q	α/°	β/°	γ/°	V_{xx}	V_{yy}	V_{zz}	C_Q/MHz	η_Q
Pseudowollastonite ($e<\chi> = 5.73$ eV, $e^2\eta = 29.1$ eV, I = 50.5%)[a]									
O1	-1.08	113	37	64	-3.97	-3.38	7.38	2.9	0.08
O2	-1.08	63	35	34	-3.59	-3.09	6.68	2.6	0.08
O3	-1.07	180	37	180	-3.54	-4.28	7.81	3.1	0.09
O4	-1.08	56	37	130	-4.24	-3.46	7.70	3.0	0.10
O5	-1.07	106	36	108	-4.36	-3.77	8.12	3.2	0.07
O6	-1.08	0	35	0	-3.19	-3.70	6.89	2.7	0.07
O7	-1.09	9	35	3	-3.28	-3.74	7.02	2.7	0.06
O8	-1.09	71	37	110	-4.39	-3.50	7.89	3.1	0.11
O9	-1.07	107	36	108	-4.22	-3.65	7.86	3.1	0.07
O10	-1.08	122	36	129	-3.86	-4.63	8.49	3.3	0.09
O11	-1.09	103	36	74	-4.28	-3.80	8.07	3.2	0.06
O12	-1.09	78	35	57	-3.87	-3.27	7.14	2.8	0.08
O13	-1.07	93	2	93	-5.20	7.40	-2.20	2.9	0.41
O14	-1.07	89	1	89	7.38	-5.16	-2.21	2.9	0.40
O15	-1.07	76	3	80	7.57	-5.25	-2.32	3.0	0.39
O16	-1.07	88	2	89	7.55	-5.21	-2.34	2.9	0.38
O17	-1.07	0	3	28	7.40	-5.15	-2.26	2.9	0.39
O18	-1.07	88	2	88	-5.13	7.43	-2.30	2.9	0.38
O14	-1.07	89	1	89	7.38	-5.16	-2.21	2.9	0.40
Parawollastonite ($e<\chi> = 5.77$ eV, $e^2\eta = 29.5$ eV, I = 51.0%)[b]									
O1	-1.09	81	8	81	-4.96	-2.92	7.88	3.1	0.26
O2	-1.09	16	40	20	7.14	-3.73	-3.40	2.8	0.05
O3	-1.16	59	30	64	-2.71	-3.58	6.30	2.5	0.14
O4	-1.16	125	33	118	-2.43	-3.58	6.01	2.3	0.19
O5	-1.11	105	3	105	8.47	-0.82	-7.65	3.3	0.81
O6	-1.08	83	3	83	8.80	-1.44	-7.36	3.4	0.67
O7	-1.04	94	36	68	-8.44	9.68	-1.25	3.8	0.74
O8	-1.03	94	36	67	-8.47	9.53	-1.06	3.7	0.78
O9	-1.02	21	45	21	-4.65	11.0	-6.35	4.3	0.15

[a] q(Si) = +1.32...+1.35, q(Ca) = +1.86...+1.93, f(O) ~ 0.00, f(Si) ~ 0.30, f(Ca) ~ 0.70
$(C_Q/MHz, \eta_Q)_{exp}$: "terminal O" = (2.2, 0.1), "bridging O" = (3.8, 0.2)

[b] (q, f): Si1 = (1.42, 0.34), Si2 = (1.39, 0.34), Si3 = (1.37, 0.31), Ca1 = (1.82, 0.61), Ca2 = (1.92, 0.72), Ca3 = (1.88, 0.70).
$(C_Q/MHz, \eta_Q)_{exp}$: "terminal O" = (2.2, 0.1), "bridging O" = (3.8, 0.2)

both kind of oxygen atoms, with "bridging" sites having a higher C_Q value (3.9 ± 0.4 MHz) than the "terminal" ones (2.9 ± 0.6 MHz) in agreement with experiments *(37)*.

A neat increase in C_Q in going from cyclo- or inosilicates to tectosilicates is predicted from the crystal structure of low-cristobalite *(45)* or of a fully dealuminated zeolite-Y *(46)* (table V). Comparison with experimental values *(47)* show a very good agreement for η_Q and a 2-3 MHz overestimate of C_Q. We have also been able to generate a unit-cell having the composition $Na_{13}Al_{13}Si_{35}O_{96}$ modeling a zeolite Na-Y having an Si/Al ratio of 2.61 *(46)*, in order to investigate the effect of the Si → (Na, Al) substitution on efg tensors. Table V shows that on going from a pure silica structure to one having Si-O-Al bridges we expect on average ΔC_Q = -2 MHz and $\Delta\eta_Q$ = +0.1. Experiments giving ΔC_Q = -1.5 MHz and $\Delta\eta_Q$ = +0.1 *(48)*, This very satisfactory last result demonstrates the ability of the PCM to modelize on the same grounds very large and complex system as well as small isolated molecules.

Prediction of NMR shielding tensors through the PCM

Let us now have a look at the possibility to modelize NMR shielding tensors using the same theoretical frame. In an appropriate PAS, the nine-component shielding tensor $\sigma_{\alpha\beta}$ which describes the very small magnetic fields arising from electronic motions around the nucleus can be diagonalized leading to *(49)*:

$$|\sigma_{zz} - \sigma_{iso}| \geq |\sigma_{yy} - \sigma_{iso}| \geq |\sigma_{xx} - \sigma_{iso}| \text{ with } \sigma_{iso} = \frac{\sigma_{xx} + \sigma_{yy} + \sigma_{zz}}{3} \qquad (7)$$

$$\Delta\sigma = \sigma_{zz} - \sigma_{iso} \quad \text{and} \quad \eta_{CS} = \frac{\sigma_{xx} - \sigma_{yy}}{\sigma_{zz} - \sigma_{iso}} \qquad (8)$$

With these conventions, the shielding anisotropy $\Delta\sigma$ is either positive or negative, while the asymmetry η_{CS} can be 0 (axial symmetry) and at most +1.

The diamagnetic contribution. The nuclear magnetic shielding tensor (NMST) $\sigma_{\alpha\beta}$ is most frequently expressed through perturbation theory as a sum of a positive diamagnetic first-order term $\sigma_{\alpha\beta}{}^d$ and a negative paramagnetic second-order term $\sigma_{\alpha\beta}{}^p$. With the origin set at the nucleus under consideration, the diamagnetic term can be written as an atomic contribution $\sigma_{at}(q)$, plus a dipole contribution arising if the point charges are not fully centred on the n-th nucleus but displaced by a distance $<\varrho>_n$, plus a quadrupole contribution arising from the spatial extent of the charge present on the n-th nucleus *(50)*. Neglecting the dipole contribution which is of the order of a few ppm leads to *(50)*:

$$\sigma^d_{xx} = \sigma^d_{at}(q) + \left(\frac{\mu_0}{4\pi}\right) \times \left(\frac{e^2}{2m_e}\right) \times \left[\sum_n{}' \frac{P_n}{r_n^3}(y_n^2 + z_n^2) + \sum_n{}' \frac{<\varrho^2/3>_n}{r_n^3}\left\{2 - \frac{3(y_n^2 + z_n^2)}{r_n^2}\right\}\right] \qquad (9)$$

Table V. PCM modelization of ^{17}O efg tensors in low-cristobalite and zeolite-Y Parametrization was $r_{at}(Na)$ = 99 pm, $r_{at}(Si)$ = $r_{at}(Al)$ = 100 pm, $e\chi°(Si)$ = $e\chi°(Al)$ = 8.575 eV. Units for V_{aa} are V.Å^{-2} and $|1 - \gamma_\infty|$ = 6.3 for computing C_Q values.

Site	q	α/°	β/°	γ/°	V_{xx}	V_{yy}	V_{zz}	C_Q/MHz	η_Q
low-cristobalite (Z = 4, $e\langle\chi\rangle$ = 10.6 eV, $e^2\eta$ = 12.9 eV, I = 48.9%)[a]									
O	-0.98	100	36	95	-11.76	20.61	-8.85	8.0	0.14
Siliceous zeolite-Y (Z = 92, $e\langle\chi\rangle$ = 11.8 eV, $e^2\eta$ = 177.9 eV, I = 47.7%)[b]									
O1	-1.00	119	45	90	-10.98	18.15	-7.15	7.1	0.21
O2	-0.96	0	35	0	-12.01	20.79	-8.79	8.1	0.15
O3	-0.95	0	28	0	-10.72	19.56	-8.84	7.6	0.10
O4	-0.92	147	45	127	-7.53	18.67	-11.15	7.3	0.19
Zeolite Na-Y $Na_{13}Al_{13}Si_{35}O_{96}$ ($e\langle\chi\rangle$ = 10.1 eV, $e^2\eta$ = 3.5 eV, I = 44.9%)[c]									
O1	-0.93	119	45	90	-9.31	15.00	-5.69	5.9	0.24
O2	-0.98	90	23	60	13.30	-9.73	-3.59	5.2	0.46
O3	-0.95	90	44	180	12.88	-7.90	-4.97	5.0	0.23
O4	-0.93	99	18	77	15.61	-7.27	-8.33	6.1	0.07

[a] (q, f)[Si] = (+1.95, +0.80), f(O) = +0.10, $(C_Q/MHz, \eta_Q)_{exp}$ = (5.3, 0.13)

[b] (q, f)[Si] = (+1.91, +0.68), f(O1) = +0.02, f(O2) = +0.12, f(O3) = +0.23, f(O4) = +0.26, $(C_Q/MHz, \eta_Q)_{static}$ = (5.7, 0.1), $(C_Q/MHz, \eta_Q)_{MAS}$ = (5.2, 0.2)

[c] (q, f): Na1 = (0.95, 0.00), Na2 = (0.80, 0.00), Na3 = (1.05, 0.02), Si = (1.63, 0.01).

$$\sigma^{d}_{xy} = -\left(\frac{\mu_0}{4\pi}\right)\times\left(\frac{e^2}{2m_e}\right)\times\left[\sum_n{}' \frac{P_n x_n y_n}{r_n^3} + \sum_n{}' \frac{3x_n y_n < \varrho^2/3 >_n}{r_n^5}\right] \quad (10)$$

with corresponding expressions for other components. In (9-10), P_n is the electronic population around the n-th nucleus having coordinates (x_n, y_n, z_n) and situated at a distance r_n from the shielded nucleus which should be omitted from the summation. The constants μ_0, e and m_e have their usual meaning with $(\mu_0/4\pi)(e^2/2m_e) = 14.09$ ppm.Å. The term $<\varrho^2>_n$ represents the averaged squared electronic distance from the nth nucleus which can be taken proportional to the average diamagnetic susceptibility of that atom *(51)*. This term is managed by our software using tabulated values *(52)* while the term $\sigma_{at}{}^d(q)$ is obtained through a polynomial approximation of the values computed through Hartree-Fock-Slater wave functions *(53)*. The partial charge correction have been included with the hope to compute more realistic σ^d values as previously suggested *(54)*.

Using (9-10), the full diamagnetic shielding tensor can be readily evaluated in any coordinate system and further diagonalized leading to a set of three Euler angles (α^d, β^d, γ^d) and three principal values ($\sigma_{iso}{}^d$, $\Delta\sigma^d$, $\eta_{CS}{}^d$). A simpler expression is however obtained, if only the isotropic average is considered owing to the traceless nature of the quadrupole contribution:

$$\sigma^{d}_{iso}/\text{ppm} = \sigma^{d}_{at}(q)/\text{ppm} + 9.39\sum_n{}' \frac{P_n}{r_n(\text{Å})} \quad (11)$$

Having applied equations (11-13) to more than one hundred different crystal structures, we have found empirically that reasonable diamagnetic terms are obtained when $P_n = N_n - q_n$, difference between the total number of valence electrons N_n and the partial charge q_n and when the summation is limited to r_n distances less than 220 pm. This procedure allows to have $P_n \sim Z_n$ for light atoms ($Z < 10$) as originally suggested *(50)* and avoids to add diamagnetic terms over 1000 ppm when heavy elements are present.

The paramagnetic contribution. The evaluation of the second-order paramagnetic contribution is not as straightforward as it requires the use an LCAO-MO formalism associated to an average excitation energy (AEE) approximation. In fact, as long as overlap terms are neglected, one is completely free to use a GIAO basis *(55)* or a natural gauge origin *(56)* to evaluate this paramagnetic term *(57)*:

$$\sigma^{p}_{xx} = -\left(\frac{4\mu_0\mu_B^2}{4\pi a_0^3}\right)\times\frac{<(a_0/r)^3>}{\Delta E}\times\left[P_{yy} + P_{zz} - P_{yy}P_{zz} + P_{yz}P_{zy} + \sum_n{}'\left(-2P_{yy_n}P_{zz_n} + 2P_{yz_n}P_{zy_n}\right)\right] \quad (12)$$

and corresponding expressions for $\sigma_{yy}{}^p$ and $\sigma_{zz}{}^p$. In (12) ΔE is the AEE, $(4\mu_0\mu_B{}^2/4\pi a_0{}^3) = 1449$ ppm.eV, $<(a_0/r)^3>$ is the average of the electron-nucleus distance r computed over p-orbitals and $P_{\mu\nu}$ the charge density and bond order matrix ($P_{\mu\nu} = 2\Sigma_j{}^{occ}\, c_{\mu j}\times c_{\nu j}$) of the LCAO-MO development. The last summation is restricted to neighbouring atoms having p-orbitals involved in both σ– and π-bonding with the shielded nucleus *(58)*. Further approximations are that cross-terms P_{yz} are generally small, and that P_{ii} terms are gross orbital populations for the shielded nucleus and π-bond orders for other nuclei.

Now, by its very definition, the global hardness η is a measure of the HOMO-LUMO gap of a compound. Consequently, it seems reasonable to assume that the AEE ΔE should scale with the global hardness. The $\langle (a_0/r)^3 \rangle$ term describes the p-orbital expansion or contraction as electrons are added or removed from the shielded nucleus. Using effective atomic numbers Z_{np}^* and Slater rules:

$$\left\langle (a_0/r)^3 \right\rangle_{np} = \frac{1}{3}\left(\frac{Z_{np}^*}{n}\right)^3 \approx \frac{\left(Z_{np}^{0*} + 0.35\times q\right)^3}{3n^3} \approx \frac{\left(Z_{np}^{0*}\right)^3}{3n^3} + \frac{0.35}{n}\times q \qquad (13)$$

For oxygen, $Z_{2p}^{0*} = 4.453$ *(59)*, i.e. $\langle (a_0/r)^3 \rangle_{2p} \sim R° + \varkappa\times q \sim 3.7 + 0.04\times q$. Optical spectra data gives R° = 4.3 *(60)*, SCF wave functions R° = 5.03 and ϰ = 1.00 *(61)* and analytical Hartree-Fock wave functions R° = 4.97 and ϰ = 0.9 *(62)*. Consequently the radial term $\langle (a_0/r)^3 \rangle$ may safely assumed to be a linear function of the partial charge q, with two adjustable constants R° and ϰ which may be optimized, but should remain close to 5 for R° and close to 1 for ϰ. The charge density and bond order matrix is the most difficult term to manage. A first simplification is obtained by considering the average value, $\sigma_{iso}^p = (\sigma_{xx}^p + \sigma_{yy}^p + \sigma_{zz}^p)/3$, and the so-called population unbalance P_u measuring the deviation of the electronic cloud from sphericity:

$$\sigma_{iso}^p / \text{ppm} = 966 \times \frac{R° + \varkappa \times q}{\Delta E(\text{eV})} \times P_u \qquad (14)$$

$$P_u = (P_{xx} + P_{yy} + P_{zz}) - ½(P_{xx}P_{zz} + P_{yy}P_{xx} + P_{yy}P_{zz}) + ½(P_{xy}P_{yx} + P_{xz}P_{zx} + P_{zy}P_{yz}) \qquad (15)$$

Isotropic orbital populations ($P_{xx} = P_{yy} = P_{zz} = P$), negligible cross-terms ($P_{xy} = P_{xz} = P_{yz} = 0$) and charge conservation then requires that:

$$P_{ss} + P_{xx} + P_{yy} + P_{zz} = N - q \qquad (16)$$

with P_{ss} gross s-orbital population and N total number of valence electrons. Then P_u should scale as $(3/2)[2P - P^2]$ or with $P = [N - (q + P_{ss})]/3$:

$$P_u = \frac{3}{2}\times\left[1 - \left(\frac{q-\Delta}{3}\right)^2\right] \quad \text{if} \quad \Delta = N - P_{ss} - 3 \qquad (17)$$

Equation (17) predicts a very characteristic parabolic variation of the shielding with the partial charge q, centred around q = 0 for carbon (N = 4, $P_{ss} \sim 1$) or nitrogen (N = 5, $P_{ss} \sim 2$), around q = +1 for oxygen (N = 6, $P_{ss} \sim 2$) and around q = +2 for fluorine (N = 7, $P_{ss} \sim 2$). However, a quick look at a typical MO diagram shows that this experimentally observed parabolic variation may also have another origin.

Let $\Delta\varepsilon$ be the difference in energy between two overlapping atomic orbitals. According to second-order perturbation theory, chemical bonding opens an energy gap ΔE which scales like $\Delta\varepsilon + S^2/\Delta\varepsilon$, where S^2 describes the radial expansion and angular variation of the overlapping orbitals. Now, a poor energy match (high $\Delta\varepsilon$) means a high AEE ($\sim \Delta\varepsilon$) through ionic bonding owing to a large electronegativity difference Conversely, a good energy match ($\Delta\varepsilon \sim 0$) means that the AEE $\sim S^2/\Delta\varepsilon$ should also remain high through covalent bonding as now electronegativities are rather close. Then the AEE is expected to vary in a non linear way with the ionicity or partial charge q, reaching a minimum value when $\Delta\varepsilon \sim S$. However, if $\Delta\varepsilon > S$ or $\Delta\varepsilon < S$, linear approximations AEE $\sim E° + E_1 \times q$ becomes valid and neglecting all terms in q^2 we get $<(a_0/r)^3>/\Delta E \sim R°/E° + (\varkappa/E° - R°E_1/E°^2)\times q \sim R°/E°$, a mere multiplicative constant if by chance $R°\times E_1 \sim \varkappa \times E°$. This very lucky compensation between gap opening and orbital contraction is responsible for the relative success of the so-called "constant AEE" approximation at moderate polarity.

Equations (14-17) allows us to modelize isotropic shielding constants, but it tells us nothing about shielding anisotropies. Let us now assume that the shielded nucleus is surrounded by n neighbouring atoms, each carrying a partial charge q_n. Knowing the structure, it is straightforward to compute the following quantities: $\Delta q_x = \Sigma_n q_n \times (3x_n^2 - r_n^2)/r_n^2$, $\Delta q_y = \Sigma_n q_n \times (3y_n^2 - r_n^2)/r_n^2$ and $\Delta q_z = \Sigma_n q_n \times (3z_n^2 - r_n^2)/r_n^2$. Now, with $P_{ii} \sim [N - (q + P_{ss})]/3 + \Delta q_i$. an anisotropic NMST can be easily obtained using (12). From a practical viewpoint, the s-orbital population may be obtained from $P_{ss} = [\chi°(s) - <\chi>]/\eta$, where $\chi°(s)$ is the electronegativity of a pure s-orbital, $<\chi>$ the mean electronegativity of the compound and η the chemical hardness of the target atom. Knowing P_{ss} and the Δq_i, the P_{ii} can be evaluated and if some overflow ($P_{ii} > 2$) occur, we may then assume that $P_{ii} = 2$. This levelling produces a charge deficit relative to equation (16), and neutrality is restored by adding excess electrons first into the s-orbital if $P_{ss} < 2$, or to the least populated p-orbital otherwise. After this charge adjustment, the two contributions σ^d and σ^p are added, giving the total NMST relative to the bare nucleus. Comparison with experimental values is then possible only if chemical shifts are systematically converted to an absolute scale. For the ^{17}O nucleus, the absolute shielding constant of carbon monoxide have been deduced from its ^{17}O spin-rotation constant: $\sigma(C^{17}O) = -42.3 \pm 17.2$ ppm *(63)*. Knowing this value, all ^{17}O chemical shifts referenced against water can be converted into absolute shielding constants according to $\sigma = 307.9 - \delta$ and an agreement between theory an experiment of the order of 20 ppm should be considered as excellent.

Crystalline MO oxides. Let us first check if the AEE scales like the global hardness η by considering ten MO oxides having well-known crystal structures *(64-70)* (table VI). The use of the coordinence-adjusted ionic crystal radii for alkaline-earth elements causes no problems, whereas other set of radii leads to negative f(O) values. Figure 1 shows a plot of the isotropic ^{17}O chemical shift measured through solid-state NMR *(71)(72)* against the reciprocal of the global hardness $e^2\eta$ given in table VI. The observed variation is quite regular and explain the strong deshielding

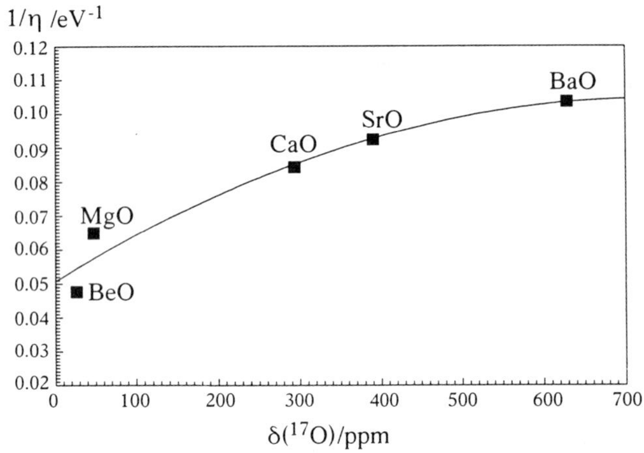

Figure 1. Variation of the ^{17}O NMR isotropic chemical shift of crystalline alkaline-earth MO oxides with the reciprocal of the crystal global hardness η.

observed on going from Be to Ba. The deviation from linearity for BeO and MgO may be attributed to the strong charge variation q(O) occurring when Be → Mg meaning that all terms $<(a_0/r)^3>$, P_u and AEE are changing. On the contrary, along the Mg →Ca → Sr → Ba. sequence q(O) remains approximately constant and accordingly a nearly perfect linear correlation (r = 0.99) is obtained for oxides having the same NaCl structural type. In a more quantitative way, let us assume that R° = 5.0 and $\varkappa$ = 1.0 in order to reduce the number of adjustable parameters. Knowing the partial charge q and the isotropic shift, we are able to compute σ^d, P_u and $<(a_0/r)^3> = R° + \varkappa \times q$ and get a ΔE value from (14). Plotting the global hardness against this ΔE value leads to a straight line: $\eta = 7.04 + 0.587 \times \Delta E$ (r = 0.996) and figure 2 shows the highly linear correlation (r = 0.999) obtained between theory and experiment when this linear (η, AEE) scaling is used for computing the chemical shifts directly from the crystalline structure. Table VI also shows that for other MO oxides, we are clearly in a situation where ΔE is almost constant. This nearly constancy of the AEE is confirmed when, as before, ΔE values are computed from the observed isotropic shift. Consequently, the observed deshielding for the sequence Zn → Cd → Hg or Sn → Pb should be attributed to the increase of $<(a_0/r)^3>$ and P_u terms with q(O). This table also shows that each series has is own $\eta = f(\Delta E)$ relationship, a fact obviously related to the normalization of the Madelung matrix.

Equation (14) is also able to explain some apparent puzzling situations in ^{17}O NMR chemical shifts variations. For example, the linear correlation displayed in fig. 5 of ref. *(41)* shows that an increase of the cation radius along the series Mn → Ru → Cr → Os → Tc → Re → V → Mo → W is associated to a shielding of the ^{17}O nucleus, whereas the linear correlation displayed in fig. 2 of ref. *(71)* shows that an increase of the cation radius along the series Be → Mg → Ca → Sr → Ba leads on the contrary to a deshielding which have been correlated to the cube of the ionic radius of the metallic cation. Now, with small cations q(O) is not very negative, meaning that $<(a_0/r)^3>$ and P_u terms of (14) are high. Increasing the cation radius makes q(O) more negative, meaning a reduced paramagnetic contribution, which explains the observed shielding in the first correlation. For the second one, let us recall that the energy of electrons confined into a box of size L is always quantified with an HOMO-LUMO gap $\Delta E = E_{n+1} - E_n = (2n+1)h^2/8m_eL^2$. Now, with the alkaline-earth series, the size of the box obviously increases with the *volume* ($\propto r^3$) occupied by the cation meaning a reduced gap ΔE and hence a strong deshielding according to (14). The pitfalls of purely empirical correlations is also evidenced in fig.2 of ref. *(71)* where another linear correlation is displayed for the Zn → Cd → Hg series. It should be clear by now that if the first correlation for the alkaline-earth series is basically correct, the second one is just an illegal extrapolation of the first, obtained by assigning a much too large radius to the Hg^{2+} cation, completely incompatible with its twofold coordination in the crystalline oxide. From our model, we know that the global hardness is almost constant for this series, suggesting that the observed deshielding is caused by the higher covalency of the Hg-O bond relative to Zn-O and not by a decrease of the AEE.

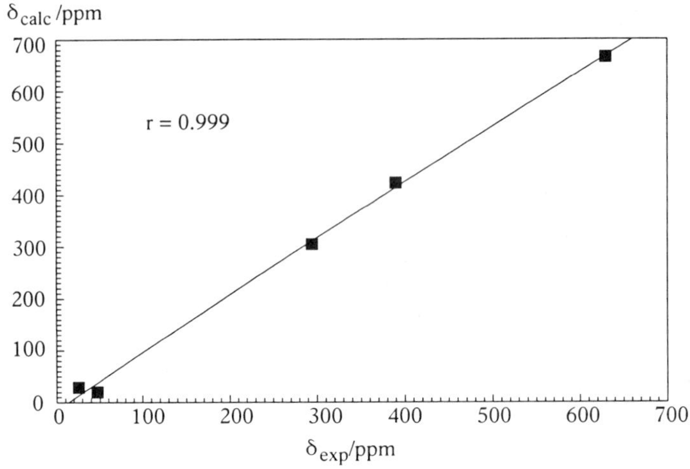

Figure 2. Linear correlation observed between experimental and computed ^{17}O NMR isotropic chemical shifts for crystalline alkaline-earth MO oxides.

Table VI. Breaking down the observed ^{17}O isotropic chemical shifts in MO oxides into diamagnetic (σ_d), radial ($R = \langle(a_0/r)^3\rangle$), population unbalance (P_u) and AEE terms. The s-mixing coefficients for computing free atomic electronegativities $\chi°$ were s = 1 for alkaline-earth elements and 0.785 (Zn), 0.701 (Cd), 0.653 (Hg), 0.377 (Sn), 0.363 (Pb).

Value	BeO	MgO	CaO	SrO	BaO	ZnO	CdO	HgO	SnO	α-PbO
ref.[d]	*(64)*	*(65)*	*(65)*	*(65)*	*(65)*	*(66)*	*(67)*	*(68)*	*(69)*	*(70)*
$e\chi°_M$ [e]	6.39	5.84	4.48	4.20	4.18	6.39	5.57	6.47	7.80	8.65
r_{at}(M)	41[a]	86[a]	120[a]	132[a]	149[a]	107[b]	118[b]	113[b]	124[c]	122[c]
$e\langle\chi\rangle$[e]	14.9	9.5	6.0	5.3	4.7	7.8	6.9	8.3	8.4	9.8
q(O)	-0.41	-0.77	-0.95	-0.94	-0.94	-0.92	-0.90	-0.77	-0.68	-0.56
I /%	20.5	38.4	47.3	47.2	47.1	45.8	44.9	38.7	34.0	28.1
$e^2\eta$[e]	21.0	15.4	11.8	10.8	9.6	12.2	12.0	12.4	11.6	11.6
f(O)	0.68	0.29	0.10	0.08	0.04	0.11	0.10	0.15	0.06	0.11
δ_{iso}[f,g]	26	47	294	390	629	-18	60	121	251	294
σ_d[f]	495	569	418	418	418	497	418	418	418	418
R	4.59	4.23	4.05	4.06	4.06	4.08	4.10	4.23	4.32	4.44
P_u	.7756	.6529	.5791	.5805	.5807	.5960	.6000	.6970	.6861	.7298
AEE[e]	24.2	13.0	8.4	6.8	4.6	20.6	21.0	18.5	11.6	11.9

[a] Value is the crystalline ionic radius for coordinence 6 or 4 (Be)
[b] Value corresponds to the s-orbital radius
[c] Value corresponnds to the p-orbital radius
[d] Crystal structure data [e] Value in eV [f] Value in ppm
[g] Values from ref. *(71)* except for Sn and Pb which are from ref. *(72)*

The decavanate ion $[V_{10}O_{28}]^{6-}$..We now investigate compounds displaying several oxygen sites within the same molecular or crystalline structure. In this case the AEE should be approximately constant and the shifts variations should reflect the combined effect of the diamagnetic, radial and population unbalance terms. The polyoxovanadate $[V_{10}O_{28}]^{6-}$ was chosen as it displays in its most symmetrical form *(73)* up to nine oxygen sites with very different coordination numbers: μ_6-O, μ_3-O, μ_2-O and μ_1-O, and as selective decoupling NMR experiments have yielded unambiguous structural assignments for all resonances of this ion in solution *(74)*. Results obtained on the naked ion (z = -6) using structural data of ref. *(73)* are gathered in table VII. The first point to be noted is the clear distinction made by the model for the different kind of oxygen sites. The most negatively charged atom is found to be the inaccessible μ_6-O, followed by the μ_3-O which should then interact preferentially with surrounding protons as confirmed by NMR techniques *(74)* or single-crystal X-ray diffraction *(75)*. Then comes the three kind of μ_2-oxo atoms with negative charge decreasing in the order O7 < O4 ~ O5 < O3, to be compared with O4 ~ O5 < O7 < O3 derived from NMR data. The same problem arises for terminal atoms with the order O9 < O6 ~ O8 from the charge and O6 ~ O8 < O9 from NMR data. Figure 3 shows the overall good correlation obtained using (14) to compute the shifts from the charges q(O), proving that our approach is physically sounded. However, we have not been able to find a set of charges reversing the order between (O7, O4-O5) and (O9, O6-O8). As these atoms are the best candidates for π-bonding with metal orbitals and as we predict a spanning over 520 ppm for the NMR resonances against ca 1100 ppm observed experimentally, our problem clearly comes form the complete neglect of π-bonding. Accordingly, assuming 50% s-orbital mixing for σ-bonding along the x-direction (P_{xxn} = -0.5), a full π-bond order between p_z-orbitals (P_{zzn} = 1) and P_{yyn} = 0, (13) predicts an increase of about 2×1449×3.75×0.5/3.9 ~ 1380 ppm for σ_{yy}, with no changes in σ_{xx} or σ_{zz}. Isotropic values of terminal atoms are then expected to increase of at least 1380/3 = 460 ppm, leading to a total shift range of about 1000 ppm in good accord with experiments. Moreover as the V-O9 distance is slightly lower than V-O6 or V-O8, the π-bond order should be higher with O9 than with O6 or O8 supporting the observed sequence. Similarly, V-O7-V bridges appear to be much more symmetrical and shorter ($2\times d_{V\text{-}O}$ = 183 pm) than V-O4-V or V-O5-V bridges ($d_{V\text{-}O}$ = 181 and 187 pm), meaning higher π-bond orders and thus more for deshielding for O7 relative to O4 or O5.

Computation of full shielding tensors. As experimental data on ^{17}O chemical shifts tensors are rather poor, we will consider just two examples for the calculation of full NMST from partial charges distribution (table VIII). For the free water molecule the molecule was assumed to lie in the xz plane with its C_2-axis pointing along the z-direction and an effective AEE of 56.1 eV was selected, in order to reproduce the experimental isotropic shift δ_{iso} = -36.1 ppm *(63)* using our model values R° = 5.0 and $\varkappa$ = 1.0. Comparison with shielding tensors computed using coupled HF perturbation theory *(76)(77)*, shows that we get a good order of

Table VII. Prediction of ^{17}O NMR chemical shifts assuming $\Delta E = \eta$ for the decavanate ion $[V_{10}O_{28}]^{6-}$. Experimental values and attributions are from ref *(74)*. Atomic parametrization was $r_{at}(V) = 135$ pm and $e\chi°(V) = 3.633$ eV leading to $e^2\eta = 3.9$ eV. The computed diamagnetic isotropic shielding constant was found to be $\sigma^d = 418.1 \pm 0.1$ ppm for all oxygen sites.

Site	q	f	δ_{exp}/ppm	$<(a_0/r)^3>$	P_u	δ_{calc}/ppm
O1 (μ_6)	1.86	-0.01	62	3.14	.0937	-4
O2 (μ_3)	-1.54	0.01	400	3.46	.3093	255
O7 (μ_2)	-1.50	0.02	789	3.50	.3345	286
O5 (μ_2)	-1.44	0.02	765	3.56	.3724	334
O4 (μ_3)	-1.44	0.02	765	3.56	.3738	336
O3 (μ_2)	-1.41	0.02	897	3.59	.3946	362
O9 (μ_1)	-1.28	0.05	1151	3.72	.4366	470
O8 (μ_1)	-1.23	0.04	1138	3.77	.4454	510
O6 (μ_1)	-1.23	0.04	1138	3.77	.4462	511

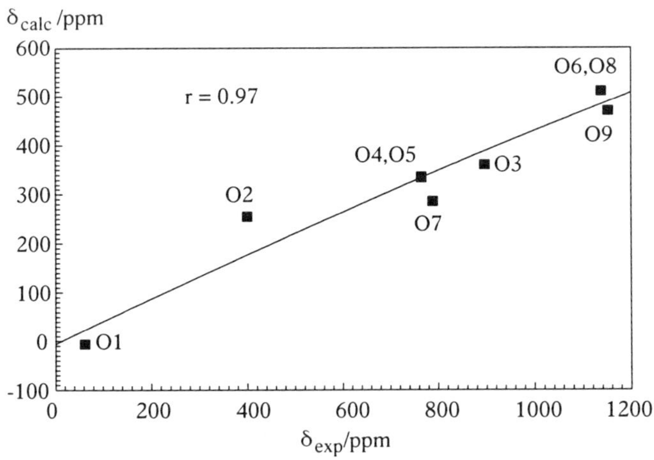

Figure 3. Linear correlation observed between experimental and computed ^{17}O NMR isotropic chemical shifts for the decavanadate ion $[V_{10}O_{28}]^{6-}$.

Table VIII. Absolute ^{17}O NMR shielding tensors (values in ppm) for the free water molecule (AEE = 56.1 eV) and a crystalline tungstate K_2WO_4 (AEE = η) assuming that $<(a_0/r)^3> = 5.0 + q(O)$. Parametrization for K_2WO_4 was $r_{at}(W)$ = 135 pm, $r_{at}(K)$ = 133 pm, eχ°(W) = 4.40 eV and eχ°(K) = 2.42 eV.

Parameter	H_2O	K_2WO_4 (e<χ> = 4.1 eV, $e^2\eta$ = 7.6 eV, I = 52.4%)		
Site	O	O1	O2	O3
q	-0.42	-1.21	-1.19	-1.17
f	0.67	0.02	0.02	0.01
σ_{xx}^d	436.5	502	492	540
σ_{yy}^d	441.5	607	555	627
σ_{zz}^d	438.5	523	482	511
σ_{iso}^d	438.8	544	510	560
$\Delta\sigma^d$	2.7	94	69	101
η_{CS}^d	0.74	0.34	0.22	0.43
P_{ss}	1.99	2.0	2.0	2.0
P_{xx}	1.11	1.21	1.33	2.0
P_{yy}	1.90	2.0	2.0	1.48
P_{zz}	1.43	2.0	1.86	1.69
P_u	.8240	.5276	.5109	.4998
$<(a_0/r)^3>$	4.58	3.79	3.81	3.83
σ_{xx}^p	-72.8	0	-103	-488
σ_{yy}^p	-112.7	-571	-522	-226
σ_{zz}^p	-106.6	-571	-489	-380
σ_{iso}^p	-97.4	-381	-371	-365
$\Delta\sigma^p$	24.6	571	403	208
η_{CS}^p	0.25	0.0	0.12	0.78
σ_{xx}	364.4	502	390	52
σ_{xx}^a	359.9/292	-256	-259	-253
σ_{yy}	333.5	36	33	401
σ_{yy}^a	323.6/247	-222	-210	-189
σ_{zz}	334.2	-48	-7	132
σ_{zz}^a	307/208	91	106	100
σ_{iso}	344.0	163	139	195
σ_{iso}^a	330.2/249	-129	-121	-114
$\Delta\sigma$	31	339	251	206
$\Delta\sigma^a$	44.6/64.5	220	227	214
η_{CS}	0.03	0.25	0.16	0.39
η_{CS}^a	0.56/0.60	0.16	0.22	0.30

[a] Values from coupled HF pertubation theory *(76-77)* for H_2O and from solid-state NMR for K_2WO_4 *(41)*.

magnitude of the anisotropy and the right direction of highest shielding. The bad value found for the asymmetry parameter shows however that work remain to be done to refine the model. Computing shielding tensors for crystalline compounds is also possible as shown in the case of potassium tungstate K_2WO_4 *(78)* whose NMR data have been recorded at 67.8 MHz *(41)*. The three oxygen atoms are found to bear very similar partial charges, in agreement with the very small range of isotropic chemical shifts measured by NMR. As the intensity of the signal coming from site 1 was twice that coming from either site 2 or 3, it was logically deduced that site 1 corresponds to the two crystallographically equivalent oxygens named O3 in table VIII. Being the site having the lowest electronic density it must then be the most deshielded in perfect agreement with experimental observations. In view of the partial charges, it thus seems logical to assign site 2 to oxygen atom O2 and site 3 to oxygen atom O1. Concerning the three shielding tensors, the following conclusions may be drawn:

i) The diamagnetic values are not unrealistic giving quite reasonable anisotropies of about 100 ppm.

ii) The P_{ii} values indicates that our software is not well-suited for handling overflow situations. It is highly probable that σ-overflow occurring in one p-orbital should be corrected through π-bonding and not by a crude and arbitrary levelling at $P_{ii} = 2.0$. Work is in progress in this direction.

iii) As the P_{ii} values are wrong, the paramagnetic values are not realistic, explaining that we are unable to reproduce even the isotropic sequence.

iv) If the absolute values of the total shielding tensor are not reliable, parameters involving differences and ratio of these absolute values like $\Delta\sigma$ or η_{CS} are not unreasonable if compared with experimental values. This allows to make at least qualitative estimation of these parameters. These results are very encouraging, in view of the simplicity of the method and the very rough approximations involved. Refining the model in the very next future should then lead to a much more reliable estimations of NMR parameters directly from the structure.

Conclusion

In conclusion, we have presented a completely novel approach for computing both NMR shielding tensors and efg tensors. It consists in first computing a partial atomic charge distribution from known crystalline or molecular structures, and then in using this distribution to approximate the various parameters governing the magnetic shielding: diamagnetic contribution, radial expansion or contraction of p-orbitals, population unbalance in these p-orbitals and average excitation energy. Besides the molecular or crystalline structure one has to provide a set of radii for evaluating the chemical hardness, and some mixing coefficients to evaluate the starting electronegativities. Having this atomic parametrization in hand, all further calculations are done through a set of programmed functions receiving in input the structure and the charge distribution computed directly from it. The only parameters which can be changed by the user are the efg value of the free atom R° and the coefficient $\varkappa$ describing how this efg changes with the partial charge q. In this study, we have not tried to optimize these two parameters to demonstrate that the knowledge of their detailed numerical values is not very critical. By using rough estimates R° ~ 5 and $\varkappa$ ~ 1 we cannot pretend to reach the same accuracy as others methods, but we are still able to understand experimental trends is a very satisfactory way. Results have been obtained for the ^{17}O nucleus on a wide range of compounds (small molecules, simple crystalline structures, polyatomic ions, and huge and/or complex crystalline structures involving disordered positions and/or partial substitutions). The only kind of compounds which have not been yet tested are glasses and liquids. In fact, the equations of the model are also very well suited to treat such cases. This just requires the coupling of the software with the output of any molecular modelization software able to produce atomic coordinates. The importance of taking into account π–bonding effects for light nuclei have also been clearly demonstrated, but work remain to be done in order to include them with the minimum computational effort. The developed formalism is obviously not limited to the ^{17}O nucleus, and current work is under progress to apply it to other nuclei such as ^{29}Si, ^{19}F, $^{14,15}N$, ^{27}Al, or ^{31}P.

Literature Cited

1. McFarlane, W. and McFarlane, H.C.E.; in *"Multinuclear NMR"*, Mason, J. ed.; Plenum Press, New-York, **1987**, pp. 403.
2. Parr, R.G.; Donnelly, R.A.; Levy, L. and Palke, W.E.; *J. Chem. Phys.* **1978**, *68*, 3801
3. Mulliken, R.S. *J. Chem. Phys.*, **1934**, *2*, 782-793.
4. Sanderson, R.T. *Science*, **1951**, *114*, 670-672.
5. Chen, E.C.M.; Wentworth, W.E. and Alaya, J.A. *J. Chem. Phys.*, **1977**, *67*, 2642.
6. Poole, R.T.; Williams, D.R.; Riley, J.D.; Jenkin, J.G.; Liesegang, J. and Leckey, R.C.G.; *Chem. Phys. Letters*, **1975**, *36*, 401.
7. Pearson, R.G.; *Coordination Chemistry Reviews*, **1990**, *100*, 403.
8. Komorowski, L.; *Chem. Phys. Letts*, **1987**, *134*, 526.
9. Parr, R.G. and Chattaraj, P.K.; *J. Amer. Chem. Soc.*, **1991**, *113*, 1854.
10. Davis, D.W.; Shirley, D.A. and Thomas, T.D.; *J. Amer. Chem. Soc.*, **1972**, *94*, 6565.
11. Pearson, R.G.; *J. Amer. Chem. Soc.*, **1985**, *107*, 6801.
12. Mortier, W.J.; Ghosh, S.K. and Shankar, S.; *J. Amer. Chem. Soc.*, **1986**, *108*, 4315.
13. Parry, D.E.; *J. Chem. Soc. Faraday Trans. 2*, **1974**, *70*, 337.
14. Van Genechten, K.A.; Mortier, W.J. and Geerlings, P. *J. Chem. Phys.*, **1987**, *86*, 5063.
15. Henry, M. *Top. Mol. Organ. Eng.*, **1997**, *15*, 273.
16. Parr, R.G. and Yang, W.; *J.Amer. Chem. Soc.*, **1984**, *106*, 4049.
17. Baekelandt, B.G.; Mortier, W.J. and Schoonheydt, R.A.; *Structure and Bonding*, **1993**, *80*, 187.
18. Fischer, R. and Ludwiczek, H.; *Monatsh. Chem.*, **1975**, *106*, 223.
19. Bratsch, S.G.; *J. Chem. Educ.*, **1988**, *65*, 34.
20. Basch, H.; Viste A. and Gray, H.B.; *Theoret. Chim. Acta (Berl.)*, **1965**, *3*, 458.
21. Waber, J.T. and Cromer, D.T.; *J. Chem. Phys.*, **1965**, *42*, 4116.
22. Slater, J.C.; *J. Chem. Phys.*, **1964**, *41*, 3199.
23. Lucken, E.A.C.; *"Nuclear quadrupole coupling constants"*, Academic Press, **1969**.

24. Paschalis, E. and Weiss, A.; *Theoret. Chim. Acta (Berl.)*, **1969**, *13*, 381.
25. Brun, E.; Derighetti, B.; Hundt, E.E. and Niebuhr, H.H.; *Phys. Letts*, **1970**, *31A*, 416.
26. Lewis, J.; Schwarzenbach, D. and Flack, H.D.; *Acta Cryst.*, **1982**, *A38*, 733.
27. Verhoeven, J.; Dynamus, A. and Bluyssen, H.; *J. Chem. Phys.*, **1969**, *50*, 3330.
28. Sternberg, U.; *Solid State Nucl. Magn. Reson.*, **1993**, *2*, 181.
29. Leadbetter, A.J.; Ward, R.C.; Clark, J.W.; Tucker, P.A.; Matsuo, T. and Suga, H.; *J. Chem. Phys.*, **1985**, *82*, 424.
30. Spiess, H.W.; Garett, B.B.; Sheline, S.K. and Rabideau, S.W.; *J. Chem. Phys.*, **1969**, *51*, 1201.
31. Kamb, B.; *Acta Cryst.*, **1964**, *17*, 1437.
32. Kamb, B.; Hamilton, W.C.; LaPlaca, S.J. and Prakash, A.; *J. Chem. Phys.*, **1971**, *55*, 1934.
33. Edmonds, D.T.; Goren, S.D.; Mackay, A.L. and White, A.A.L..; *J. Magn. Reson.*, **1976**, *23*, 505.
34. Edmonds, D.T.; Goren, S.D. and White, A.A.L..; *J. Magn. Reson.*, **1977**, *27*, 35.
35. Kuhs, W.F.; Finney, J.L.; Vettier, C. and Bliss, D.V.; *J. Chem. Phys.*, **1984**, *81*, 3613.
36. Jorgensen, J.D.; Beyerlein, R.A.; Watanabe, N. and Worlton, T.G.; *J. Chem. Phys.*, **1984**, *81*, 3211.

37. Mueller, K.T.; Baltisberger, J.H.; Wooten, E.W. and Pines, A.; *J. Phys. Chem.*, **1992**, *96*, 7001.
38. Sasaki, S.; Fujino, K.; Takeuchi, Y. and Sadanaga, R.; *Acta Cryst.*, **1980**, *A36*, 904.
39. Shannon, R.D.; *Acta Cryst.*, **1976**, *A32*, 751.
40. van der Wal, R.J.; Vos, A. and Kirfel, A.; *Acta Cryst.*, **1987**, *B43*, 132.
41. Schramm, S. and Oldfield, E.; *J. Amer. Chem. Soc.*, **1984**, *106*, 2502.
42. Yamanaka, T. and Mori, H.; *Acta Cryst.*, **1981**, *B37*, 1010.
43. Hesse, K.F.; *Z. Kristallogr.* **1984**, *168*, 93.
44. Timken, H.K.C.; Schramm, S.E.; Kirkpatrick, R.J. and Oldfield, E.; *J. Phys. Chem.*, **1987**, *91*, 1054.
45. Dollase, W.A.; *Z. Kristallogr.* **1965**, *121*, 369.
46. Hriljac, J.A.; Eddy, M.M.; Cheetham, A.K.; Donohue, J.A. and Ray, G.J.; *J. Solid State Chem.*, **1993**, *106*, 66.
47. Spearing, D.R.; Farnan, I. and Stebbins, J.F., *Phys. Chem. Min.*, **1992**, *19*, 307.
48. Timken, H.K.C., Turner, G.L., Gilson, J.P., Welsh, L.B. and Oldfield, E.; *J. Amer. Chem. Soc.*, **1986**, *108*, 7231.
49. Shore, J.S.; Wang, S.H.; Taylor, R.E.; Bell, A.T. and Pines A., *J. Chem. Phys.*, **1996**, *105*, 9412.
50. Gierke, T.D. and Flygare, W.H.; *J. Amer. Chem. Soc.*, **1972**, *94*, 7277.
51. Gierke, T.D., Tigelaar, H.L. and Flygare, W.H.; *J. Amer. Chem. Soc.*, **1972**, *94*, 330.
52. Malli, G. and Froese, C.; *Int. J. Quantum. Chem.*, **1967**, *1S*, 99.
53. Saxena, K.M.S. and Narasimhan, P.T.; *Int. J. Quantum. Chem.*, **1967**, *1*, 731.
54. Moniz, W.B. and Poranski Jr., C.F.; *J. Magn. Reson.*, **1973**, *11*, 62.
55. Pople, J.A.; *J. Chem. Phys.*, **1962**, *37*, 53.
56. Karplus, M. and Das, T.P.; *J. Chem. Phys.*, **1961**, *34*, 1683.
57. Zilm, K.W. and Duchamp, J.C.; in *Nuclear Magnetic Shieldings and Molecular Structure*, Tossel, J.A. ed., Kluwer Academic Publishers, **1993**, pp. 315.
58. Pople, J.A.; *Mol. Phys.*, **1963**, *7*, 301.
59. Clementi, E. and Raimondi, D.L.; *J. Chem. Phys.*, **1963**, *38*, 2686.

60. Barnes, R.G. and Smith, W.V.; *Phys. Rev.*, **1954**, *93*, 95.
61. Velenik, A. and Lynden-Bell, R.M.; *Mol. Phys.*, **1970**, *19*, 371-381.
62. Malli, G. and Fraga, S.; *Theoret. Chim. Acta*, **1966**, *6*, 54-63.
63. Wasylishen, R.E.; Mooibroek, S. and Macdonald, J.B.; *J. Chem. Phys.*, **1984**, *81*, 1057.
64. Sabine, T.M. and Hogg, S.; *Acta Cryst.*, **1969**, *B25*, 2254.
65. Vidal-Valat, G.; Vidal, J.P. and Kurki-Surno, K.; *Acta Cryst.*, **1978**, *A34*, 594.
66. Albertsson, J., Abrahams, S.C. and Kvick, A.; *Acta Cryst.*, **1989**, *B34*, 34.
67. Davey, W.P. and Hoffman, E.O.; *Phys. Rev.*, **1920**, *15*, 333.
68. Aurivillius, K.; *Acta Chem. Scand.*, **1964**, *18*, 1305.
69. Pannetier, J. and Denes, G.; *Acta Cryst.*, **1980**, *B36*, 2763.
70. Leciejewicz, J.; *Acta Cryst.*, **1961**, *14*, 1304.
71. Turner, G.L.; Chung, S.E. and Oldfield, E.; *J. Magn. Reson.*, **1985**, *64*, 316.
72. Bastow, T.J.; and Stuart, S.N.; *Chem. Phys.*, **1990**, *143*, 459.
73. Punte, G.; Rivero, B.E.; Rigotti, G. and Navaza, A.; *Acta Cryst.*, **1988**, *C44*, 216.
74. Besecker, C.J.; Klemperer, W.G.; Maltbie, D.J. and Wright, D.A.; *Inorg. Chem.*, **1985**, *24*, 1027.
75. Day, V.W.; Klemperer, W.G. and Maltbie, D.J.; *J. Amer. Chem. Soc.*, **1987**, *109*, 2991.
76. Arrighini, G.P.; Maestro, M. and Moccia, R.; *Chem. Phys. Letters*, **1970**, *7*, 351.
77. Appleman, B.R.; Tokuhiro, T.; Fraenkel, G. and Kern, C.W.; *J. Chem. Phys.*, **1974**, *60*, 2574.
78. Koster, A.S.,; Kools, F.X.N.M. and Rieck, G.D., *Acta Cryst.*, **1969**, *B25*, 1704.

Chapter 21

Local and Long-Range Effects on NMR Shieldings in Main-Group Metal Oxides and Nitrides

J. A. Tossell

Department of Chemistry and Biochemistry, University of Maryland, College Park, MD 20742

The NMR shieldings of cationic main group elements, e.g. B, Al and Si, in their oxides and nitrides, are determined almost entirely by the composition and geometry of their 1st and 2nd coordination spheres. Longer range effects related to ring size do occur, but are small in magnitude. By contrast, recent computational studies on B oxides and Si nitrides indicate that O and N NMR shieldings are strongly modified by atoms in the 3rd and higher coordination spheres. Such long-range effects complicate the calculation of O and N NMR shieldings in solids, but they also indicate that the shieldings of such elements can be effective probes of the mid- and long-range structures of amorphous oxides and nitrides. Similar long-range effects on N shieldings will be shown for phosphorus nitrides and for hexagonal BN.

Solid-state NMR spectroscopy is a powerful technique for the determination of structure in amorphous solids and for species in solution. Extensive studies of NMR shieldings of electropositive or cationic atoms, such as Si, have established that shieldings are strongly influenced by the composition of the 1st coordination sphere and also show significant, consistent and easily measurable trends related to the identify and geometry of 2nd coordination sphere atoms. Atoms in the 3rd and more distant coordinate spheres generally have little effect upon the shielding. Experimental trends in Si shielding observed experimentally arise from variations in the coordination number (i.e. the number of atoms in the 1st coordination sphere), the extent of polymerization of the silicate tetrahedra, the degree of replacement of one net-work forming cation by another (e.g. coupled Na^{+}, Al^{+3} for Si^{+4} substitution), the size of the rings of tetrahedra present and the Si-O-Si angles (*1,2*). Similar trends are seen in gas-phase molecules, species in aqueous solution and in both crystalline and amorphous solids. Polarized double-zeta basis set Hartree-Fock level calculations using small molecular cluster models reproduce these trends semiquantitatively, as we will show.

Calculated absolute shieldings for Si in such model systems also agree well with those observed in solids. As seen in Table 1, shieldings calculated at the modest 6-311(2d) basis set (*3*) level using the GIAO method (*4*) as implemented in the software GAUSSIAN94 (*5*), for equilibrium distances calculated at the 6-31G* SCF level using the GAMESS software (*6*), are in good agreement with absolute shieldings (*7*) for both the gas-phase molecule SiF_4 and the solid-state species SiO_4^{-4}. By contrast, the calculated absolute N shielding (*8*) for $(SiH_3)_3N$ is in good agreement

Table 1. Comparison of shieldings calculated at the 6-311(2d) GIAO level and experimental absolute shieldings for some Si and N compounds.

Si species	**calc. shielding (ppm)**	**exp. absolute shielding[b,c] (ppm)**
Si species		
SiF_4	500.5	482
SiO_4^{-4}	456.8	439.5
N species[a]		
$(SiH_3)_3N$	297.6	281
Si_3N_4	206.1, $Si_9N_9H_{21}$ model, 6-31G*	170 N(1) 152 N(2)

a. ref. 12
b. Si absolute shieldings, ref. 7
c. N absolute shieldings, ref. 8

with the experimental value for the gas-phase molecule (*9*), but is much larger than the N shieldings observed for either of the 3-coordinate N sites in crystalline α-Si_3N_4 (*10*). Only when much larger cluster models are considered, e.g. $Si_9N_9H_{21}$, do we obtain absolute shieldings which approach experiment. We previously reported this result (*11,12*) for Si_3N_4 and C_3N_4 models. Our conclusions about the long-range nature of N shieldings in solids have recently been confirmed for C_3N_4 (*13*) by other researchers using a very different quantum mechanical approach (density functional theory with pseudopotentials and a plane-wave basis).

Computational Methods

The basic computational method is that of coupled Hartree-Fock perturbation theory (*14*). At present we prefer the GIAO implementation mentioned above because of its computational efficiency and ease of use, but we have previously used a common gauge-origin method as implemented in the software SYSMO (*15*) as well as the random-phase approximation, localized orbital (RPA LORG) approach as implemented in the software RPAC (*16*).

Trends in NMR Shieldings for ^{29}Si and Other Cationic Species

If trends in NMR shieldings are determined only by the local environment than it is still possible to develop a comprehensive picture of mid-range order within a non-crystalline material using a multi-nuclear approach. This can be illustrated in Figure 1, taken from ref. 17, which shows a snapshot from a molecular dynamics simulation of an alkali silicate glass. If we imagine spheres drawn about each of the nuclei, Si, O and K, to identify the range of atoms determining each nuclides' NMR shielding, such spheres will overlap and we have enough information to characterize mid-range as well as short-range order.

We will now consider the important trends in NMR shielding for electropositive atoms, starting with that involving coordination number for compounds of Si, Na and Al. Although Si is four-coordinate in most all of its oxidic compounds at atmospheric pressure, it can transform to 5- or 6-coordination at high-pressure (*18*). As shown in Table 2, calculated shieldings for 4-, 5- and 6-coordinate hydroxides at the 6-31G* GIAO level show the same increase in shielding with coordination number as observed experimentally. Earlier studies (*19*) using the common-origin coupled Hartree-Fock approach showed the right trend, but the change is shielding was exaggerated because of inadequately compensated diamagnetic contributions from the ligand cores. Similar trends in shielding with coordination number are seen for other nuclides, such as ^{23}Na (*20*), which shows a shielding increase of around 5 ppm as the CN increases by 1, for CN between 5 and 10. As seen in Table 3, shieldings calculated at the 6-31G* GIAO level for the $Na(OH_2)_n^+$, n=2, 4, 5, 6 and 8 series show a trend very similar to experiment for CN's between 4 and 8. All these hydrated species are deshielded with respect to free Na^+, with the deshielding a sensitive function of the CN and the Na-O distance. An accurate evaluation of shielding as a function of identity and number of nearest neighbour atoms also sometimes makes it possible to identify species in amorphous systems. For example we have identified a species $AlF_3(O\text{-}Si)_2$ in F-bearing aluminosilicate glasses (*21*) based upon a comparison of our calculated NMR shieldings and Al nuclear quadrupole coupling constants with experimental values (*22*).

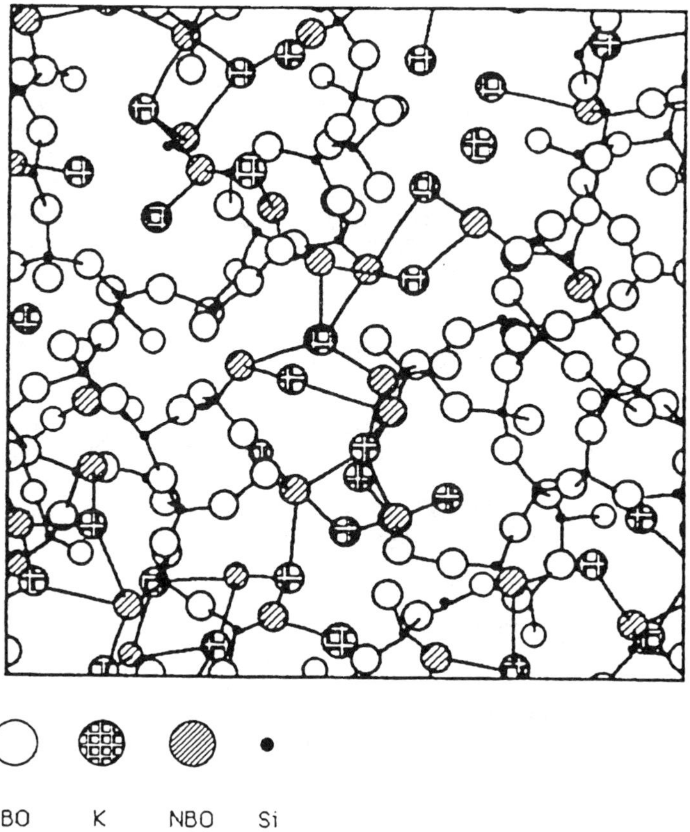

Figure 1 The atomic scale structure of the potassium silicate glass containing 20 mol % K_2O (Reproduced with permission from reference 17. Copyright 1991 American Institute of Physics.)

Table 2. Calculated shieldings for different coordination number Si hydroxides (6-31G* GIAO NMR and optimized geometries) compared with experimental values (ref. 18).

Molecule	calculated σ(ppm)	exp. σ(ppm)
$Si(OH)_4$	510.2	-106
$Si(OH)_5^{-1}$	557.2	-147
$Si(OH)_6^{-2}$	604.6	-198

Table 3. Calculated ^{23}Na NMR shieldings for Na^+, $Na(OH_2)n^+$, n=2, 4, 5, 6 and 8 and $Na(OH_2)_5O(SiH_3)_2^{+1}$ (6-31G* GIAO at 6-31G* geometries)

molecule	σ(ppm)
free Na^+	623.2
$Na(OH_2)_2^+$	604.2
$Na(OH_2)_4^+$	581.8
$Na(OH_2)_5^+$	583.3
$Na(OH_2)_6^+$	588.9
$Na(OH_2)_8^+$	598.1
$Na(OH_2)_5O(SiH_3)_2^+$	591.2

An even more important trend in ^{29}Si NMR shieldings is the increase in shielding as the degree of polymerization increases (*23*), often shown as a plot of Q^n (where n is the number of O atoms of the tetrahedron shared with other tetrahedrally coordinated atoms such as Si or Al) vs. NMR chemical shift. This trend has been reproduced by RPA LORG calculations comparing SiH_3OH and $(SiH_3)_2O$, which show an increase in ^{29}Si shielding due to formation of the Si-O-Si bridge, which becomes larger as the Si-O-Si angle increases (*24,25*). Such an increase in shielding with angle is seen in general in a wide variety of polymerized silicates (*26*). It is also observed that the ^{29}Si shielding increases as the size of the silicate ring increases, for both solids and molecules (*27*). This trend is reproduced by calculations on $[H_2SiO]_n$, n=2,3, oligomers (*25*). Such trends are also seen for more realistic models, such as in the comparison of $Si(OH)_4$ vs. the single 4-ring compound $Si_4O_{12}H_8$ at the 6-31G* GIAO level (*28*), where an increase of about 21 ppm is seen. The effect of replacing Si^{+4} by Na^+, Al^{3+} may be assessed by comparing $Si_4O_{12}H_8$ and $Si_2Al_2O_{12}H_8Na_2$, which show a shielding differnce of about 13 ppm, consistent with experiment (*29*). Similar trends are calculated to occur for 4-coordinate Al in polymerized species like $Al_nO_n(OH)_{2n}^{-n}$, whose calculated properties are shown in Table 4, and which serve as models for species with Al-O-Al bridges in recently synthesized aluminosiloxanes (*30*). Polymerization of the monomeric unit $Al(OH)_4^-$ increases the ^{27}Al shielding, as does an increase in n (ring size) or the addition of counterions such as Li^+ or H^+. A similar effect is seen for 4-coordinate Be^{+2} (*31*). As shown in Table 5, hydrolysis of $Be(OH_2)_4^{+2}$ (i.e. formation of $Be(OH_2)_3OH^{+1}$) deshields the ^9Be by almost 2 ppm, while trimerization of this species to give $Be_3(OH)_3(OH_2)_6^{+3}$ increases the shielding by about 1.2 ppm, giving a 0.8 ppm difference between the parent ion and the hydrolyzed trimer, in good agreement with the experimental difference of 0.61 ppm (*32*). These differences do not change if we consider hydrated species, e.g. $Be(OH_2)_4...6H_2O^{+2}$. The calculated changes in ^1H spectra for these species also agree with experiment (33) and our calculations indicate that the ^{17}O spectra will also be diagnostic of speciation.

It is important to note that for ^{17}O the nuclear quadrupole coupling constant is also an important NMR parameter, which can be used to characterize both the identity of the tetrahedrally coordinated atoms, T, in a T-O-T' linkage, and the value of the T-O-T' angle. Earlier calculations (*24*) showed a decrease in the O NQCC as the Si-O-Si angle decreases, a trend verified by experiment (34, 35). Later calculations(*36*) indicating a progessive decrease in the O NQCC for Si-O-Si, Si-O-Al and Al-O-Al angles were also verified.

It is important to note that polymerization of tetrahedral units does not always lead to an increase in the shielding of the tetrahedral cation. For example, at the 6-31G* GIAO level $B(OH)_3$ and $[B(OH)_2]_2O$ give essentially the same ^{11}B NMR shielding. However, the ^{11}B NMR shielding in borates is diagnostic of structure, particularly for B atoms in single 3-rings or boroxol, $B_3O_3(O-)_3$, rings, as discussed previously (*37, 38*). Comparison with experimental values (*39,40*) indicates that the characteristic "boroxol" value of the ^{11}B shielding is actually found in a number of boron oxide species containing in part 4-coordinate B, which can be distinguised by their Raman spectra. In "drum" aluminosilicate molecules (*41*) the shielding of ^{29}Si is also found to be smaller than that in the corresponding monomeric unit, a result reproduced by calculations at the 6-31G* GIAO level (*42*).

Table 4. Calculated ^{27}Al and ^{17}O NMR shieldings and electric field gradients at bridging and nonbridging O's for aluminates and aluminosilicates.

molecule	σ**Al**	σ**O**	$\lvert q_{Obr} \rvert$
$Al(OH)_4^-$	549.0	325.8	-
$Al_4O_4(OH)_8^{-4}$	567.3	320.9 br, 304.9, 305.3, nbr	0.48
$Al_6O_6(OH)_{12}^{-6}$	572.3	323.4 br, 294.2, 300.2, 300.4, 301.0 nbr	0.57
$Al_4O_4(OH)_8Li_4$	570.0	305.0 br, 329.9 nbr	0.36
$Al_4O_4(OH)_8H_4$	572.7	310.2 br, 331.9 nbr	1.29
$Si_2Al_2O_4(OH)_8^{-2}$	571.2	289.0 br, 298.2, 323.0 nbr	0.75

Table 5. ^9Be,^{17}O and ^1H NMR shieldings (in ppm) calculated for Be aquo and hydroxo complexes using the 6-31G* basis set, the GIAO method and 6-31G* optimized geometries.

molecule	σ**Be**	σ**O**	σ**H**
$Be(OH_2)_4^{+2}$	119.98 115.33[a] 112.66[b]	313.4 - 321.7 (H_2O)	27.2 - 27.5 (H_2O) av. 27.3
$Be(OH_2)_3OH^{+1}$	118.01 113.32[a] 111.21[b]	312.0 - 315.0 (H_2O) 325.1 (OH)	28.3 - 28.9 (H_2O) 31.1 (OH) av. 29.0
$Be_2(OH)_2(OH_2)_4^{+2}$	119.10	318.3 (OH) 328.7 (H_2O)	28.2 (H_2O) 30.1 (OH) av. 28.8
$Be_3(OH)_3(OH_2)_6^{+3}$	119.21	321.9 (OH) 315.7 (H_2O)	28.1 (H_2O) 30.7 (OH) av. 29.0
$Be(OH_2)_4...6H_2O^{+2}$	119.41	310.2 - 317.4 (H_2O)	23.4 - 29.4 (H_2O)
$Be(OH_2)_3OH...6H_2O^{+1}$	117.78	300.8 - 317.2 (H_2O) 329.6 (OH)	24.6 - 29.4 (H_2O) 30.3 (OH)
$Be_2(OH)(OH_2)_6^{+3}$	120.06	319.1 (OH) 300.7 - 302.7 (H_2O)	27.2 - 28.3 (H_2O) 30.2 (OH) av. 28.1
$Be(OH)_4^{-2}$	117.61	313.3 - 314.0 (OH)	35.4 (OH)
H_2O		328.0	31.8
OH^-		321.1	40.3

a. 6-311G(2d,p)
b. BLYP

Shieldings for Anionic Species, O And N

We first noted long-distance effects upon the shielding of anionic species in our calculation of the ^{17}O NMR of some large borate molecular clusters (*38*). Although the calculated ^{11}B shieldings seemed to depend only on the local environment, the ^{17}O NMR shieldings seemed to depend on longer range effects, with the O atoms at the middle of the molecular cluster model, e.g. $B_9O_{15}H_3$, systematically deshielded compared to those on the outside. The calculated values for O shieldings, compared with the experimental ^{17}O NMR results, forced us to conclude that B_2O_3 glass actuallly possessed a good deal of long-range order, in agreement with the conclusion of the experimentalists (*39*). Such effects appeared even more strongly in our study of the ^{15}N shielding of Si_3N_4 (*11,12*) as discussed in the introduction. Although the simplest model molecule $(SiH_3)_3N$ gave a N NMR shielding consistent with the experimental value (*9*) for the gas-phase molecule its shielding was very much larger than the values (170 ppm for the N(1) sites and 152 ppm for the N(2) sites, based on the absolute N shielding scale in ref.8) observed in crytalline Si_3N_4 (*10*). The first study (*11*), using the RPA LORG method for $[SiH_3]_2N[SiH_2(NH_2)]$, showed a shielding of the central N which was strongly dependent upon the conformation of the molecule. Results from the later study (*12*) using larger molecular cluster models and the 6-31G* GIAO method are shown in Table 6. Pictures of the crystal structure of α-Si_3N_4 and of molecular models for it are shown in Figure 2. Only for the largest model, $Si_9N_9H_{21}$, do the calculated values begin to match experiment. This indicates that N atoms in the 3rd coordination sphere significantly deshield the central N. This effect is even greater when O rather than N is the 3rd nearest neighbor, but is smaller for C 3rd nearest neighbors. Similar effects, but of smaller magnitude, are seen for C-N compounds. Such effects can be interpreted qualitatively as arising from the lone pairs of electrons on the N atoms in such compounds, but it is important to note that other properties dependent upon the nature of the N lone pair, such as the electric field gradient at the N nucleus, do not show such large effects of molecular size (*12*).

It is well known that N shieldings show strong solvent effects (*43*), but such effects are roughly an order of magnitude smaller than the solid-state effects seen in Si_3N_4. Recently such effects have been confirmed computationally for C_3N_4 using a very much different method, applicable to both solids and molecules (*13*). Solid C_3N_4 has beeen of much recent interest since the prediction (*44*) that in the α-Si_3N_4 it would be harder than diamond. Later calculations and experimental data are described in ref. 13, where calculated C and N NMR shieldings for various crystalline forms of C_3N_4 and those of related molecules are also given. These authors discuss the sensitivity of N NMR shifts to their environment. The most relevant result to compare to our work is their calculated deshielding of N in the α-C_3N_4 of about 50 ppm compared to $(CH_3)_3N$. This molecule to solid deshielding is of the same magnitude that we calculate.

We have begun to apply the same techniques to the recently characterized crystalline P_3N_5 (*45*). The experimental abolute NMR shieldings for this compound are 182 for the 2-coordinate N's and 132 for the 3-coordinate N's. As shown in Table 7, calculated shieldings at the central N for various model molecules representing 2-coordinate and 3-coordinate sites differ greatly as the size of the model molecule increases. Pictures of the molecules studied are shown in Figure 3. For example, in going from the simplest 3-coordinate model $(PH_3)_3N^{+3}$ to $[P(NH_2)_3]_3N^{+3}$ the

Table 6. N NMR shieldings evaluated at the GIAO SCF level using 6-31G* basis sets and 6-31G* geometries for Si compounds (6-31G(2d) geometries for C compounds).

molecule		symmetry	σ^N (ppm)
$(SiH_3)_3N$		C_{3v}	288.1
$(SiH_3)_2N(SiH_2NH_2)$		C_1	256.8
$(SiH_2NH_2)_3N$		C_{3v}	247.7
$Si_6N_5H_{15}$		C_{3v} C_3	244.7 230.1
$Si_9N_9H_{21}$		D_{3h}	206.1 (inner) 237.1 (outer)
$(CH_3)_3N$	nonplanar, planar,	C_3 C_{3h}	263.2 307.2
$C_6N_5H_{15}$		C_{3v}	259.3
$C_9N_9H_{21}$		D_{3h}	235.9 (inner) 248.7 (outer)
$(SiH_2OH)_3N$		C_{3v}	221.6
$(SiH_2CH_3)_3N$		C_{3v}	260.8

Table 7. Experimental and calculated N NMR shieldings in models for P_3N_5 (6-31G* optimized geometries and 6-31G* or 6-311(2d,p) GIAO NMR).

solid/molecule	σ(ppm)	
experimental P_3N_5 shieldings		
two-coordinate N	182	
three-coordinate N	132	
calculated shieldings	6-31G*	6-311(2d,p)
two-coordinate N		
$(PH_3)_2N^{+1}$	319.1	307.0
$[(NH_2)_3P]_2N^{+1}$	233.4	214.7
$P_3N_3H_6$	255.8	232.8
$P_3N_9H_{12}$	231.9	211.9
three-coordinate N		
$(PH_3)_3N^{+3}$	281.9	268.0
$[P(NH_2)_3]_3N^{+3}$	157.7	139.7

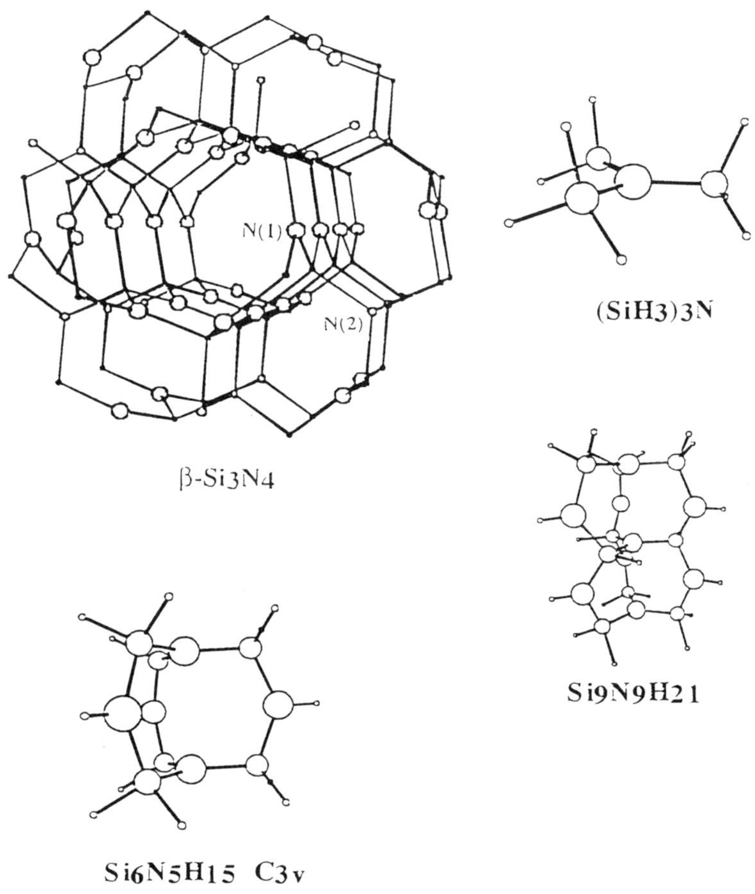

Figure 2 The crystal structure of α-Si_3N_4 and molecular models for the N(2) site.

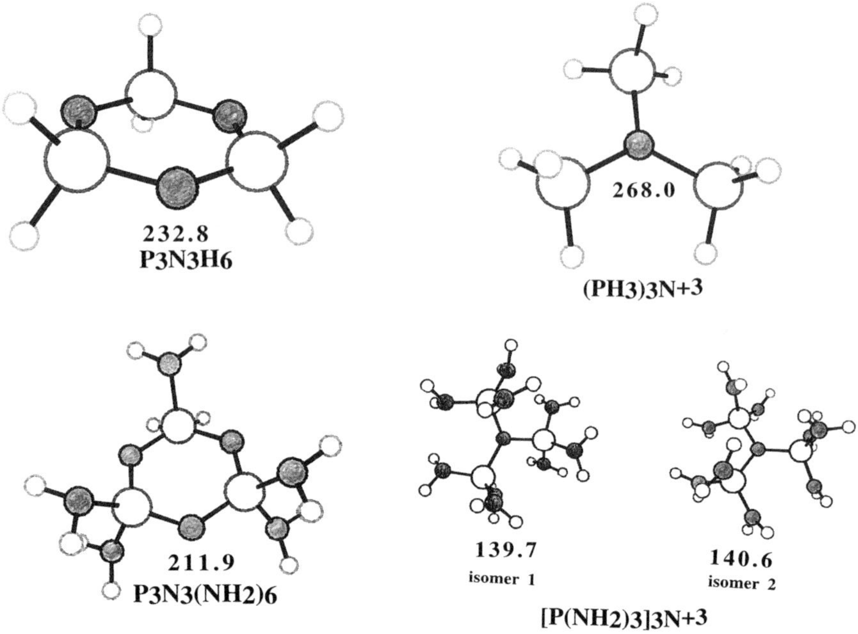

Figure 3 Molecular models for two-coordinate and three-coordinate N sites in P_3N_5, with formulas and calculated central N shieldings.

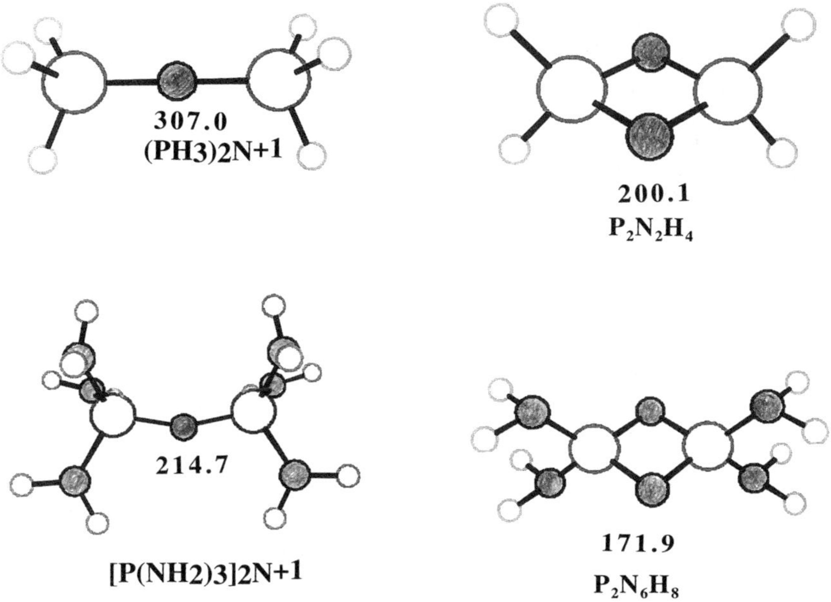

Figure 3 *Continued.*

central N shielding droups by about 120-130 ppm. For the largest 2-coordinate and 3-coordinate models at the 6-311(2d,p) basis set level agreement with experiment is good.

Preliminary studies of cluster models for hexagonal crystallne BN yield the N shieldings shown in Figure 4. A very interesting point is that the N's in the $B_{12}N_{12}H_{12}$ model are all strongly shielded with respect to the smallest model, $(BH_2)_3N$. Thus the molecule to solid change is not invariably deshielding.

The same effects we have seen for N shieldings appear as well in O shieldings, but with a somewhat smaller magnitude. For example. 6-31G* GIAO calculations give the central O in $[Si(OH)_3]_2O$ as deshielded by about 62 ppm compared to $(SiH_3)_2O$. There is some experimental evidence which can be interpreted as supporting this effect. Certainly both bridging and nonbriding O's in silicates show rather wide ranges of NMR shieldings (*35*). A recent study on nanophase MgO (*46*) also indicates that changes in the 3rd coordination sphere can signifcantly modify ^{17}O NMR shieldings. The molecule to solid effect was probably not noticed earlier because ^{17}O is quadrupolar and O often occupies low-CN, low-symmetry sites with large quadrupolar coupling constants. Thus, ^{17}O solid-state NMR shieldings are rather difficult to study so the number of shieldings determined is not very large (compared for example to ^{29}Si shieldings). Further theoretical and experimental studies of this effect on O shieldings are warranted.

Conclusion

We have seen that for cationic species trends in NMR shieldings are fairly local and can be modeled accurately using molecular cluster models and Hartree-Fock level methodologies. We can reproduce the shielding trends arising from changes in coordination number, degree of polymerization, replacement of one 2nd coordination sphere cation by another, T-O-T' angle and ring size. For anionic species the cluster model is more problematic (even when the anionic species is at the center of the cluster). Long-distance effects, particularly from the 3rd coordination shell, can strongly perturb the shieldings. Thus anion shieldings in molecules and solids often differ greatly. Although the molecule to solid effect is usually deshielding, we have discovered at least one case (BN) were the shielding increases in the solid. Calculations on C_3N_4 by others using a new solid-state approach to NMR shieldings show the same effect. Certainly this effect makes solid-state (and possibly solution) O and N shieldings more difficult to interpret and to model, but it also suggests that the shieldings of these elements can potentially give valuable information on the long-range structure of amorphous solids.

Acknowledgments

This work was supported by DOE Office of Basic Energy Sciences, Geosciences Program, Grant DE-FG02-94ER14467.

Literature Cited

(1) (a) Fyfe, C. A. *Solid State NMR for Chemists*, C.F.C. Press, Guelph, Canada, **1983** (b) Engelhardt, G. and Michel, D. High-resolution Solid-state NMR of Silicates and Zeolites, Wiley, **1986.**

(2) Kirkpatrick, R. J., 341, in *Reviews in Mineralogy*, Vol. 18, *Spectroscopic Methods in Mineralogy and Geology*, Hawthorine. F. C. ed., Mineralogical Society of America, Washington, D. C. ,**1988.**

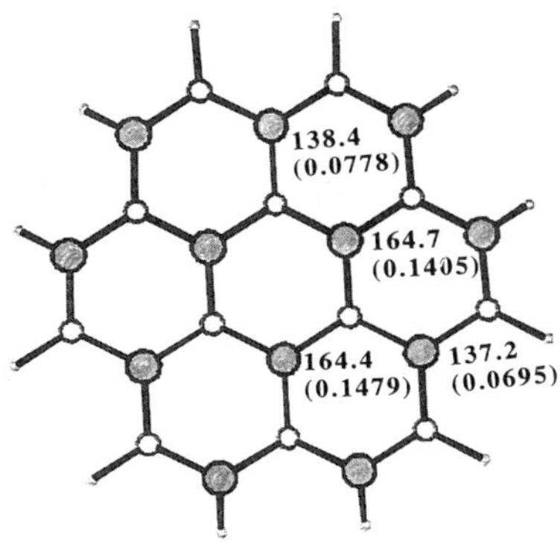

B12N12H12

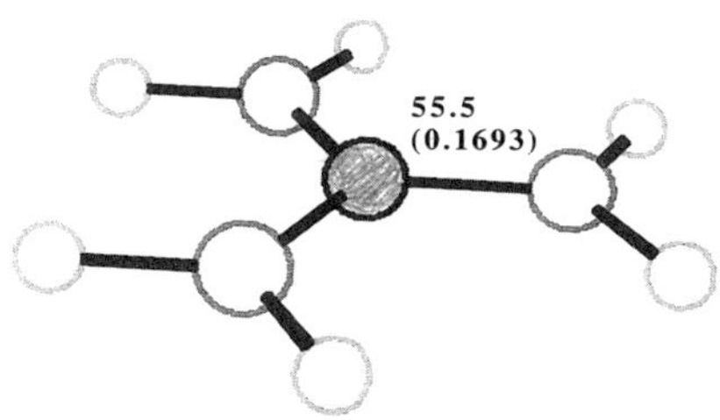

(BH2)3N

Figure 4 $B_{12}N_{12}H_{12}$ model for solid hexagonal BN and molecular $(BH_2)_3N$ at optimized geometries, with calculated N shieldings.

(3) Hehre, W. J., Radom, L., Schleyer, P. v. R. and Pople, J. A. *Ab Initio Molecular Orbital Theory*, Wiley, **1986.**
(4) Hinton, J. F., Guthrie, P. L., Pulay, P. and Wolinski, K., *J. Mag. Res.*, **1993**, *A103*, 188.
(5) Frisch, M. J., Trucks, G. W., Schlegel, H. B., Gill, P. M. W., Johnson, B. G., Robb, M. A., Cheeseman, J. R., Keith, T., Petersson, G,. A., Mongomerty, J. A., Raghavachari, K., Al-Laham, M. A., Zakrzewski, V. G., Ortiz, J. V., Foresman, J. B., Peng, C. Y., Ayala, P. Y., Chen, W., Wong, M. W., Andres, J. L., Replogle, E. S., Gomperts, R., Martin, R. L., Fox, D. J., Binkley, J. S., Defrees, D. J., Baker, J., Stewart, J. P., Head-Gordon, M., Gonzalez, C. and Pople, J. A. GAUSSIAN94, Rev. B.3, Gaussian, Inc., Pittsburgh, PA **1994.**
(6) Schmidt, M., W., Baldridge, K. K., Boatz, J. A., Elbert, S. T., Gordon, M. S., Jensen, J. H., Koseki, S., Matsunaga, N., Nguyen, K. A., Su, S. J., Windus, T. L., Dupuis, M., and Montgomery, J. A.., *J. Comput. Chem.*, **1993**, *14*, 1347.
(7) Jameson, C. J. and Jameson, A. K., *Chem. Phys. Lett.*, **1988**, *149*, 300.
(8) Jameson, C. J., Jameson, A. K., Oppusunggu, D., Willie, S., Burrell, P. M. and Mason, J., *J. Chem. Phys.*, 1981, *74*, 81.
(9) Anderson, D. W. W., Bentham, J. E. and Rankin, D. W. H., *J. Chem. Soc. Dalton Trans.*, **1973**, 1215.
(10) Harris, R. K., Leach, M. J. and Thompson, D. P., *Chem. Mater.*, **1990**, *2*, 320.
(11) Tossell, J. A., *Int. J. Quant. Chem. Symp.* **1995**, *29*, 443.
(12) Tossell, J. A, ,*J. Mag. Res.*, **1997**, *127*, 49.
(13) Yoon, Y.-G., Pfrommer. B. G., Mauri, F. and Louie, S. G., *Phys. Rev. Lett.*, **1998**, *80*, 3388.
(14) Lipscomb, W. N., *Advan. Magn. Reson.*, **1966**, *2*, 137.
(15) Lazzeretti, P. and Zanasi, R., *J. Chem. Phys.*, **1980**, *72*, 6768.
(16) Hansen, Aa. E. and Boumain, T. D., *J. Chem. Phys.*, **1985**, *82*, 5035.
(17) Huang, C. and Cormack, A. N., *J. Chem. Phys.*, **1991**, *95*, 3634.
(18) Xue, X., Stebbins, J. F., Kanzaki, M., McMillan, P. F. and Poe, B., *Amer. Mineral.*, **1991,** *76*, 8.
(19) Tossell, J. A., *J. Non-Cryst. Solids,* **1990**, *120*, 13.
(20) Xue, X. and Stebbins, J. F., *Phys. Chem. Mineral.*, **1993**, *20*, 297.
(21) Tossell, J. A., *Amer. Mineral.*, **1993**, 78, 16.
(22) Kohn, S. C., Dupree, R., Mortuza, M. G. and Henderson, C. M. B., *Amer. Mineral.*, **1991**, 76, 309.
(23) Lippmaa. E., Magi, M., Samoson, A., Engelhardt, G. and Grimmer, A.-R., *J. Amer. Chem. Soc.*, **1980**, *102*, 4889.
(24) Tossell, J. A. and Lazzeretti, P., *Phys. Chem. Mineral.*, **1988**, *15*, 564.
(25) Tossell, J. A., *Chem. Mater.*, **1994**, *6*, 239.
(26) Oestrike, R., Yang, W.-H., Kirkpatrick, R. J ., Hervig, R. L., Navrotsky, A. and Montez, B., Geochim. Cosmochim. Acta, **1987**, *51*, 2199.
(27) Marsmann, H. ^{29}Si NMR spectroscopic results, in *^{17}O and ^{29}Si NMR: NMR Basic Principles and Progress*, Diehl, P., Fluck, E. and Kosfeld, R. Ed., Springer, Berlin, vol. 17, **1981.**
(28) Tossell, J. A. and Saghi-Szabo, G., *Geochim. Cosmochim. Acta*, **1997**, *61*, 1171.
(29) Lippmaa, E., Magi, M., Samoson, A., Tarmak, M. and Engelhardt, G., *J. Amer. Chem. Soc.*, 1981, *103*, 4992.
(30) Veith, M., Jarczyk, M. and Huch, V., *Angew. Chem. Int. Ed.*, **1998**, *37*, 105.
(31) Tossell, J. A., *J. Mag. Resn.*, **1998**, *135*, 203.
(32) Cecconi, F., Ghilardi, C. A., Midollini, S. and Orlandini, A., *Inorg. Chem.*, **1998**, *37*, 146.
(33) Akitt, J. W. and Duncan, R. H., *J. Chem. Soc., Faraday Trans. I,* **1980**, *76*, 2212.
(34) Farnan, I., Grandinetti, P. J., Baltisberger, J. H., Stebbins, J. F., Werner, U., Eastman, M. A, and Pines, A., *Nature*, **1992,** *358*, 31.

(35) Xue, X., Stebbins, J. F. and Kanzaki, M., *Amer. Mineral.*, **1994**, *79*, 31.
(36) Tossell, J. A., *J. Phys. Chem.,* **1996**, *100*, 14828.
(37) Tossell, J. A., *J. Non-Cryst. Solids*, **1995**, *183*, 307.
(38) Tossell, J. A., *J. Non-Cryst. Solids*, **1997**, *215*, 236.
(39) Youngman, R. E. and Zwanziger, J. W., *J. Non-Cryst. Solids*, **1994**, *168*, 293.
(40) Youngman, R. E. and Zwanziger, J. W., *J. Phys. Chem.*, **1996**, *100*, 16720.
(41) Voigt, A., Murugavel, R., Ritter U., and Roesky, H. W., *J. Organomet. Chem.*, **1996**, *521*, 279.
(42) Tossell, J. A., *Inorg. Chem.*, **1998**, *37*, 2223.
(43) Witanowski, M., Biedrzycka, Z., Sicinska, W., Grabowski, Z. and Webb, G. A., *J. Mag. Resn.*, **1991**, *124*, 127.
(44) Cohen, M. L., *Phys. Rev. B*, **1985**, *32*, 7988.
(45) Schnick, W., Lucke, J. and Krumeich, F., *Chem. Mater.*, **1996**, *8*, 281.
(46) Chadwick, A. V., Poplett, I. J. F., Maitland, D. T. S. and Smith, M. E., *Chem. Mater.*, **1998**, *10*, 864.

Chapter 22

Ab Initio Calculations of ^{31}P NMR Chemical Shielding Anisotropy Tensors in Phosphates: The Effect of Geometry on Shielding

Todd M. Alam

Department of Aging and Reliability in Bulk Materials, Sandia National Laboratories, Albuquerque, NM 87185–1407

Molecular geometry variations have a pronounced influence on the magnitude and orientation of the phosphorous (^{31}P) nuclear magnetic resonance (NMR) chemical shielding anisotropy (CSA) tensor (σ) in phosphates. *Ab initio* calculations of the ^{31}P σ tensor as a function of internal bond angles and bond lengths for simple phosphate systems are described. The CSA tensors were calculated using GAUSSIAN 94 implementation of the gauge-including atomic orbital (GIAO) method, at the Hartree-Fock (HF) level. Both the isotropic value of the chemical shielding and the principal components of the anisotropic tensor were determined. It is demonstrated that differences in the behavior of the isotropic values and the principal components of the CSA tensor can provide a powerful tool to investigate structural variations.

The ability to calculate and correlate material properties directly with molecular structure remains an important objective in material science. The correlation of the chemical shift or magnetic shielding in nuclear magnetic resonance (NMR) spectroscopy with bond angles, bond length, coordination number and interactions with neighboring atoms has seen a long and rich history in a wide variety of applications. Recently there has been renewed interest in the development of more accurate correlations between NMR observables and the local structure in amorphous materials, including polymers (*1*), composites, mineral oxides (*2,3*) and glasses.(*4*) Because the use of standard scattering techniques can prove difficult in amorphous systems, NMR spectroscopy has become a powerful tool to probe the local and medium range order in these amorphous materials.

The determination of the NMR chemical shift or chemical shielding anisotropy (CSA) tensor from quantum mechanical methods remains a crucial component in the future development of structural correlations. The prediction of NMR parameters using *ab initio* techniques has under gone major advancements in the last decade (*5*), including solutions to the gauge dependence of the chemical shift (*6*), improvements in the basis sets, and the extension to higher levels of theory including density functional theory (DFT) techniques.(*7,8*) Advances in computational speed have also relaxed many of the restrictions involved in previous investigations of large sized molecular clusters. With these advances there have been an increasing number of *ab initio* investigations for a wide variety of nuclei, including ^{11}B (*6*), ^{13}C (*1,9,10*), ^{15}N (*9,10*), ^{17}O (*9-11*), ^{29}Si (*3,12-16*), ^{31}P (*10,17-21*), ^{77}Se and ^{113}Cd. (*10,22*)

A complication encountered in amorphous inorganic materials is the formation of three-dimensional networks, compared to the one-dimensional chains common in synthetic organic polymers. This network formation can lead to a very large number of structural variables that may influence the observed NMR CSA tensor. *Due to this increased complexity it therefore becomes crucial to separate and distinguish the effects of different structural variations on the observed NMR spectra.*

In our laboratory we are particularly interested in obtaining a measure of structure and distribution for ultraphosphate glasses. There have been a limited number of calculations for phosphorous (^{31}P) systems, including semi-empirical correlations between structure and the isotropic chemical shift in phosphate glasses.(*18-21*) To aid in the development of future NMR-structure correlations in phosphate glasses, *ab initio* NMR calculations of the ^{31}P CSA tensor in simple phosphate systems were initiated and are presented here. The influence of bond lengths, bond angles and torsional angles on the CSA tensor are discussed. In addition, analysis of both the isotropic values and the principal components of the anisotropic shielding tensor are presented, providing a method of distinguishing different structural changes.

Theoretical Method and Model Structures

The chemical shift anisotropy (CSA) tensors were calculated using the parallel version of the GAUSSIAN 94 software package (*23*) on a multi-node SGI ONYX processor. The gauge-including atomic orbital (GIAO) method (*24*) at both the Hartree-Fock (HF) and density functional theory (DFT) level of theory were employed for the calculation of the shielding tensors. Four uniform basis sets were previously investigated with various combinations of polarization and diffuse functions: STO-3G, 3-21G, 6-31G and 6-311G.(*25*) The majority of the shielding results presented here were obtained using HF methods at the 6-311++G(2d,2p) level. All geometry re-optimizations employed the DFT B3LYP hybrid functional, utilizing Becke's exchange functional and the LYP correlation functional.(*7,26,27*) A detailed study of basis set dependence of ^{31}P CSA tensor in phosphates, including a comparison between HF and DFT methods, has previously been reported (*25*), and will not be detailed here.

Not all 9 components of σ are typically reported when describing the chemical shielding tensor. Instead, the 3 principal components (or eigen values) in a principal axis system (PAS) are reported. The CSA tensor can also be described by three additional parameters; 1) the isotropic value (or trace), σ_{iso}, of the shielding tensor and is defined as

$$\sigma_{iso} = \frac{1}{3}(\sigma_{11} + \sigma_{22} + \sigma_{33}), \qquad (1)$$

along with 2) the anisotropy ($\Delta\sigma$), defined by

$$\Delta\sigma = \sigma_{33} - \frac{1}{2}(\sigma_{22} + \sigma_{11}), \qquad (2)$$

and, 3) the shielding asymmetry parameter (η) given by

$$\eta = \frac{(\sigma_{22} - \sigma_{11})}{(\sigma_{33} - \sigma_{iso})}. \qquad (3)$$

The identity of the principal components in σ are defined by

$$|\sigma_{33} - \sigma_{iso}| \geq |\sigma_{11} - \sigma_{iso}| \geq |\sigma_{22} - \sigma_{iso}|. \qquad (4)$$

Experimental geometries were utilized when available, otherwise geometries optimized at the B3LYP level using a 6-311++G(2d,2p) basis set are employed. In some instances the NMR CSA tensor calculations were performed on both experimental and computational optimized geometries (as noted in the text). The angles and distances for the investigated compounds are summarized in Table I. For the calculations of CSA tensor variations due to changes in geometry, angles and distances were not re-optimized for each new configuration unless otherwise noted. If geometry re-optimizations were performed, the geometry of the new configuration was optimized using a B3LYP/ 6-311++G(2p,2d) level of theory.

For PH_3, geometry optimization at the B3LYP 6-311++G(2d,2p) level resulted in $r_{PH} = 1.418$ Å and $\theta_{HPH} = 93.4°$, and are equivalent to the experimentally determined geometry. For the PO_4^{3-} anion an optimized structure was utilized, and has slightly longer P-O bond lengths than earlier utilized.(*28*) For phosphoric acid, H_3PO_4 there are rather substantial differences between the experimental geometry (*29,30*) and the geometry obtained from computational optimization. The *ab initio* optimized results are very similar to earlier investigations utilizing a 6-31G(d) basis set.(*31*) The major difference observed between experimental and the computer optimized structures were the relative differences between the short (P=O) and the longer P-OH bond distances. Previously it has been shown that P-O bond distances and bond order play a large role in the magnitude of both the isotropic chemical shift and $\Delta\sigma$.(*18,19,32-34*) Because of this influence, calculations of the NMR CSA tensor were performed on both the experimental and optimized structures (*vide infra*). The agreement between the optimized structure reported here for the eclipsed pyrophosphate cluster, $H_4P_2O_7$, and recent *ab initio* calculations are very good.(*35,36*) A larger discrepancy was noted when compared to an earlier *ab initio* investigation (*31*), and can be attributed to the complexity of the method and basis set employed for the calculations.

Table I. Molecular Geometries used for *Ab Initio* NMR Calculations

Compound					
	r_{PH} (Å)	θ_{HPH}			
PH_3 [a]	1.415	93.3			
	r_{PO} (Å)	$\theta_{OPO(1)}$ [d]	r_{OH} (Å)	θ_{POH}	θ_{POP} [g]
PO_4^{3-} [b]	1.591	109.5	--	--	--
H_3PO_4 [c]	1.496(P=O) 1.546(P-OH)	112.7	0.991	117.0	
H_3PO_4 [b]	1.467(P=O) 1.596(P-OH)	116.3	0.963	111.7	
$H_4P_2O_7$ [b] (eclipsed)	1.461(P=O) 1.592(P-OH) 1.607(P-O-P)	113.0[e] 103.1[f]	0.963	111.1	140.3
$H_4P_2O_7$ [b] (staggered)	1.463(P=O) 1.587(P-OH) 1.614(P-O-P)	114.4[e] 102.2[f]	0.963	111.8	132.2

[a] Experimental geometry (*37*). [b] Optimized geometry, B3LYP 6-311++G(2d,2p). [c] Experimental geometry from neutron diffraction (*29,30*). [d] Corresponds to the O-P-O(1) bond angles where in H_3PO_4 O(1) is the terminal P=O, while in $H_4P_2O_7$ O(1) is the bridging oxygen, O(2) is the terminal P=O and O(3) the OH oxygens. [e] Represents the O(1)-P-O(2) bond angle. [f] Represents the O(1)-P-O(3) bond angle. [g] The bridging P-O-P bond angle.

Calculation of Chemical Shielding Anisotropy Tensors

The chemical shielding anisotropy (CSA) tensors for the simple phosphate structures listed in Table I, were calculated using *ab initio* methods. The isotropic tensor value (σ_{iso}), the three principal components (σ_{ii}), the CSA anisotropy ($\Delta\sigma$) and asymmetry parameter (η) were evaluated and are given in Table II.

The compound PH_3 represents the standard for phosphorous NMR shielding calculations. Experimental absolute shielding values of 594 ± 10 ppm with $\Delta\delta$ = -50 ± 15 ppm have been reported.(*38*) These values agree well with the calculations reported here, as well as previous theoretical investigations.(*20,21,39-42*) Results for the highly symmetric PO_4^{3-} indicated a large deshielding with respect to PH_3, σ_{iso}

of ~ 314 ppm, and is different than previous reports.(*17,28*) For H_3PO_4 substantial differences are observed for the different structures (experimental versus computer optimized). The isotropic shielding values between the two investigated structures are very similar, but the principal components are very sensitive to structural changes, with variations on the order of ~20 to 30 ppm. These Hartree-Fock results for H_3PO_4 (experimental neutron geometry) agree well with the σ_{iso} = 353.9 ppm reported previously by Kutzelnigg and co-workers using the IGLO method.(*6*) They also report an anisotropy of $\Delta\sigma$ = 182.7 ppm, which differs from the 143 ppm reported in Table II. These chemical shielding anisotropies are all larger than the $\Delta\sigma$ ~ 65 ppm observed for frozen phosphoric acid (Alam, personal communications).(*34*) For the pyrophosphate cluster the eclipsed and staggered conformation produced similar CSA tensors. It is interesting to note that the phosphorous atoms are inequivalent in the staggered confirmation.

Table II. *Ab Initio* NMR CSA Tensor for Simple Phosphate Clusters[a]

Compound[b]	σ_{iso}(ppm)[c]	σ_{33}(ppm)[c]	σ_{22}(ppm)[c]	σ_{11}(ppm)[c]	$\Delta\sigma$(ppm)[c]	η[c]
PH_3 [d]	590.1	561.9	604.2	604.2	-42.3	0.0
PO_4^{3-} [e]	313.6	313.6	313.6	313.6	0.0	0.0
H_3PO_4 [d]	351.5	447.0	303.9	303.9	143.1	0.0
H_3PO_4 [e]	347.3	477.8	282.1	282.1	195.7	0.0
$H_4P_2O_7$ [e] (Eclipsed)	358.9	484.1	300.0	292.4	187.9	0.06
$H_4P_2O_7$ [e,f] (Staggered)	354.2	461.5	308.0	293.0	161.0	0.14
	355.3	480.9	297.1	287.8	188.6	0.07

[a] NMR shielding tensors calculated using Hartree-Fock method the 6-311++G(2d,2p) basis set. [b] Geometry of investigated compounds listed in Table I. [c] CSA tensor defined using Equations 1-4. [d] Experimental structure. [e] Optimized structure. [f] In the staggered confirmation the phosphorous are inequivalent.

Influence of Geometry on Chemical Shielding Anisotropy Tensor

Variations in geometry are expected to have direct influence on the ^{31}P chemical shielding tensor in a variety of compounds, but these effects have seen limited *ab initio* investigations.(*10,20,21,40,41,43*) The geometry effects for PH_3 and simple phosphate molecules are described below. With increasing molecular complexity there are additional degrees of freedom that may influence the NMR shielding tensor, but computer simulations allow individual effects to be easily separated. One should recall that the absolute values depend on the level of theory or basis set utilized, but it is the sign and trends observed here that are important.

PH_3 The standard PH_3 molecule provides a simple example of molecular geometry effects on the CSA tensor. If PH_3 maintains C_{3v} there are only two variables that described the geometry, the P-H bond length (r_{PH}) and the HPH bond angle (θ_{HPH}). One can also consider distortions to the symmetry due to non-equivalent differences in bond lengths. Changes in the isotropic and principal components of the CSA tensor with variation in r_{PH} are shown in Figure 1. A linear decrease in σ_{iso}, and all three principal components σ_{ii}, is observed with increasing P-H bond length. A change of approximately –480 ppm/Å is observed for σ_{iso}, and significantly larger than the derivative reported previously for PH_3.(*41*) For PH_3 molecules maintaining C_{3v} symmetry, the principal components σ_{11} and σ_{22} are equal and reveal a much larger variation with changes in r_{PH} than the σ_{33} component of the shielding tensor. If only a single P-H bond length is varied from the equilibrium 1.45 Å length, the C_{3v} symmetry is destroyed, producing inequivalent σ_{11} and σ_{22} tensor components. The magnitude of variation with increasing r_{PH} is significantly smaller than that observed when all three P-H bond lengths were varied (as would be predicted), with the σ_{22} component having a larger variation than the σ_{11} component.

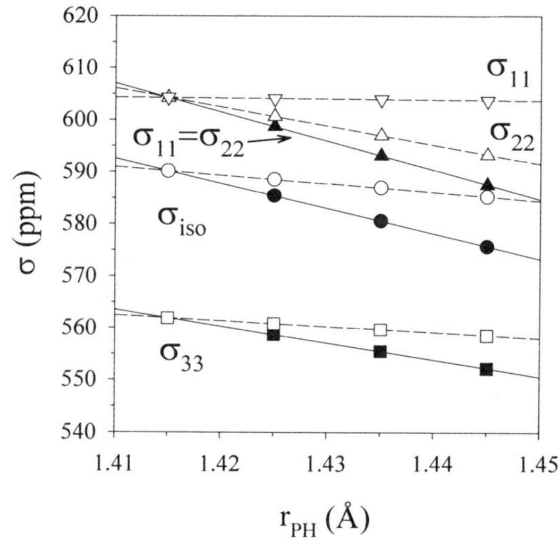

Figure 1: CSA tensor (σ) variation with P-H bond distance (r_{PH}) in PH_3. The isotropic (σ_{iso}) value and principal components (σ_{33}, σ_{22}, σ_{11}) of the CSA tensor are shown. Solid symbols are for variation of all three P-H bond lengths, while open symbols are for variation of a single P-H bond length.

Variations in the ^{31}P shielding tensor due to changes in HPH bond angles were also investigated for PH_3. The changes from a pyramidal geometry to the planar form produce large variations in the different components of the shielding tensor, and have been investigated previously.(*41*) In Figure 2a, the isotropic and principal components of the tensor are shown as a function of HPH angle, in which the PH bond lengths were re-optimized for each new geometry. Very different behavior is observed for the different principal components of the shielding tensor.

The σ_{33} tensor element shows a linear decrease in shielding with increasing θ_{HPH} angle, while the σ_{11} and σ_{22} tensor elements show an initial decrease in the shielding, with a minimum at $\theta_{HPH} \sim 105°$, followed by a steep increase in shielding with higher θ_{HPH} bond angles.

One question that arises in these types of investigations is what amount of the observed variation is due to changes in the θ_{HPH} bond angle, and what effects are due to changes in the r_{PH} bond lengths that result during re-optimization. To distinguish these two influences the changes in the CSA tensor with θ_{HPH} keeping r_{PH} fixed at the equilibrium values of 1.415 Å, are shown in Figure 2b. Comparison to Figure 2a, clearly demonstrates that the bond angle variation produces the majority of the CSA tensor variation. The minor differences observed to result from variations in the bond length, and are most pronounced near $\theta_{HPH} \sim 120°$ where the isotropic and σ_{11} and σ_{22} tensor elements are larger in the re-optimized geometry (Figure 2a) than the geometry with fixed P-H bond lengths (Figure 2b). During re-optimization r_{PH} was found to decrease with increasing θ_{HPH} angle, and from Figure 1 this change in known to produce an increase in magnitude of the shielding tensor.

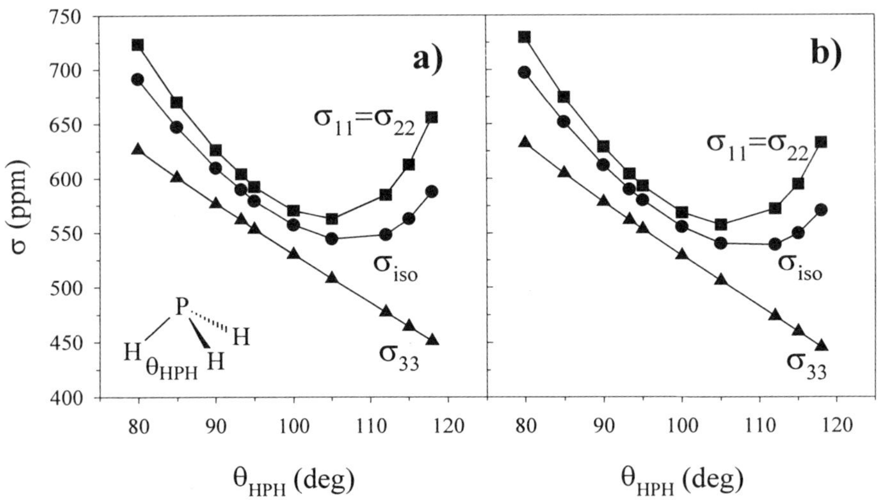

Figure 2: CSA tensor (σ) variation with HPH bond angle (θ_{HPH}) in PH_3; a) geometry optimization of P-H bond distance performed at each new configuration, b) P-H bond distance fixed at experimental equilibrium distance of 1.415 Å.

These effects of changes in geometry on the shielding tensor in PH_3 provide the framework for discussion of trends observed in the simple phosphate systems described below. In theory, the difference in the variation for the isotropic and individual principal components of the chemical shielding tensor can be used to distinguish between different geometry variations, and should be considered in future experimental and *ab initio* investigations of phosphorous compounds.

PO_4^{3-} The variation of the σ_{iso} due to changes in the P-O bond length for PO_4^{3-} is shown in Figure 3. An approximately linear decrease (–625 ppm/Å) with increasing r_{PO} is observed. When only one P-O bond length is increased a much smaller decrease in the shielding is observed (Figure 3). The lengthening of a single P-O bond also removes the high degree of symmetry in PO_4^{3-} creating non-equivalent principal components of the shielding tensor. The magnitude of the inequivalence is very small, such that the individual tensor components are not shown in Figure 3 for clarity.

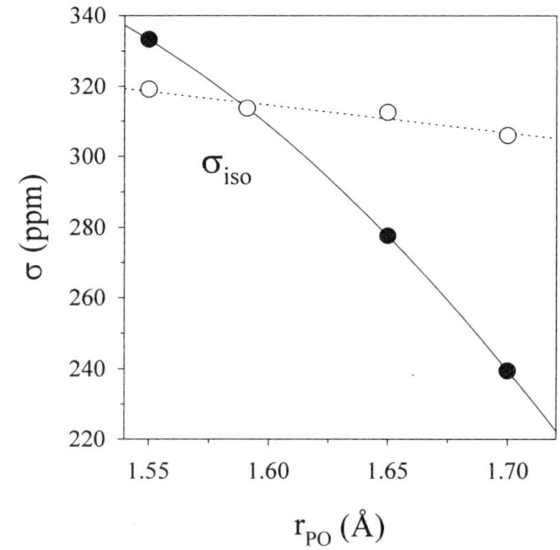

Figure 3: CSA tensor (σ) variation with P-O bond distance (r_{PO}) in PO_4^{3-}. Only the isotropic (σ_{iso}) value of the CSA tensor is shown. Solid symbols are for variation of all three P-O bond lengths, while open symbols are for variation of a single P-O bond length.

H_3PO_4 The variations of the shielding tensor with changes in P-O bond lengths for phosphoric acid, H_3PO_4, are shown in Figure 4. These calculations utilize the computationally optimized geometry for the equilibrium structure as presented in Table I. With increasing length of the terminal oxygen bond, $r_{PO(1)}$, a linear decrease in σ_{iso} and all three principal components σ_{ii} was observed (Figure 4a). The trend is similar to that observed in both PH_3 and PO_4^{3-}, and is consistent with the idea of decreasing bond order effects on the phosphorous shielding tensor. The decrease in the isotropic chemical shielding is approximately -278 ppm/Å and is much smaller than that observed for PH_3 or PO_4^{3-}, but represents the variation of only a single bond length.

Increases in the P-OH bond length, $r_{PO(2)}$ produces a decrease in the isotropic shielding component of approximately -158 ppm/Å, as well as the σ_{11} and σ_{22} components of the shielding tensor (Figure 4b). A slight *increase* in the σ_{33} shielding component is observed with increasing $r_{PO(2)}$ (Figure 3b). Recall that σ_{33} is aligned along the terminal P=O bond in H_3PO_4, such that increases in the P-OH bond length provides a small increase in shielding along this direction. Variation of a single P-OH bond length produces a smaller variation in the shielding tensor values (Figure 3b), as well as removing the C_{3v} symmetry. This loss of symmetry produces inequivalent σ_{11} and σ_{22}, though in H_3PO_4 this difference remains minor.

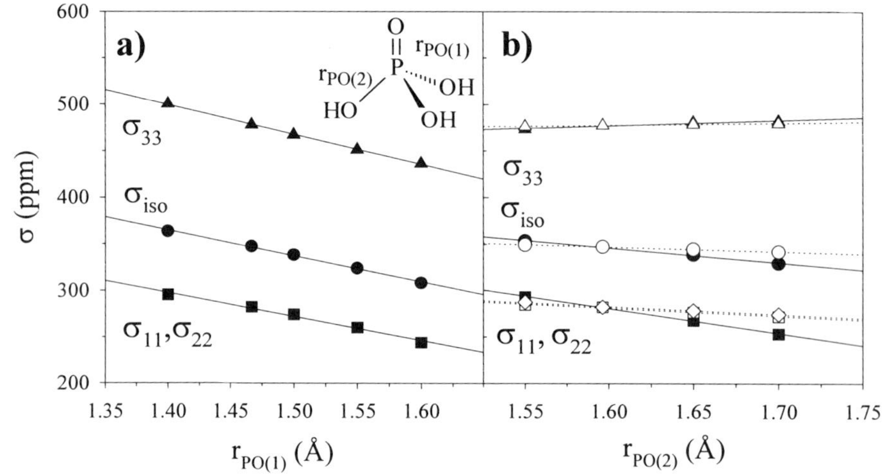

Figure 4: CSA tensor (σ) variation with P-O bond distance (r_{PO}) in H_3PO_4. The isotropic (σ_{iso}) value and principal components (σ_{33}, σ_{22}, σ_{11}) of the CSA tensor are shown. a) Variation due to changes in the terminal P=O bond length, $r_{PO(1)}$. b) Variation due to changes in the P-OH bond length $r_{PO(2)}$. Solid symbols are for variation of all three P-OH bond lengths, while open symbols are for variation of a single P-OH bond length.

The variation of the CSA tensor with changing OPO(1) bond angle is shown in Figure 5. For this calculation the O(1) oxygen refers to the terminal P=O bond, such that this angle is different than the angle description in PH_3. The isotropic component and the σ_{33} tensor element show increasing shielding with increasing $\theta_{POP(1)}$, while the σ_{11} and σ_{22} decrease. *This again demonstrates that angle variations can be distinguished from bond length variation by investigation of the individual principal tensor components.* It is also interesting to note that for H_3PO_4 all the variations have approximately a linear dependence, compared to the very non-linear behavior noted in PH_3.

Hydrogen Bonding Effects *Ab initio* calculations represent a single molecule in vacuum, while experimentally molecules in solids are far from isolated. One of the early observations for phosphoric acid, H_3PO_4, was that the $\Delta\sigma$ observed experimentally (~ 65 ppm) is much smaller than values obtained from simulation (Table II). Whether this is a result of a deficiency in the level of theory used, or represents intra- or inter-molecular effects on the experimental values has yet to be reconciled.

Analysis of the neutron diffraction data for H_3PO_4 (*29*) clearly shows that there is significant hydrogen bonding within the crystal. For example the terminal oxygen has two hydrogen bonds with adjacent molecules. These hydrogen bonds are all approximately 1.55 to 1.60 Å in length. To address the impact of hydrogen bonding on the calculation CSA tensors, a simple model was investigated where a single hydrogen is brought into close contact with the terminal oxygen of the modeled phosphoric acid. The variation of the isotropic and individual principal components of the CSA tensor is shown in Figure 5b. In the limit of very close contact (~ 1 Å) the CSA anisotropy vanishes, as is expected for a symmetric molecule (recall that the other three OH distances are ~0.96 Å, as given in Table I). As the hydrogen bond length is increased the isotropic value and σ_{11} and σ_{22} show a linear decrease in magnitude, while the σ_{33} component (parallel to the hydrogen bond) reveals a linear increase. Similarly the anisotropy of the shielding tensor increases dramatically with increasing hydrogen bond length. This clearly show that inter-molecular effects, such as hydrogen bonding, need to be considered when using *ab initio* techniques to evaluate NMR CSA tensors.

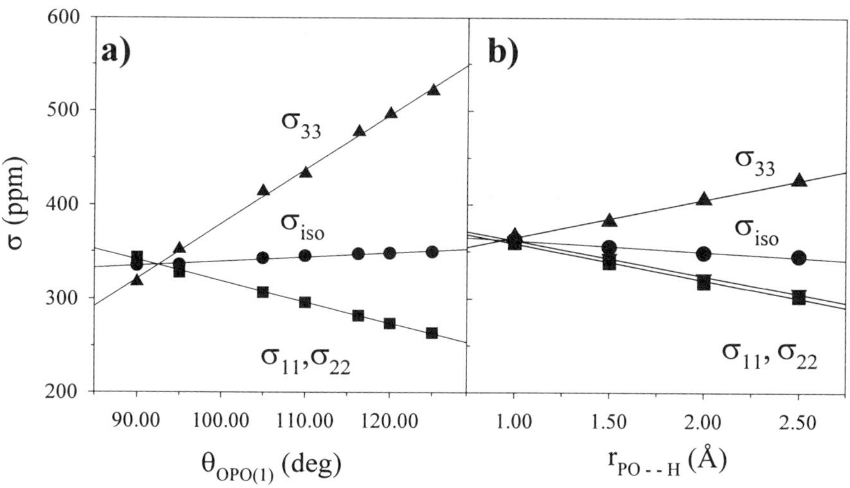

Figure 5: CSA tensor (σ) variation in H_3PO_4. a) Variation as a function of the OPO(1) bond angle, where O(1) designates the terminal P=O oxygen. b) Variation with changes in the hydrogen bond length, $r_{PO—H}$. The single variable hydrogen bond is associated with the terminal P=O oxygen at a PO-H angle of 175°.

$H_4P_2O_7$ For the pyrophosphate cluster, variations in the CSA tensor due to changes in the bridging P-O-P bond angle, θ_{POP}, and the O(1)-P-O-P torsional angle were investigated. In Figure 6 the isotropic and principal components of the shielding tensor as a function of θ_{POP} are shown. The variation of the ^{31}P chemical shift as a function of bridging angle has been investigated previously for phosphate systems using semi-empirical methods.(*18,19*) The results reported here provide a comparison to those investigations, demonstrating the future utility of *ab initio* calculations. For Figure 6a, all of the P-O bond distances were re-optimized at each θ_{POP}, while in Figure 6b the P-O bond distances were maintained at the equilibrium length.

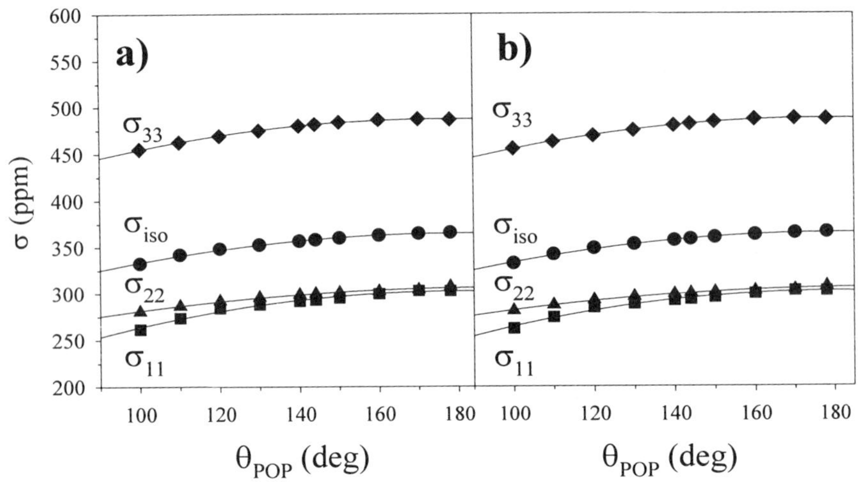

Figure 6: Variation of the CSA tensor (σ) with changes in the bridging P-O-P bond angle θ_{POP} for the pyrophosphate cluster, $H_4P_2O_7$. In a) the P-O bond lengths were re-optimized for each configuration, while in b) the P-O bond lengths are maintained at the equilibrium value.

In both cases the tensor components show a slight increase with increasing θ_{POP}, leveling near ~ 180°. This corresponds to a decrease in the chemical shift of ~ 30 ppm between 90° and 180° degrees. This is slightly larger than the angle dependence reported by Sternberg and co-workers (*18*), but is smaller than the change reported for the H_3POPH_3 dimer.(*43*) The similarity between Figure 6a and 6b might suggest that changes in the P-O bond lengths have a minor impact on the CSA tensor. Inspection of the P-O bond lengths for re-optimization at each θ_{POP} angle is shown in Figure 7a. Interestingly only the bridging P-O bond distance ($r_{PO(1)}$) varies appreciably with changes in the bridging angle. This variation in the bond length is consistent with the correlation between bond order and changes in the bridging POP bond angle reported by Lager and Gibbs.(*33*)

Variation in the CSA tensor with changes in the OPO(1)P torsional angle ϕ (see inset Figure 7a) are shown in Figure 7b. The CSA variations observed for torsional angle changes are small compared to the variation observed for changes in the POP bridging angle (Figure 6). Note in particular that the isotropic chemical shielding is almost constant over the entire ϕ range investigated. The principal components show a slightly larger variation, with $\Delta\sigma$ increasing with higher values of ϕ. It should be noted that the phosphorous nuclei in these pyrophosphates are inequivalent, and that the results shown in Figure 7b correspond to the P nuclei between the oxygen defining the torsional angle. The neighboring P nuclei experiences very small changes as a function of ϕ, but these results are not discussed here. The CSA tensor variation shown in Figure 7b are much smaller than those noted by Losso and co-workers (*19*). In their semi-empirical investigation they reported the torsional angle effect of the chemical shift of meta-phosphate groups, which may not be directly comparable to the pyrophosphate groups as investigated here.

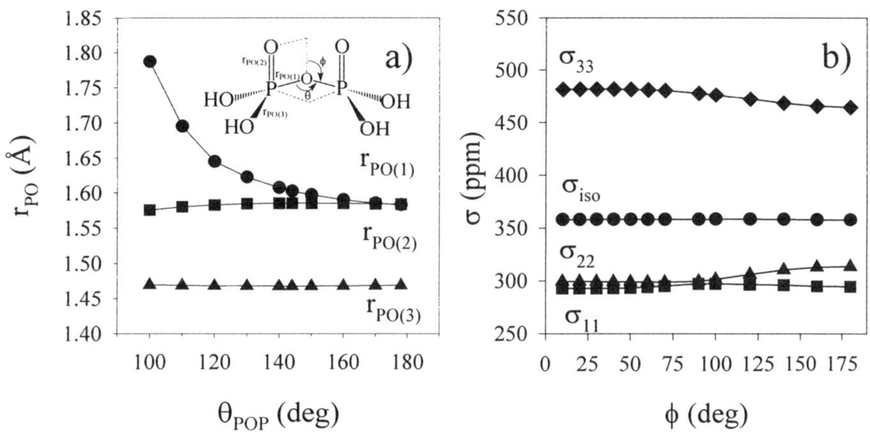

Figure 7: a) Variation of the different P-O bond lengths (r_{PO}) in $H_4P_2O_7$ as a function of bridging angle θ_{POP}. b) Variation of the CSA tensor (σ) as a function of the torsional angle ϕ.

The use of *ab initio* calculations also allows details about subtle variations in the CSA tensor to be examined and explained. An excellent example is the determination of the CSA tensor orientation within the molecular framework. For the pyrophosphate cluster, $H_4P_2O_7$, the largest principal component σ_{33} lies approximately parallel (~10°) to the terminal P=O bond, while the orientation of the σ_{11} and σ_{22} tensor elements are a function of the θ_{POP} and ϕ angles. Inspection of Figure 7b shows the σ_{11} tensor element increases up to ϕ ~ 100° then decreases as ϕ approaches 180°. In contrast the σ_{22} tensor element is relatively constant and only begins to increase in magnitude at ϕ ~ 100°. This change in behavior is a result of the tensor definition used in Equation 4, in which the magnitude of the principal component determines the identity. While there are small changes in the orientation of the σ_{11} and σ_{22} tensor elements with the θ_{POP} and ϕ angles, there is a distinct discontinuity observed at $\phi \sim 100^{o}$. From the shielding calculation it becomes obvious that at $\phi \sim 100^{o}$ the identity of the σ_{11} and σ_{22} tensor elements switch as a result of the definition in Equation 4. The ability to readily calculate and compare CSA tensor orientation with experimental results will prove an important aspect of future a*b initio* NMR investigations.

Summary

These investigations demonstrate that *ab initio* calculations of the NMR ^{31}P shielding tensor provide a powerful tool for investigations of simple phosphate systems. Investigations into larger clusters and the effects of cations on the CSA tensor are presently being pursued. From inspection of the entire CSA tensor, different structural changes can be distinguished, and should be determined in future experimental and theoretical investigations. In simple pyrophosphate systems, changes in the ^{31}P CSA tensor values appear to be dominated by angle variation, rather than distance variations within the molecule. In this investigations it is shown that changes in the bridging P-O-P angle produce the largest influence on the value of the CSA tensor for a wide range of P-O-P angles. While the majority of $P_2O_7^{2-}$ structures reported to date have P-O-P bonds angles ranging from 135° to 155°, several unique phases (including some demonstrating negative thermal expansion coefficients) and high temperature forms containing linear P-O-P bond angle have recently been investigated using novel two-dimensional exchange ^{31}P MAS NMR experiments.(*44,45*) It is clear that ^{31}P NMR can still provide a wealth of information about these types of materials. The application of *ab initio* techniques to these unique and distinct phases plus other phosphate systems will prove to be an exciting avenue of future research.

Acknowledgments. Sandia is a multiprogram laboratory operated by Sandia Corporation, a Lockheed Martin Company, for the United States Department of Energy under Contract DE-AC04-94AL85000. Research sponsored in part by the Phillips Laboratory, Air Force Materiel Command, USAF, under cooperative agreement number F29601-93-2-0001. The views and conclusions contained in this document are those of the author(s) and should not be interpreted as necessarily representing the official policies or endorsements, either expressed or implied, of Phillips Laboratory or the U.S. Government.

Literature Cited

1. Born, R.; Spiess, H. W. *Ab Initio Calculations of Conformational Effects on ^{13}C NMR of Amorphous Polymers*; Born, R.; Spiess, H. W., Ed.; Springer Verlag: New York, 1997; Vol. 35, pp 126.
2. Tossell, J. A. In *Nuclear Magnetic Shieldings and Molecular Structure*; Tossell, J. A., Kluwler Academic Publishers: Dordrecht, The Netherlands, 1993, pp 279-296.
3. Dupree, R.; Kohn, S. C.; Henderson, C. M. B.; Bell, A. M. T. In *Nuclear Magnetic Shieldings and Molecular Structure*; Tossell, J. A., Kluwler Academic Publishers: Dordrecht, The Netherlands, 1993, pp 421-435.
4. Phillips, B. L.; Kirkpatrick, R. J.; Xiao, Y.; Thompson, J. G. In *Nuclear Magnetic Shieldings and Molecular Structure*; Tossell, J. A., Kluwler Academic Publishers: Dordrecht, The Netherlands, 1993, pp 203-220.
5. de Dios, A. C., *J. Prog. NMR Spec.* **1996**, *29*, 229-278.
6. Kutzelnigg, W.; U., F.; Schindler, M. *The IGLO-Mehtod: Ab-initio Calculations and Interpretation of NMR Chemical Shifts and Magnetic Susceptibilities*; Kutzelnigg, W.; U., F.; Schindler, M., Ed.; Springer-Verlag: New York, 1991; Vol. 23, pp 262.
7. Becke, A. D., *J. Chem. Phys.* **1993**, *98*, 5648-5652.
8. Rauhut, G.; Puyear, S.; Wolinski, K.; Pulay, P., *J. Phys. Chem.* **1996**, *100*, 6310-6316.
9. Cheeseman, J. R.; Trucks, G. W.; Keith, T. A.; Frisch, M. J., *J. Chem. Phys.* **1996**, *104*, 5497-5509.
10. de Dios, A. C.; Jameson, C. J., *Annual Reports on NMR Spectroscopy* **1994**, *29*, 1-69.
11. Tossell, J. A.; Lazzeretti, P., *Phys. Chem. Minerals* **1988**, *15*, 564-569.
12. Sternberg, U.; Priess, W., *J. Magn. Reson., Ser. A* **1993**, *102*, 160-165.
13. Wolff, R.; Vogel, C.; Radeglia, R. In *Nuclear Magnetic Shieldings and Molecular Strucuture*; Tosell, J. A., Kluwer Academic Publishers: Dordrecht, The Netherlands, 1993, pp 385-400.
14. Wolff, R.; Radeglia, R., *J. Molecular Structure (Theochem)* **1994**, *313*, 111-119.
15. Cremer, D.; Olsson, L.; Ottosson, H., *J. Molecular Structure (Theochem)* **1994**, *313*, 91-109.
16. Hinton, J. F.; Guthrie, P. L.; Pulay, P.; Wolinski, K., *J. Magn. Reson., Ser. A* **1993**, *103*, 188-190.
17. Bouman, T. D.; Hansen, A. E., *Chem. Phys. Lett.* **1990**, *175*, 292-299.
18. Sternberg, U.; Pietrowski, F.; Priess, W., *Zeitschrift Physik. Chemie Neue Folge* **1990**, *168*, 115-128.
19. Losso, P.; Schnabel, B.; Jager, C.; Sternberg, U.; Stachel, D.; Smith, D. O., *J. Non-Cryst. Solids* **1992**, *143*, 265-273.
20. Jameson, C. J.; de Dios, A. C.; Jameson, A. K., *J. Chem. Phys.* **1991**, *95*, 9042-9053.
21. Jameson, C. J.; de Dios, A.; Jameson, A. K., *Chem. Phys. Lett.* **1990**, *167*, 575-582.

22. Ellis, P. D.; Odom, J. D.; Lipton, A. S.; Chen, Q.; Gulick, J. M. In *Nuclear Magnetic Shieldings and Molecular Structure*; Tossell, J. A., Kluwer Academic Publishers: Dordrecht, The Netherlands, 1993, pp 539-556.
23. Frisch, M. J.; Trucks, G. W.; Schlegel, H. B.; Gill, P. M. W.; Johnson, B. G.; Robb, M. A.; Cheeseman, J. R.; Keith, T. A.; Petersson, G. A.; Montgomery, J. A.; Raghavachari, K.; Al-Laham, M. A.; Zakrezewski, V. G.; Ortiz, J. V.; Foresman, J. B.; Cioslowski, J.; Stefanov, B. B.; Nanayakkara, A.; Challacombe, M.; Peng, C. Y.; Ayala, P. Y.; Chen, W.; Wong, M. W.; Andres, J. L.; Replogle, E. S.; Gomperts, R.; Martin, R. L.; Fox, D. J.; Binkley, J. S.; Defrees, D. J.; Baker, J.; Stewart, J. P.; Head-Gordon, M.; Gonzalez, C.; Pople, J. A. *Gaussian 94 (Revision D.3)*; D.3 ed.; Frisch, M. J.; Trucks, G. W.; Schlegel, H. B.; Gill, P. M. W.; Johnson, B. G.; Robb, M. A.; Cheeseman, J. R.; Keith, T. A.; Petersson, G. A.; Montgomery, J. A.; Raghavachari, K.; Al-Laham, M. A.; Zakrezewski, V. G.; Ortiz, J. V.; Foresman, J. B.; Cioslowski, J.; Stefanov, B. B.; Nanayakkara, A.; Challacombe, M.; Peng, C. Y.; Ayala, P. Y.; Chen, W.; Wong, M. W.; Andres, J. L.; Replogle, E. S.; Gomperts, R.; Martin, R. L.; Fox, D. J.; Binkley, J. S.; Defrees, D. J.; Baker, J.; Stewart, J. P.; Head-Gordon, M.; Gonzalez, C.; Pople, J. A., Ed.; Gaussian, Inc.: Pittsburg, PA, 1995.
24. Ditchfield, R.; Hehre, W. J.; Pople, J. A., *J. Chem. Phys.* **1972**, *54*, 724-728.
25. Alam, T. M., Sandia National Laboratories, **1998**, SAND98-2053.
26. Becke, A. D., *Phys. Rev. A.* **1988**, *38*, 3098-3100.
27. Lee, C.; Yang, W.; Parr, R. G., *Phys. Rev. B* **1988**, *37*, 785-789.
28. Tossell, J. A.; Lazzeretti, P., *Chem. Phys. Lett.* **1987**, *140*, 37-40.
29. Blessing, R. H., *Acta Cryst.* **1988**, *B44*, 334-340.
30. Cole, F. E. ; Cole, F. E., Ed.; Washinton State University: Pullman, WA, 1966.
31. O'Keefee, M.; Domenges, B.; Gibbs, G. V., *J. Phys. Chem.* **1985**, *89*, 2304-2309.
32. Un, S.; Klein, M. P., *J. Am. Chem. Soc.* **1989**, *111*, 5119-5124.
33. Lager, G. A.; Gibbs, G. V., *American Mineralogist* **1973**, *58*, 756-764.
34. Vila, A. J.; Lagier, C. M.; Wagner, G.; Olivieri, A. C., *J. Chem. Soc. Chem. Comm.* **1991**, 683-685.
35. Uchino, T.; Ogata, Y., *J. Non-Cryst. Solids* **1995**, *191*, 56-70.
36. Uchino, T.; Ogata, Y., *J. Non-Cryst. Solids* **1995**, *181*, 175-188.
37. Lazzeretti, P.; Zanasi, R., *J. Chem. Phys.* **1980**, *12*.
38. Gierke, T. D.; Flygare, W. H., *J. Am. Chem. Soc.* **1972**, *94*, 7277-7283.
39. Keil, F.; Ahirichs, R., *J. Chem. Phys.* **1979**, *71*, 2671-2675.
40. Fleischer, U.; Schindler, M.; Kutzelnigg, *J. Chem. Phys.* **1987**, *86*, 6337-6347.
41. Chesnut, D. B.; Foley, C. K., *J. Chem. Phys.* **1986**, *85*, 2814-2820.
42. Gauss, J., *Ber. Bunsenges. Phys. Chem.* **1995**, *99*, 1001-1008.
43. Lindsay, C. G.; Tossell, J. A., *Phys. Chem. Minerals* **1991**, *18*, 191-198.
44. Meier, B. H.; Iuliucci, R. J., *J. Am Chem. Soc.* **1998**, *120*, 9059-9062.
45. Feike, M.; Graf, R.; Schnell, I.; Jäger, C.; Spiess, H. W., *J. Am. Chem. Soc.* **1996**, *118*, 9631-9634.

Chapter 23

Application of Nuclear Shielding Surfaces to the Fundamental Understanding of Adsorption and Diffusion in Microporous Solids

Cynthia J. Jameson[1], A. Keith Jameson[2], Angel C. de Dios[3], Rex E. Gerald II[4], Hyung-Mi Lim[5], and Pavel Kostikin[1]

[1]Department of Chemistry, University of Illinois at Chicago, Chicago, IL 60607
[2]Department of Chemistry, Loyola University, Chicago, IL 60626
[3]Department of Chemistry, Georgetown University, 37th and O Streets, NW, Washington, DC 20057
[4]Argonne National Laboratory, Argonne, IL 60439
[5]Chonnam National University, Chonnam, South Korea

The intermolecular chemical shift provides a means by which very detailed questions can be answered towards a fundamental understanding of two basic processes in the many technological applications of microporous materials: adsorption and diffusion. We demonstrate that the detailed equilibrium properties of physisorbed small molecules and the rates of cage-to-cage migration in zeolites can be observed and quantitatively reproduced by theoretical calculations. Some of the questions which have been answered include: When molecules are adsorbed in a microporous solid at a given loading, what fraction of the cavities have exactly *n* molecules? How many molecules of type 2 can be found in those cavities that have exactly *n* molecules of type 1? How do the type, size, and locations of cations affect the distribution of molecules within a microporous solid? How is the NMR chemical shift affected by confinement with *n* molecules of another type in a cavity? A zeolite cage provides a model system for the understanding of solvent effects on chemical shifts, for the theoretical investigation of the temperature dependence of medium chemical shifts under constant volume conditions in which very detailed comparisons with experiment can be carried out.

Microporous solids, zeolites in particular, are widely used in heterogeneous catalytic processes, separations, oil recovery, and other industrial processes (*1*). A microscopic understanding of elementary processes at surfaces, such as adsorption and diffusion is an important fundamental problem and may assist in interpreting more complicated surface chemistry. A simple rare-gas physisorption system is a good starting point to investigate the distribution and dynamic behavior of adsorbed species. We use a combination of NMR spectroscopy and computer simulations in fundamental studies of adsorption and diffusion. ^{129}Xe NMR is widely used in the characterization of microporous materials. In zeolites (*2-5*), polymers (*6-8*), coals (*9*), clays (*10*) and other porous materials (*11*), the mobility of the Xe atoms is such that only one Xe signal is usually observed in the NMR spectrum. The average chemical shift of the single peak under fast exchange and its dependence on the temperature and the average Xe occupancy are all that can be observed in nearly all microporous solids. In these cases the Xe chemical shift is known empirically to depend on zeolite pore and channel dimensions (*12-15*), location of cations (*16*), co-adsorbed molecules (*17,18*), dispersed metal atoms (*1,4,5*) and paramagnetic ions (*19,20*), on water content (*21*), blockage of pores by coking (*22*), domains of different composition or crystallinity (*23*), etc.

The Model System

We choose to carry out systematic studies in specifically those zeolites (type *A*) of known structure where the position of every Si and Al atom and also every charge-balancing cation is known, where the actual equilibrium distribution of cages containing specifically known numbers of Xe atoms is directly observed, where the average chemical shift of ^{129}Xe in cages containing a specific number of Xe atoms is individually observable, as well as its temperature dependence, and where the temperature dependence of the average Xe occupancy (the adsorption isotherm) can also be observed at the same time. For our model system we have chosen zeolite Na*A*. This is an aluminosilicate of formula $Na_{12}[(SiO_2)_{12}(AlO_2)_{12}]$ whose crystal structure is well characterized. The framework structure provides a simple cubic arrangement of contiguous large cages (alpha cages). A Na^+ ion in each of the 6 windows to the cages keep the Xe atoms adsorbed inside for sufficiently long residence times such that each Xe_n is observed as a distinct signal in the high resolution NMR spectrum. By making measurements in such a system, for example, pure Xe in Na*A*, or a mixture of Xe and other sorbate molecules in competitive adsorption in zeolite Na*A*, we expect to be able to obtain directly from experiment, information about adsorption and diffusion not otherwise available. These observations provide direct tests of those details of computer simulations that are usually averaged over. These provide more critical and detailed constraints on the computer modeling than otherwise.

Observations

The most interesting and challenging experimental data to be interpreted are the chemical shifts observed in the Xe, Xe_2, Xe_3, ..., Xe_8 clusters trapped in the alpha cages of the zeolite Na*A* (*24,25*). Here, what is observed is not an average chemical shift under fast exchange of the Xe with a very large number of cavities containing a variable number of Xe atoms, but rather the individual chemical shifts of trapped clusters of Xe atoms, Xe_1, Xe_2, Xe_3,... up to Xe_8. The chemical shifts have been measured as a function of temperature (*24*) and these pose a dynamical averaging challenge for the theory. The shielding of the ^{129}Xe in the Xe_n clusters changes according to n; the most deshielded nuclei are in the Xe_8 cluster, but the individual cluster shifts do not change with loading. The cluster shift is in the direction given by $\sigma_1(T)$ observed in the gas phase, however, the incremental change is not a fixed amount but increases slightly in going from 1 to 6. Furthermore, there is a big change in shielding in going from Xe_6 to Xe_7 and then again in going from Xe_7 to Xe_8. These individual Xe_n peaks are also observed experimentally in zeolite K*A* (*26*)and in zeolite Ag*A* (*27*). The net effect of the larger cation size on the Xe_n clusters (found experimentally) is to increase the chemical shift and enhance the increments $[\delta(Xe_n) - \delta(Xe_{n-1})]$ (compared to zeolite Na*A*). Under magic angle spinning, it has been possible to narrow the peaks corresponding to the Xe_n clusters in zeolite cages oriented in various directions relative to the magnetic field (*28*) . With very narrow lines it has been possible to distinguish by their different intermolecular shifts the Xe_n signals in the distinguishable cages that arise upon substitution of Ca^{2+} ions for Na^+ ions in the same zeolite: The progression of Xe_n signals associated with each cage type can be observed. (Incidentally, they have somewhat different distributions, as can be observed from the Xe_n intensities.)(*28*) This is a dramatic illustration of the sensitivity of the ^{129}Xe shielding to the environment.

In studies of competitive adsorption, the usually measured quantity is the overall composition of the adsorbed phase for a given composition of the bulk phase in equilibrium with it. It has been found that chemical shifts can provide a more detailed description. In a mixture of Xe and Kr in Na*A* zeolite it was possible to observe the individual signals from Xe_nKr mixed clusters as well as the Xe_n clusters under magic angle spinning (*28*). The absolute ^{129}Xe chemical shifts of the Xe_nKr mixed clusters and the increments between Xe_nKr and the Xe_{n+1} in various Xe-Kr mixtures in Na*A* zeolite, are shown in Table I.

Table I. ^{129}Xe chemical shifts of mixed clusters Xe_nKr in the alpha cages of zeolite Na*A* (ppm)

	$\delta(Xe_nKr)$		$\delta(Xe_nKr) - \delta(Xe_n)$	
Cluster	*OBSD*[a,b]	*GCMC*[a,b]	*OBSD*	*GCMC*[b]
Xe_1Kr	84.7	85.7	9.9	8.6
Xe_2Kr	103.3	102.6	11.0	9.9
Xe_3Kr	124.5	121.3	12.8	11.6
Xe_4Kr	148.9	144.3	15.7	14.4
Xe_5Kr	174.7	172.0	16.3	16.8
Xe_6Kr	209.9	209.1	26.5	25.1

[a] relative to an isolated Xe atom, Ref. (*28*). [b] Ref. (*29*).

In the competitive adsorption of Xe and Ar in Na*A* zeolite, the Ar atoms are in fast exchange, so that individual mixed clusters like Xe_nAr_m can not be observed. Here, the Xe_n peaks are observed just as in pure Xe in this zeolite, except that the individual chemical shifts are observed to be dependent on the loadings of Ar (primarily) and Xe (secondarily) (*30*). Here the total intermolecular chemical shift measured relative to the isolated Xe atom, the ^{129}Xe chemical shift for Xe_n in an alpha cage with an average number of Ar atoms under fast exchange. The differences in shielding, $[\langle\sigma(Xe_nAr_{ave})\rangle - \langle\sigma(Xe_n)\rangle]$, obtained directly from experiment, are a direct measure of the intermolecular effects of Ar, and are therefore a direct measure of the average number of Ar atoms in the cage with Xe_n. Similarly, for mixtures of Xe with CH_4, N_2, CO_2 or CO, the differences in the shielding of the Xe_n peak in the mixture sample and the corresponding Xe_n peak in the pure Xe sample is a direct measure of the average number of molecules of, respectively, CH_4, N_2, CO_2, or CO, in the same cage as the *n* Xe atoms. To extract this average number for every Xe_n in any such coadsorbed mixture in zeolite Na*A*, at any mole fraction of the constituent gases in the bulk gas in equilibrium with the molecules adsorbed inside the zeolite, we require the Xe-CH_4, XeN_2, Xe-CO_2, and Xe-CO intermolecular shielding surface.

Intermolecular Shielding Surfaces

An intermolecular shielding surface is a mathematical function describing the dependence of the nuclear shielding as a function of the configuration of two interacting molecules. Some shielding surfaces have been calculated for interacting rare gas pairs (*31*), and rare gas atoms interacting with cations, diatomic molecules, or H_2O (*32*), and for dimers and n-mers of H_2O molecules (*33,34*).

Shielding Surfaces for Two Interacting Molecules. In the Maryland meeting Jameson and de Dios reported the first ab initio calculations of the rare gas pair intermolecular shielding surfaces for Ar_2, ArNe, Ne_2, and NeHe (*35*). With these shielding surfaces it was possible to calculate the second virial coefficient for nuclear shielding as a function of temperature, using the well established intermolecular potential functions $V(R)$ for the pair.

$$\sigma_1(T) = \int_0^{\infty} 4\pi R^2 \mathrm{d}R\, [\sigma(R) - \sigma(\infty)] \exp\, [-V(R)/k_BT].$$

It had been found that the shielding functions are nearly conformal in the range $\infty > R > r_0$.and that the scaling factors relating intermolecular shielding functions of rare gas pairs are $\langle a_0^3/r^3\rangle_1\ \alpha_1(0)\ U_1U_2/(U_1+U_2)$, with the R scaling according to r_0. The values of $\langle a_0^3/r^3\rangle_{np}$ for the free atom, the static electric dipole polarizability $\alpha(0)$ and the ionization potentials, U are well known for the rare gas atoms, which permit the scaling of the Ar-Ar shielding function to any of the rare gas pairs. The calculation of $\sigma_1(T)$ for Xe-Xe, Xe-Kr, and Xe-Ar could therefore be carried out and the results could be compared with the experimentally measured density coefficients of nuclear shielding $\sigma_1(T)$ for these pairs. The ab initio calculations of the shielding function at the RPA level was found to be very close to that obtained at the level of SOPPA, using LORG and SOLO methods (localized orbitals/local origins and second order localized orbitals/local origins) respectively (*31*). These earlier calculations are validated by more recent results. It has been found, more recently, that the Ne-Ne intermolecular shielding function calculated at the SCF-GIAO level is nearly indistinguishable from that calculated at the CCSD(T) level for the range 2.2 - 4.0 Å where r_0 = 3.07 Å (Bühl, M.; Kaupp, M.; Malkin, V. G.; Malkina, O. L. *J. Comp. Chem.*, **1999**, in press.). It appears that the difference between the Hartree-Fock and the highest level of electron correlation for the intermolecular shielding function of the rare gas pair is indeed small. Also, it has been found (de Dios and Jameson, to be published) that the ab initio Xe-Xe intermolecular shielding function is very close to the Xe-Xe function obtained by scaling the ab initio Ar-Ar shielding function in the range $4r_0$ Å > R > $0.9r_0$. The new Xe-Xe shielding function was calculated using a large Xe atom basis set (240 basis functions), large enough that the counterpoise correction to the intermolecular shielding is negligibly small.

For Xe interacting with molecules N_2, CO and CO_2, the ^{129}Xe shielding functions have been calculated (*36*) as a function of two variables, the center of mass distance and the angle θ that the linear molecule axis makes with the center-of-mass-distance vector. Here, too, the counterpoise correction to the shielding function is very small, indicating that the Xe basis set used is large enough. These intermolecular shielding functions have been used to calculate the second virial coefficients of ^{129}Xe shielding for the pairs Xe-N_2, Xe-CO, and Xe-CO_2. When using presently available interaction potential functions, V(R,θ), the comparisons of calculated and experimental $\sigma_1(T)$ values reveal good agreement with the temperature behavior of all three pairs, and good agreement with trends in relative magnitudes of these three compared with each other and with Xe-Xe, Xe-Kr and Xe-Ar values, although the recovery of absolute magnitudes is not complete. Improved potentials are clearly necessary.

To describe the numerical results of the ab initio shielding calculations, a choice of functional form is required to represent the shielding surface in analytic form. For the rare gas pair, it was found that the inverse R dependence of σ (R) has a behavior c;lose to $R^{-6.7}$, $R^{-6.8}$, and $R^{-7.4}$. Therefore, a function of the form

$$\sigma(R) - \sigma(\infty) = \sum_n a_n R^{-n}$$

with n = even only, 6 up to 12 or 14, would appear to be an adequate description. The coefficients a_n are obtained by least squares fitting to the ab initio values at various R. On the other hand, for the Xe-linear molecule interaction, a_n may be taken to be a function of the angle θ, reflecting the symmetry of the system, such as

$$a_n = \sum_{\lambda=0,even} a_{n\lambda} P_\lambda(\cos\theta)$$

with λ = 0, even only, up to 6 or 8 for Xe-N_2 and Xe-CO_2, or λ = 0, 1, 2, 3, 4... for Xe-CO. The coefficients $a_{n\lambda}$ are found by least squares fitting to the ab initio values of σ(R,θ) that have been calculated at intervals of 15°. An alternative description is

$$\sigma(R) - \sigma(\infty) = \sigma(R_{Xe\text{-}O}) + \sigma(R_{Xe\text{-}O'}) + \sigma(R_{Xe\text{-}C})$$

$$\sigma(R_{Xe\text{-}O}) = \sum_n a_n^{(Xe\text{-}O)} R_{Xe\text{-}O}^{-n}$$

$$\sigma(R_{Xe\text{-}C}) = \sum_n a_n^{(Xe\text{-}C)} R_{Xe\text{-}C}^{-n}$$

where a pairwise additive form is used. For more complex interacting molecules, where it is not feasible to find a simple description in terms of Legendre expansions, the pairwise additive form is the functional form of choice.

Shielding Surfaces for a Molecule Adsorbed on a Zeolite Fragment. Since the Xe is adsorbed onto an inner surface of the zeolite, it would be ideal to include the entire zeolite cage framework atoms and ions in the calculations. However, this size calculation is not yet feasible, thus, only a fragment of the cage is used. For this purpose, a fragment terminated with $OSi(OH)_3$ groups rather than the more commonly used $OSiH_3$ groups is used. All atoms are placed at the atomic coordinates obtained from high resolution x-ray or neutron scattering. Counterpoise corrections are required for each ab initio calculation of the supermolecule with various positions of the adsorbed Xe atom. This is crucial here, because the zeolite framework atoms and ions provide a large number of additional orbitals in which the electrons of the rare gas atom may be found, effectively increasing the basis set size in going from isolated rare gas atom to the supermolecule. A pairwise additive functional form is practical here, in terms of the distance between the rare gas atom and each O atom and each Na^+ ion. These calculations have been reported by Jameson and Lim for a zeolite with a 1:1 ratio of Si to Al in the framework and various charge-balancing cations.(*37*).

Computational Methods

The grand canonical ensemble is appropriate for adsorption systems, in which the adsorbed phase is in equilibrium with the gas at some specified temperature. The use of a computer simulation allows us to calculate average macroscopic properties directly without having to explicitly calculate the partition function. The grand canonical Monte Carlo (GCMC) method as applied in this work has been described in detail earlier (*38*) . The aspects involving binary fluid mixtures have been described previously in our Xe-Ar work (*30*).

Cut-and-shifted effective potentials describe the interaction between the adsorbed fluid and the zeolite. The Xe-O and Xe-Na parameters are unchanged from our previous work on Xe in Na*A* (*38*) and Xe-Ar mixtures in Na*A* (*30*). The Kr-O and Kr-Na potential parameters were obtained as described in Ref. (*29*). The CH_4-O and CH_4-Na potential surfaces were described by the pairwise Lennard-Jones parameters given by Woods and Rowlinson (*39*).

The Xe-Xe potential used is of the Maitland-Smith form, as described in our previous simulations of Xe in Na*A* (*38*), fitted to the best available Xe-Xe potential of Aziz and Slaman (*40*). Likewise the Xe-Kr and the Kr-Kr potentials were taken from the best available potentials of Aziz *et al.* (*41, 42*), and fitted to the Maitland-Smith form. The Xe-CH_4 and the CH_4-CH_4 potential functions used were of the atomic site-site form based on the best available potentials of Liuti et al and Righini et al. respectively (*43, 44*).

The simulation box is a unit cell of zeolite Na*A*, with the atomic coordinates, including the Na cations, taken from the x-ray single crystal refinement of the dehydrated zeolite by Pluth and Smith (*45*). Periodic boundary conditions were imposed using the minimum image convention, consistent with the cut-and-shifted potentials employed (*46*). The core of the program effects the creation/destruction and displacement of one molecule at a time and calculates the associated energy change, ΔU in each case. This is used to continuously update the total configurational energy of the system, without having to recalculate every interaction at every step. The displacement step uses the adsorbed phase composition in the choice of either fluid. The creation/destruction step begins with the decision to either create or destroy a particle. If the decision is to destroy a particle (that is, the particle is assumed to go into the gas phase) the choice between destroying a molecule of Xe or CH_4 is made proportionately to the gas phase composition, i.e., ρ_{Xe} and ρ_{CH4} . ;likewise, the choice of creating a molecule of Xe or CH_4 (i.e., remove it from the bulk gas and place it in the zeolite) is made according to the gas phase composition. It is necessary to know *a priori* the ratio of the gas densities in equilibrium with the adsorbed phase before the simulation starts. This is easily done by first calculating the chemical potential of the Xe and the CH_4 appropriate to the temperature and densities in the gas mixture, using the virial coefficients (*30*). These chemical potentials, the temperature, and the mole fraction of Xe in the gas are the parameters of a GCMC simulation.

In our approach, the ^{129}Xe shielding, like the energy, is taken to be expressible as a sum of pairwise contributions, using atom-atom shielding functions that are likewise cut and shifted. The Xe-O and Xe-Na shielding functions are based on *ab initio* calculations on model systems and are the same as was used in our previous work (*30,38*). The Xe-Xe and the Xe-Kr shielding functions are based on *ab initio* calculations and are taken from Ref. (*32*). The Xe-CH_4 shielding functions are based on ab initio calculations at three symmetrical orientations of the CH_4 molecule relative to the rare gas taken at various distances and fitted to shielding contributions from H and C centers (*47*). These have been shown to reproduce reasonably well the temperature dependent density coefficients of the ^{129}Xe chemical shifts for pure xenon gas and for Xe in Kr, and Xe in CH_4 gas mixtures. At each step in the simulation at which the atom-atom interaction energy is calculated, the shielding contribution is calculated too, when one of the atoms involved is a xenon atom. Since the shielding functions go to large negative values at close approach, it is quite important to have Xe-Xe, Xe-Kr and Xe-CH_4 potentials that have the correct behavior at these short distances, especially close to r_0. This is the reason for using as accurate two-body potentials as possible for Xe-Xe and Xe-Kr or Xe-CH_4. The Maitland-Smith form provides a superior fit to the best rare gas pair potentials and is just as inexpensive as the Lennard-Jones form in computational overhead. The Maitland Smith form is also used for the Xe-C and Xe-H parts of the Xe-CH_4 potential.

The results of the GCMC simulations are analyzed to provide the usual averages for each (μ_{CH4}, μ_{Xe},T), such as $\langle n \rangle_{Xe}$, $\langle m \rangle_{Kr}$, the average number of atoms per alpha cage, the xenon distribution, P_n, i.e., the fractions of the cages that are occupied by *n* Xe atoms, as well as the detailed occupancies $f(Xe_nKr_m)$, i.e., the fractions of the cages that are occupied by *n* Xe atoms and *m* Kr atoms. In addition, the one-body distribution functions, pair distribution functions, and average shieldings for individual mixed clusters Xe_nKr_m or $Xe_n(CH_4)_m$, which are independent of loading, are accumulated over all the GCMC runs. All calculations reported here were carried out on IBM RISC/6000 (models 560 and 365) workstations.

Given that the total potential energy is a sum over all the pairwise interactions at distances less than some cutoff distance, and given the interaction potential function for the Xe-other sorbate interactions, likewise taken to be pairwise additive, then the properly weighted large number of configurations (a million or so) can be summed in a Grand Canonical Monte Carlo (GCMC) scheme (*38*). At each of these configurations the shielding of every Xe nucleus is calculated. Ab initio shielding calculations for ^{39}Ar in the triangular and linear Ar_3 clusters have shown that as long as the Ar atoms do not get much closer than about $0.9r_0$, the isotropic shielding contributions can be obtained by additivity. We therefore construct the ^{129}Xe shielding surface as we go, summing up over all the atoms that are sufficiently close (less than 12 Å). The approach is to treat the whole system (any number of Xe atoms per unit cell and atoms of the zeolite framework) as a supermolecule and to calculate the shielding function at each configuration by assuming that the intermolecular shielding is pairwise additive, that is, for a given ^{129}Xe nucleus the shielding is a sum of contributions from Xe-Xe as a function of distance to all the other Xe atoms, plus Xe-O as a function of distance to all the O atoms of the zeolite framework, plus Xe-Na^+ as a function of distance to all the Na^+ counterions. The Xe-Al and Xe-Si contributions are neglected; the Xe atom can not get close to these atoms due to the O atoms bridging the Al and Si. The shielding surface is thus constructed on the fly, using the pair shielding functions $\sigma(Xe\text{-}O_{zeo})$, $\sigma(Xe\text{-}Na_{zeo})$ and $\sigma(Xe\text{-}Xe)$ derived from ab initio calculations described above (*31, 37*), changing with the numbers of neighbors as the numbers of sorbate atoms fluctuate in the grand canonical ensemble. The isotropic ^{129}Xe shielding averaged over all the Xe atoms in a cluster are separately stored for each of the eight cluster types so that the average chemical shift for each Xe_n can be obtained. This is done at each temperature. The averages at various temperatures can then be compared with the experimental chemical shifts at these temperatures (*38*).

Results

The distributions of the Xe atoms among the cages of the zeolite are provided by the relative intensities of the Xe_1, Xe_2, Xe_3, ...Xe_8 peaks seen individually in the ^{129}Xe NMR spectrum. Thus, the NMR experiment provides a direct measure of the distribution of Xe atoms among the cavities, e.g., what fraction of the zeolite cages have 5 exactly Xe atoms? This is reproduced very well by the grand canonical simulations described above.

The calculated and experimental quantities being compared are in absolute terms, that is, the shieldings are relative to the free Xe atom in both calculation and experiment. The agreement of the GCMC average with experiment is excellent, especially considering that there are no adjustable parameters once the Xe-zeolite potential parameters are adopted. That the increments between the Xe_n cluster shifts are well-reproduced, including their temperature dependence, might have been expected from the good agreement found in the gas phase chemical shifts based on the same ab initio-derived Xe-Xe intermolecular chemical shift function, since these chemical shift increments are largely due to the Xe-Xe interactions.

What about the changes in the intermolecular chemical shifts when the atoms of the zeolite cage are changed? The GCMC simulations permit examination of the separate contributions coming from the cations, the zeolite framework (effectively represented by the O atoms), and Xe-Xe contributions. The difference between the Xe-O and Xe-cation contributions in K*A* and Na*A* is interesting since the same $\sigma(Xe...O_{zeol})$ and $V(Xe\text{-}O_{zeol})$ functions were used in the GCMC simulations in both zeolites. The direct cation contribution to the chemical shift is larger for K than for Na in all cluster sizes, and this increases with cluster size. This is more pronounced in the K*A* cage due to the in-out arrangements of the K ions (*26*). The excluded volume effect due to the latter leads to smaller Xe-O contributions in K*A* and larger contributions from the Xe-Xe interactions, becoming more severe with increasing number of Xe atoms due to the more pronounced deshielding exhibited by the $\sigma(r_{Xe\text{-}Xe})$ function at shorter distances. Although the calculated numbers can and do change upon changing the parameters of the V(Xe-K) potential, the trends are all preserved.

The calculated differences between the chemical shift of Xe_n in K*A* and Na*A* are in semi-quantitative agreement with experiment, and the GCMC simulations reproduce the trends for the temperature dependence of the ^{129}Xe chemical shifts of the Xe_n clusters in K*A* (*26*) just as well as they did for Xe_n in Na*A* (*38*). The GCMC averaging for Xe_n in partially Ca^{2+} ion-exchanged Na*A* also reproduce the chemical shift increments assigned to cages having 1 or 2 Ca^{2+} ions. Polarization terms in the energy of Xe due to interactions with cations, a non-pairwise additive potential energy term, had to be included to take into account the differences in mono- versus di-valent cations (*48*).

The absolute ^{129}Xe chemical shifts of the Xe_nKr mixed clusters and the increments between Xe_nKr and the Xe_{n+1} are reproduced by GCMC averaging in various Xe-Kr mixtures in Na*A* zeolite, as shown in Table I. In Figure 1 we compare the experimental Xe_n cluster chemical shifts in 9 different samples of Xe-CH_4 mixtures in Na*A* with the average chemical shifts from the GCMC simulations with the same overall $\langle n\rangle_{Xe}$ and $\langle m\rangle_{CH4}$ as experiment. Here the total intermolecular chemical shift measured relative to the isolated Xe atom, the ^{129}Xe chemical shift for Xe_n in an alpha cage with an average number of CH_4 molecules under fast exchange (which is directly calculated in the GCMC simulation) is plotted in comparison with the experimental values. We see that in an absolute measure, the calculated chemical shifts are in very good agreement with experiment. Furthermore, the differences in shielding, $[\langle\sigma(Xe_nCH_{4\ \langle m\rangle})\rangle - \langle\sigma(Xe_n)\rangle]$, may be compared directly with experiment, are a direct measure of the intermolecular effects of CH_4, and are therefore a direct measure of the average number of CH_4 molecules in the cage with *n* Xe atoms. These calculated shielding differences are in quantitative agreement with experiment for samples with various Xe and CH_4 loading (*47*). The simulations also provide the average number of CH_4 molecules associated with each individual Xe_n. Once the average ^{129}Xe chemical shifts in the Xe_n $(CH_4)_m$ mixed cluster chemical shifts have been calculated for each *n* and *m* combination in the GCMC simulations, the Xe_n chemical shifts observed in any sample of Xe-CH_4 mixtures in zeolite Na*A* can be converted directly to the average number of CH_4 molecules in the same cage with *n* Xe atoms while the fraction of cages containing exactly *n* Xe atoms is directly given by the relative intensity of the Xe_n peak. In other words, the ^{129}Xe chemical shifts provide the detailed distribution of the Xe-CH_4 mixture among the cavities of the zeolite. Other mixtures (Xe-CO, Xe-CO_2, Xe-N_2,...) can be studied in this way.

The ^{129}Xe chemical shifts of each of the peaks associated with Xe_1, Xe_1Kr_1, Xe_2, Xe_2Kr_1, ...Xe_6Kr_1, ...Xe_8 are observed individually. On the other hand, because Ar or CH_4 are freely migrating from one cage to the next, only an average occupancy of Ar or CH_4 is seen by the Xe nuclei. Thus, only an average Xe_n NMR chemical shift is observed, differing from that of the Xe_n peak when pure xenon is trapped in the zeolite, by an amount which is directly related to the average number of Ar or CH_4 molecules (visiting sorbates) in the same cage as Xe_n. This average Xe_n chemical shift is reproduced by the Monte Carlo simulations, for all the peaks for all the samples. In other words, the simulations find the correct fraction of cages that have an exact number of Xe atoms, and for those cavities having that number of Xe atoms, the simulations find the correct average number of Ar or CH_4 in the same cavity. In the special case of Xe-Kr mixtures, the mixed Xe_nKr occupancy is directly observed since individual peaks can be observed with either zero or one Kr atom with each Xe_n..

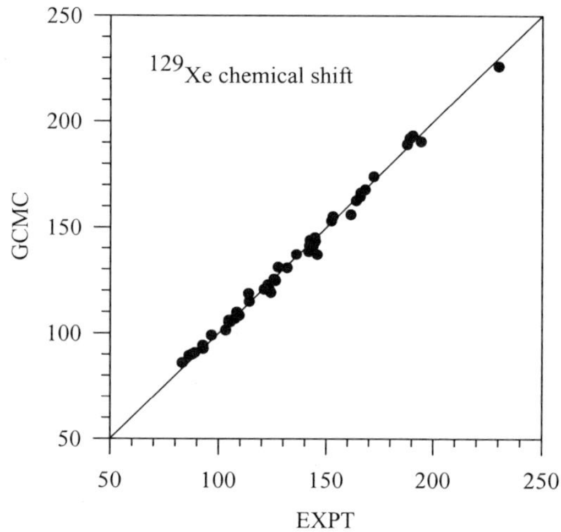

Figure 1. ^{129}Xe chemical shifts of Xe_n with CH_4 under fast exchange in zeolite Na*A*, obtained from GCMC simulations, compared with experimental values. The shifts are in ppm relative to the isolated Xe atom.

This is possible because both Xe and Kr have long residence times in a specific cavity. The Ar or CH_4 molecule is freely exchanging between cavities but the Xe_n chemical shift reports the average number of Ar or CH_4.with Xe_n, thus providing a more complete picture of distribution of mixed adsorbed species within a zeolite than has ever been obtained before.

We have investigated the effect of the cation on the chemical shifts and on the adsorption by observing Xe adsorbed in Na*A* in which all of the Na ions have been exchanged with K (zeolite K*A*) (*26*), and in Na*A* in which only some of the Na ions have been exchanged with Ca^{2+} ions (*28*). In the latter, we have been able to obtain the chemical shifts of Xe_n in cages containing no Ca (i.e., pure zeolite Na*A*), in cages containing one Ca and in cages containing two and three Ca ions. Each of these types of cages provides its own signature in the form of a progression of peaks giving the intensity distribution according to the Xe_n occupancies of cages of that particular type, and the Xe_n chemical shifts that are unique to that type of cage. The GCMC simulations support the assignments and provide a measure of the role of polarization of Xe by di and monvalent cations. We have also interpreted the ^{129}Xe NMR observations of Xe in Ag*A* by Moudrakovski, Ratcliffe and Ripmeester. (*27*) using the same GCMC simulations (*49*). The behavior of Xe_n in Ag*A* is very similar to that of Xe_n in Na*A*, and the interpretation is similarly successful. The observation of Xe_n chemical shifts in the same zeolite framework with differing cations Na^+, K^+, Ag^+, and Ca^{2+}, permit the interpretation of the physisorption of Xe in the A type zeolite in a quantitative way. Since the occupancy of each cage is directly observable, the chemical shifts are therefore clearly associated with a fixed number of Xe atoms in a cage which has the same framework atoms in the same arrangement, while at the same time permitting a variation in the type of cations present. In this way, detailed comparisons of simulation predictions of distributions and chemical shifts against experiments are possible. We have been able to understand how the type, size, and locations of cations affect the distribution of molecules within a cavity.

A more detailed picture of adsorption than ever before is made possible by the complete interpretation of the Xe_n chemical shifts, which, of course, is made possible by the use of ab initio intermolecular shielding functions. The excellent agreement of the simulation results with experiment in turn gives us confidence that the GCMC simulations are robust and the potential parameters being used are not too unreasonable. In this way, intermolecular shielding functions play a very important role in the fundamental understanding of adsorption. Without the chemical shifts, only an adsorption isotherm may be used for testing the simulations, and this quantity is a highly averaged physical quantity which only provides partitioning of the sorbate between inside and outside the zeolite. We on the other hand, have this inside-outside partitioning since we observe both the adsorbed and the bulk gas peaks, and we also have the detailed distributions of both types of sorbate molecules among the cavities of the zeolite. Temperature dependent studies have provided us with additional information about the one- and two-body distributions of the n Xe atoms within a cavity (*50*). The Xe-zeolite part of the average Xe_n chemical shift is a function of the one body distribution of the Xe in the cavity and the Xe-Xe part of the average Xe_n chemical shift is a function of the two-body Xe-Xe distribution function.

Our group and the Pines group have measured for Xe in zeolite Na*A*, via magnetization transfer and 2D-EXSY, the individual rate constants for a Xe atom leaving a cage containing *n* Xe atoms, going into an adjacent cage containing (m-1) Xe atoms, for all *n* and *m* ranging from 1 to 8 (*51, 52*). A similar measurement has been made by Ripmeester's group for Xe in zeolite Ag*A* (*27*). The set of microscopic rate constants provide a very detailed picture of the diffusion of Xe atoms within the zeolite, that is, from the rate constants, the difussion coefficient for a sample at any average loading can be obtained. It is the large chemical shifts separating the Xe_n clusters that make such direct NMR measurements of microscopic rates possible.

Conclusions

We have investigated various mixtures in zeolite Na*A* as model systems for competitive adsorption in microporous solids. This work provides a more detailed comparison of experiment with GCMC simulations than has ever been possible. When we observed experimentally the clusters of *n* Xe atoms and *m* Kr atoms in the same cage, it was the first instance where the number of molecules of the second sorbate occupying the same cage as *n* atoms of the first sorbate have actually been determined. The Xe_n and the mixed Xe_nKr_m clusters are trapped in the alpha cages of this zeolite for times sufficiently long that it is possible to observe individual peaks in the NMR spectrum for each cluster. The ^{129}Xe chemical shift of the Xe_nKr_m peak is a measure of the distribution of the *n* Xe and *m* Kr atoms within the alpha cage. The absolute chemical shifts for the Xe_n and Xe_nKr peaks observed at 300 K, spanning a 200 ppm range, are in excellent agreement with the average chemical shifts for these mixed clusters in the GCMC simulations. ^{129}Xe chemical shift increments upon addition of one Kr atom to the Xe_n cluster are increasing monotonically with cluster size. A significant conclusion, which may be very helpful in competitive adsorption cases where such detailed comparisons with experiment are not feasible, is that the magnitude of the increments can be estimated semi-quantitatively from the known gas phase shifts in mixtures of Xe and Kr and from the Xe_n cluster shift increments. Similar observations, have been possible for other mixtures such as Xe-Ar and Xe-CH_4, except that in these cases, the Ar or CH_4 are in fast echange among the cages, so that only an average number of molecules occupying the same cage as an Xe_n is observed, and the Xe_n chemical shift is a direct measure of this average number. In these cases, the chemical shift of the Xe_n for every n, in all samples containing various mole fractions of Xe and other sorbate are reproduced very well by the GCMC simulations, thus permitting the complete interpretation of distributions and chemical shifts. Similarly, the Xe-CO, Xe-N_2, Xe-CO_2 shielding functions recently calculated by us (*36*) will provide a similar opportunity for interpretation of the competitive adsorption of these mixtures in zeolite Na*A*, using the NMR measurements that we have already carried out in these systems. Furthermore, the Xe chemical shifts in any open network zeolites, where only a single adsorbed peak signal can be obtained under fast Xe exchange can also be interpreted using the same simulations (*53*).

Acknowledgments

The research described here has been supported by the National Science Foundation (Grants No. CHE-9210790 and CHE-9528066) and the University Scholar program of the University of Illinois Foundation.

Literature Cited

1. *Zeolites and Related Microporous Materials: State of the Art 1994;* Weitkamp, J.; Kärge, H. G.; Pfeifer, H.; Hölderich, W., Eds.; Studies in Surface Science and Catalysis; Elsevier: Amsterdam, 1994; Vol.84.
2. Fraissard, J.; Ito, T. *J. Chem. Phys.* **1982,** *76*, 5225.
3. Fraissard, J.; Ito, T. *Zeolites* **1988**, *8*, 350.
4. Dybowski, C.; Bansal, N.; Duncan, T. M. *Ann. Rev. Phys. Chem.* **1991**, *42,* 433.
5. Barrie, P. J.; Klinowski, J. *Prog. NMR Spectrosc.* **1992**, *24*, 91.
6. Stengle, T. R.; Williamson, K. L. *Macromolecules* **1987,** *20,* 1428.
7. Miller, J. B.; Walton, J. H.; Roland, C. M. *Macromolecules* **1993,** *26*, 5602.
8. Kentgens, A. P. M.; van Boxtel, H. A.; Verweel, R. J.; Veeman, W. S. *Macromolecules* **1991,** *24*, 3712.
9. Suh, D. J.; Park, T. J.; Ihm, S. K.; Ryoo, R. *J. Phys. Chem.* **1991,** *95*, 3767.
10. Barrie, P. J.; McCann, G. F.; Gameson, I.; Rayment, T.; Klinowski, J. *J. Phys. Chem.* **1991,** *95*, 9416.
11. Cheung, T. T. P. *J. Phys. Chem.* **1989,** *93*, 7549.
12. Derouane, E. G.; Nagy, J. B. *Chem. Phys. Lett.* **1987,** *137*, 341.
13. Demarquay, J.; Fraissard, J. *Chem. Phys. Lett.* **1987,** *136*, 314.
14. Johnson, D. W.; Griffiths, L. *Zeolites* **1987,** *7*, 484.
15. Ripmeester, J. A.; Ratcliffe, C. I.; Tse, J. S. *J. Chem. Soc., Faraday Trans. 1* **1988,** *84*, 3731.
16. Kim, J.-G.; Kompany, T.; Ryoo, R.; Ito, T.; Fraissard, J. *Zeolites* **1994,** *14*, 427.
17. Wu, J. F.; Chen, T. L.; Ma, L. J.; Lin, M. W.; Liu, S. B. *Zeolites* **1992,** *12*, 86.
18. Chmelka, B. F.; Pearson, J. G.; Liu, S. B.; Ryoo, R.; de Menorval, L. C.; Pines, A. *J. Phys. Chem.* **1991,** *95*, 303.
19. Liu, S. B.; Fung, B. M.; Yang, T. C.; Hong, E. C.; Chang, C. T.; Shih, P. C.; Tong, F. H.; Chen, T. L. *J. Phys. Chem.* **1994,** *98*, 4393.
20. Boddenberg, B.; Hartmann, M. *Chem. Phys. Lett.* **1993,** *203*, 243.
21. Fraissard, J.; Kärger, J. *Zeolites* **1989,** *9*, 351.
22. Barrage, M. C.; Bonardet, J. L.; Fraissard, J. *Catal. Lett.* **1990,** *5*, 43.
23. Maistriau, L.; Derouane, E. G.; Ito, T. *Zeolites* **1990,** *10*, 311.
24. Jameson, C. J.; Jameson, A. K.; Gerald II, R. E.; de Dios, A. C. *J. Chem. Phys.* **1992**, *96*, 1676.
25. Chmelka, B. F.; Raftery, D.; McCormick, A. V.; de Menorval, L. C.; Levine, R. D.; Pines, A. *Phys. Rev. Lett.* **1991,** *66*, 580.
26. Jameson, C. J.; Jameson, A. K.; Gerald II, R. E.; Lim, H.-M. *J. Chem. Phys.* **1995,** *103*, 8811
27. Moudrakovski, I; Ratcliffe, C. I.; Ripmeester, J. A. *J. Am. Chem. Soc.* **1998**, *120*, 3123.
28. Jameson, A. K.; Jameson, C. J.; de Dios, A. C.; Oldfield, E.; Gerald II, R. E.; Turner, G. L. *Solid State Nucl. Magn. Reson.* **1995,** *4*, 1.
29. Jameson, C. J.; Jameson, A. K.; Lim, H.-M. *J. Chem. Phys.* **1997**, 107, 4364.
30. Jameson, C. J.; Jameson, A. K.; Lim, H.-M. *J. Chem. Phys.* **1996,** *104*, 1709.
31. Jameson, C. J.; de Dios, A. C. *J. Chem. Phys.* **1992,** *97,* 417
32. Jameson, C. J.; de Dios, A. C. *J. Chem. Phys.* **1993,** *98*, 2208.

33. Malkin, V. G.; Malkina, O. L.; Steinebrunner, G.; Huber, H. *Chemistry Eur. J.*, **1996**, *2*, 452.
34. Chesnut, D. B.; Rusiloski, B. E. *J. Mol. Struct. (THEOCHEM)* **1994**, *314*, 19.
35. *Nuclear Magnetic Shieldings and Molecular Structure*; Tossel, J. A., Ed.; NATO ASI Series C; Kluwer: Dordrecht, 1993, Vol. 386.
36. de Dios, A. C.; Jameson, C. J. *J. Chem. Phys.* **1997,** *107*, 4253.
37. Jameson, C. J.; Lim, H.-M. *J. Chem. Phys.* **1995,** *103*, 3885
38. Jameson, C. J.; Jameson, A. K.; Baello, B. I.; Lim, H.-M. *J. Chem. Phys*. **1994**, *100*, 5965.
39. Woods, G. B.; Rowlinson, J. S. *J. Chem. Soc., Faraday Trans. 2* **1989,** *85*, 765.
40. Aziz, R. A.; Slaman, M. J. *Mol. Phys*. **1986,** *57*, 825.
41. Aziz, R. A.; van Dalen, A. *J. Chem. Phys.* **1983,** *78*, 2402.
42. Aziz, R. A.; Slaman, M. J. *Mol. Phys.* **1986,** *58*, 679.
43. Liuti, G.; Pirani, F.; Buck, U.; Schmidt, B. *Chem. Phys*. **1988**, *126*, 1.
44. Righini, R.; Maki, K.; Klein, M. L. *Chem. Phys. Lett*. **1981**, *80*, 301.
45. Pluth, J. J.; Smith, J. V. *J. Am. Chem. Soc*. **1980,** *102*, 4704.
46. Allen, M. P.; Tildesley, D. J. *Computer Simulation of Liquids;* Clarendon: Oxford, 1987.
47. Jameson, C. J.; Jameson, A. K.; Kostikin, P.; Baello, B. I., to be published
48. Jameson, C. J.; Lim, H.-M.; Jameson, A. K. *Solid State Nucl. Magn. Reson.* **1997,** *9*, 277.
49. Jameson, C. J.; Lim, H. M. *J. Chem. Phys.* **1997,** *107*, 4373.
50. Jameson, C. J.; Jameson, A. K.; Lim, H.-M.; Baello, B. I. *J. Chem. Phys.* **1994,** *100*, 5977.
51. Jameson, A. K.; Jameson, C. J.; Gerald II, R. E. *J. Chem. Phys*.**, 1994,** *101*, 1775.
52. Larsen, R. G.; Shore, J.; Schmidt-Rohr, K.; Emsley, L.; Long, H.; Pines, A.; Janicke, M.; Chmelka, B. F., *Chem Phys. Lett*., **1993**, *214*, 220.
53. Jameson, C. J.; Jameson, A. K.; Gerald II, R. E.; Lim, H.-M. *J. Phys. Chem.* **1997,** *101*, 8418.

Indexes

Author Index

Subject Index

B

D

E

F

G

H

O

P

Q

T

U

V

W

X

Z